INTO SPACE - A JOURNEY OF HOW HUMANS ADAPT AND LIVE IN MICROGRAVITY

Edited by **Thais Russomano**
and **Lucas Rehnberg**

Into Space - A Journey of How Humans Adapt and Live in Microgravity
http://dx.doi.org/10.5772/intechopen.70684
Edited by Thais Russomano and Lucas Rehnberg

Contributors

Robert J Reynolds, Steven Day, Alex Michael, Bratislav Stankovic, Tony Milligan, Jun Chen, Sara Langston, Carole Tafforin, Stephanie Smith, Susan Childers, Andrzej J. Paszczynski, Laura Galdamez, Nick Caplan, Andrew Winnard, Dorothee Debuse, Anil Kulkarni, Paul Sherman, John Sladky, Christian Laforsch, Miriam Knie, Bernard Wolfschoon Ribeiro, Jessica Fischer, Burkhard Schmitz, Kay Van Damme, Ruth Hemmersbach, Donat-P. Häder, Florian Peter Michael Kohn, Claudia Koch, Ramona Ritzmann

First published in London, United Kingdom, 2018 by IntechOpen
IntechOpen is the global imprint of INTECHOPEN LIMITED, registered in England and Wales, registration number: 11086078, The Shard, 25th floor, 32 London Bridge Street
London, SE19SG – United Kingdom
Printed in Croatia

British Library Cataloguing-in-Publication Data
A catalogue record for this book is available from the British Library

Additional hard copies can be obtained from orders@intechopen.com

Into Space - A Journey of How Humans Adapt and Live in Microgravity, Edited by Thais Russomano and Lucas Rehnberg

p. cm.

Print ISBN 978-1-78923-220-2
Online ISBN 978-1-78923-221-9

For EU product safety concerns:
IN TECH d.o.o., Prolaz Marije Krucifikse Kozulić 3, 51000 Rijeka, Croatia,
info@intechopen.com or visit our website at intechopen.com.

Meet the editors

Professor Thais Russomano, MD, MSc, PhD, has 30 years of experience in Aerospace Medicine, Space Physiology, Telemedicine and eHealth. She is a Brazilian MD, with an MSc degree in Aerospace Medicine (USA), a PhD degree in Space Physiology from King's College London (KCL) and a 2-year experience as a researcher at DLR, Germany. She founded the MicroG research centre, PUCRS, in Brazil, which she coordinated for 18 years. Thais is an international guest lecturer, CEO of InnovaSpace (UK) and CMO of the International Space Medicine Consortium (USA). She is a senior lecturer at CHAPS, School of Basic and Medical Biosciences, Faculty of Life Sciences and Medicine, KCL, contributing to Aviation and Space Courses. Thais holds seven patents in Space Sciences and Biomedical Engineering and is a member of academies, associations and societies, contributing to several of their scientific committees. She is a published author of books, book chapter and numerous scientific papers related to her areas of expertise.

Dr. Lucas Rehnberg, MBBS, BSc, MSc, graduated from Barts and the London School of Medicine and Dentistry in 2015 and is currently a clinical fellow in Intensive Care Medicine, University Hospital Southampton, UK. Lucas also has an undergraduate degree in Biomedical Sciences and a master's degree in Human and Applied Physiology from Kings College London (2009). Since 2011, Lucas has been a visiting research associate at the Centre of Human and Applied Physiological Sciences (CHAPS), KCL, and a consultant in Space Medicine at the MicroG research centre, PUCRS, Brazil, since 2012. More recently, he has become involved with InnovaSpace in the role of SGen Hub Coordinator. Lucas has over a decade of experience in space life sciences, from conducting research to publishing papers in the area of space medicine, as well as international presentations and teaching.

Contents

Preface IX

Chapter 1 **Reimagining Icarus: Ethics, Law and Policy Considerations for Commercial Human Spaceflight 1**
Sara M. Langston

Chapter 2 **Basic Methodology for Space Ethics 17**
Tony Milligan

Chapter 3 **From the Individual to the Cultural Space Group 31**
Carole Tafforin

Chapter 4 **Acute and Chronic Effects of Hypobaric Exposure upon the Brain 45**
Paul Sherman and John Sladky

Chapter 5 **Spaceflight Induced Changes in the Central Nervous System 71**
Alex P. Michael

Chapter 6 **The Effect of Gravity on the Nervous System 87**
Florian P.M. Kohn, Claudia Koch and Ramona Ritzmann

Chapter 7 **Spaceflight: Immune Effects and Nutritional Countermeasure 103**
Anil D Kulkarni, Marie-Francoise Doursout, Asmita Kulkarni, Alamelu Sundaresan, Takehito Miura, Koji Wakame and Hajime Fujii

Chapter 8 **Countermeasure Development for Lumbopelvic Deconditioning in Space 121**
Andrew Winnard, Dorothee Debuse and Nick Caplan

Chapter 9 **Tumor Cells in Microgravity 139**
Jun Chen

Chapter 10 **Plants in Space 153**
Bratislav Stankovic

Chapter 11 **Approaches to Assess the Suitability of Zooplankton for Bioregenerative Life Support Systems 171**
Miriam Knie, Bernard Wolfschoon Ribeiro, Jessica Fischer, Burkhard Schmitz, Kay Van Damme, Ruth Hemmersbach, Donat-P. Häder and Christian Laforsch

Chapter 12 **Are We Alone? The Search for Life on Mars and Other Planetary Bodies 209**
Stephanie A. Smith, Andrzej Paszczynski and Susan E. Childers

Chapter 13 **Exploring the Stratosphere: What We Missed by Shooting for the Moon 229**
Laura Galdamez

Chapter 14 **The Mortality of Space Explorers 253**
Robert J. Reynolds and Steven M. Day

Preface

All living organisms on Earth are subject to the influence of gravity. This gravitational force has dictated their anatomy and physiology over millions of years, including that of human beings. Therefore, if humans were "dolls" made of modelling clay, the mould would be gravity.

Removing the effect of Earth's gravitational force alters all organic functioning. One of the first responses to space flight is the inability of the vestibular system to cope with the absence of gravity and, consequently, body spatial orientation. A disease called space motion sickness, which is characterized by impairment of performance, nausea, vomiting and a diffuse malaise, occurs in astronauts and lasts for the first 72 hours of a space mission. In the weightless environment of space, a headward shift of blood and body fluids progressively happens over the course of weeks and deeply affects the cardiopulmonary and neuro-ophthalmic systems, remodelling heart and lung function as well as vascular pressures in the brain and cardiovascular system. When astronauts return to Earth, the cardiovascular system, already adapted to microgravity, causes concern by succumbing to orthostatic intolerance and decreasing exercise aerobic capacity. In space, humans develop anaemia and the immune system is depressed, showing that weightlessness affects human physiology down to a cellular, even molecular, level.

In 3 to 5 days, the body begins to adapt to the space environment, and within 6 weeks, it starts to work in accordance with its new setting. Some systems, however, do not adapt favourably. The mechanical unloading of muscles and bones in space affects the musculoskeletal system, causing significant atrophy, especially in the anti-gravitational groups of muscles in the back and legs. It is believed that the normal process of bone formation and resorption is disturbed, resulting in loss of bone mass, primarily in the lower body. The decrease in bone mass is a huge concern, as the loss can range from 1% to 2% of total bone mass per month, impacting on astronauts' health, not only in space but also particularly after their return to Earth.

At 27,000 km/h, a spacecraft completes an orbit around the Earth every 90 minutes, which affects the circadian rhythm of the astronaut. Seeing the world through a little window inside the confines of a spacecraft obviously has implications for human psychology. Emotional issues related to isolation and confinement can also have a significant impact on crewmembers, a situation that is likely to be exacerbated by the increased duration and distance from Earth that an interplanetary trip to Mars will bring. The concept of our finitude as a universal species becomes evident.

We benefit on Earth from the protection provided by the atmosphere and magnetic field of our planet, which shields us from much of the radiation present in outer space. However, when humans leave this cocoon, they become exposed to space radiation that can have deleterious effects on numerous organs and systems, even putting the life of an astronaut in danger. The International Space Station sits just within the protective magnetic field in low Earth orbit, but even so, astronauts are exposed to levels of radiation that are ten times higher than on Earth. Any future trip to Mars will be subject to even greater levels of radiation exposure, and this is a serious problem that is still to be solved.

This book presents and discusses some of these physiological, anatomical and cellular changes that happen to astronauts during short- or long-term space missions, which have been called space deconditioning. These body alterations alone or in combination, acute or chronic, can lead to in-flight undesirable health and operational consequences, especially if an emergency situation were to occur, or in the post-flight phase of a mission, when astronauts return to Earth, or even the moons or Mars, and have to readapt to a gravitational field. Other important aspects are also considered, such as the ethical and cultural issues related to manned spaceflights and the development of techniques and models related to the creation of safe extraterrestrial environments where human beings can live and work.

The book *Into Space: A Journey of How Humans Adapt and Live in Microgravity* will give an overview of the complexity of manned space flights, showing how interdisciplinary this subject is and discussing the challenges that space physiologists, physicians and scientists must face as humans seek to conquer the final frontier of space.

Professor Thais Russomano, MD, MSc, PhD
King's College London, UK

Dr. Lucas Rehnberg, MBBS, BSc, MSc
University Hospital Southampton, UK

Chapter 1

Reimagining Icarus: Ethics, Law and Policy Considerations for Commercial Human Spaceflight

Sara M. Langston

Additional information is available at the end of the chapter

http://dx.doi.org/10.5772/intechopen.74716

Abstract

Commercial human spaceflight presents an area for engaging novel human activity and objectives, to include space exploration, entertainment, transportation and extraterrestrial resource acquisition. The inherent dangers and lack of scientific and medical certainty involved however raise interrelated questions of ethics, bioethics, law and public policy. This is particularly the case with spaceflight participant (SFP) screening, selection, and commercial human spaceflight activities where regulations are currently silent or lacking. In the absence of established law, ethics can play an important role by informing industry standards, policies and best practices. Understanding the fundamental ethical values at stake in the application of new technologies and societal opportunities therefore is a significant step in establishing a practical, moral and sustainable framework for human expansion into space. As the frequency and reliability of private human spaceflight activities advances, spaceflight is likely to take on the legal and ethical vestiges of common carriers, with distinct passenger rights and higher standards of care attributed to the launch operator as a common carrier. This chapter raises some of the complex issues and challenges that face the private spaceflight industry and that merit collaborative discussion across disciplines and the global space transportation community going forward.

Keywords: ethics, bioethics, space medicine, space law, spaceflight

1. Introduction

Commercial human spaceflight presents a novel area for diverse human activity, whether conducted for exploration, entertainment, transportation or extraterrestrial resource acquisition. The inherent dangers and lack of scientific and medical certainty involved raise interrelated questions concerning ethics, bioethics and public policy, particularly in regard to spaceflight participant (SFP) screening, selection and commercial human spaceflight activities and practices. *Ethics* is the study of how things should be. It is through moral reasoning that society discerns ideal human values and what constitutes right action for governments, communities and individuals [1]. Ethical discussions concerning novel industry activities are particularly relevant to commercial human spaceflight because in developing these new technologies and spheres of human activity, existing social moral values are evaluated, developed and enacted with far reaching implications and consequences.

Ethical values were in fact acknowledged in discussions at the beginning of the Space Age, even serving a role in political agendas, and incorporated into the international legal framework. The United Nations' 1963 Declaration of Legal Principles Governing the Activities of States in the Exploration and Use of Outer Space, along with the subsequent space treaties, are principle-based instruments and emerged out of the geopolitical climate of the Cold War (from 1950s–1970s). The utopian principles espoused by the international community at this time, most importantly: that outer space is to be used for peaceful purposes only and for the benefit and in the interest of all countries, were codified in the 1967 Outer Space Treaty, which remains the foundational treaty pertaining to space activities today [2]. Overall, the legal principles espoused in these international instruments can also be seen to apply traditional ethical principles of *beneficence* (duty to do good) and *non-maleficence* (duty to do no harm) in the space environment, on celestial bodies, and to other actors in space, while establishing a practical moral framework for action.

Contemporary social values on human spaceflight and exploration also impact the developing customs and social norms on new commercial practices, actual uses of space and its resources, and public human spaceflight development. Here, a *practical ethics* approach is useful in applying ideals or ethical principles to practical engagement in life through codes of conduct and protocols, usually specific to context, discipline and industry. Doctors, lawyers, and engineers are among the professional disciplines that have long established codes of conduct. Policy and law, on the other hand, generally apply *normative ethics*, establishing moral frameworks through applicable rules, theories, principles and guidelines. *Bioethics* specifically relates to practical ethics and issues arising in medicine and biology. Bioethical issues include healthcare, patient rights, individual rights over one's own body, medical malpractice, and the use of human subjects in studies and scientific experimentation. NASA, for instance, frequently confronts bioethical concerns in its human spaceflight program, which requires evaluating and balancing the actual and prospective risk of harm to astronauts with the prospective benefits and mission objectives.

In practice, ethics, policy and law, are not always distinguishable disciplines. Law and ethics are particularly intertwined in biomedicine. The public healthcare system and medical sector moreover are governed by a combination of laws and policies on a range of procedural and substantive bioethical issues [3]. New technologies and surgical options (e.g. telemedicine) and progressively increasing human activity in extreme environments that test the limits of human endurance (e.g. space), are pushing back the boundaries of established professional, legal and community values on acceptable risk and scientific uncertainty. As a result, the new commercial space transportation industry (NewSpace) requires new informed legal and ethical approaches to human spaceflight and emerging aerospace activities.

The expansion of commercial space capabilities and actors, in particular, require a meaningful understanding of the values and implications of these activities to society in order to adopt appropriate standards and guidelines. This chapter raises primary ethical issues in commercial human spaceflight along with relevant law and policy concerns. Identifiable categories of ethics and spheres of practice in this regard include:

- Medical ethics
- Scientific research and ethics
- Ethical decision-making frameworks
- Astronaut ethics
- Technology and ethics
- Environment and ethics

2. Medical ethical implications

Space medicine and ethics questions have existed since human spaceflight programs began. The distinguishing factor with commercial spaceflight lies in the accountability and regulatory oversight that exists for government space programs and civil space agencies unlike private space research and human spaceflight enterprises. NewSpace entities and related industries bear the burden of establishing practical ethical policies, procedures and professional codes of conduct for private individuals. Apart from the professional responsibility of participating disciplinary experts (doctors, lawyers, scientists and engineers etc.), ethics-based principles and procedures are not currently identified in the commercial human spaceflight regulatory or industry decision-making frameworks.

While a notable distinction may be drawn between law and ethics the two disciplines often work hand in hand. To be clear, law is focused on external acts and consequences rather than the internal moral intentions of a person [3]. After all a divergent intention alone does not constitute an unlawful act, although intention alone can form the basis for a moral framework. In law, the action or practical steps taken in furtherance of the intention form the basis for a judicial determination of lawful/unlawful activity. In the absence of a regulatory regime governing the human aspects of commercial spaceflight activities—particularly where questions of risk, uncertainty and persons are concerned—it becomes even more crucial to establish a practical ethical framework for operation as spaceflight activities fundamentally invoke a myriad of moral issues in human interrelationships.

2.1. Medical forum shopping

Medical forum shopping is a foreseeable ethical and legal concern where regulatory standardization is lacking or is inconsistent between jurisdictions. No regulatory requirements currently exist for SFP medical screening and health selection criteria. The current regulatory perspective is that individuals have the right to participate in novel spaceflight activities at their own risk. As a result, screening, selection and training discrepancies may arise between states, launch operators and certifying medical practitioners. This lack of standardized requirements creates a potential lacuna for SFPs with specific health profiles (e.g. pacemaker recipient) to shop across states for medical certification or waive into a 'go' status in order to participate in one or more desirable human spaceflight experiences. This could lead to increased risk of harm for the individual (in-flight/post-flight), spaceflight and crew, as well as uninvolved third parties. Analogous cases of forum shopping for licensing and/or obtaining legal and jurisdictional benefits have occurred in other areas, for example, in transportation and private international law.

2.2. Physician's professional responsibility

The United States is currently the only state with national space legislation and regulations directly pertaining to commercial (private) spaceflight operators, crew and ordinary civilians as spaceflight participants. Since the flight requirements, actors and requirements are distinguishable from state-sponsored astronaut/ cosmonaut corps this chapter will focus on the pertinent commercial space regulations as relates to private and commercial human spaceflight in the U.S. In this regard, the U.S. Code of Federal Regulations (CFR), Title 14, Parts 67 and 460.5 specifically requires spaceflight crewmembers to be cleared by an Aviation Medical Examiner (AME), as required for regular air crews. The regulations are silent, however, as to SFPs. Lawmakers and industry favor a regime of informed consent and liability waivers over governance for SFPs due to the novelty and numerous uncertainties of private spaceflight. Regardless, this does not waive a physician's professional and ethical duty towards the SFP.

In the absence of regulation, the Federal Aviation Administration Office of Commercial Transportation's (FAA-AST) published Recommended Practices for Human Space Flight Occupant Safety (2014) (Recommended Practices), providing voluntary guidance for limited duration suborbital and orbital spaceflights. The Recommended Practices suggest that SFPs seek consultation from a medical professional with "appropriate aerospace knowledge and experience" [4]. No clarification or definition for this 'knowledge' or 'experience' is provided. As a result, the interpretation and implementation of this requirement fall under the purview of the medical examiner and commercial operator.

Here professional ethics would suggest that a general practitioner performing a pre-flight evaluation on a SFP collaborate with a qualified AME, whether this is required by the launch operator or not. Relevant aerospace medical knowledge is essential to provide necessary preventative and post-flight treatments tailored to the particular SFP, particularly those with any preexisting conditions, and in relation to the specific flight activity and duration. For instance, with pharmaceutical treatments, drugs metabolize differently in microgravity environments, and a person may experience related physiological events even after returning to Earth [5]. This signifies greater implications of administering over-the-counter and/or prescription drugs before, during and after spaceflight. Mitigating and preventative treatments therefore need to consider all stages of spaceflight.

In addition, health, safety and medicine are closely aligned in spaceflight, as it is in aviation. Adverse effects of over the counter and prescription drugs may impair a SFP's ability to perform required safety critical functions as well as his/her reaction time in conducting these functions or in the event of an emergency [6]. If the SFP were to become impaired in flight due to the effects of a medication or an exacerbated preexisting condition this could jeopardize the safety of the individual, as well as that of the crew, mission and potentially other third parties. The expertise of an aerospace medical practitioner and access to professional aerospace medical research networks (e.g. Civil Aerospace Medical Institute) are thus invaluable resources for certifying physicians in determining individual risks for health events and appropriate courses of medical intervention for commercial SFPs.

2.3. Conflicts of interest

It has long been acknowledged that physicians owe a duty of care to their patients—this entails both a professional ethical and legal obligation. Since health and fitness screenings of SFPs are left largely to the approval of the commercial launch operator under US law, it is essential that a physician is clear of potential conflicts of interest. For example, physicians and medical personnel contracted by commercial launch operators to screen SFPs for spaceflight could present a conflict of interest if, for instance, they receive any incentive or benefit (from anyone) for doing so. In any case, physicians are morally and legally obligated to maintain their professional integrity, standard of care and patient confidentiality regardless of third party (e.g. commercial operator, research entities) incentives or desires to clear customers for spaceflight activities, or reliance on the SFP's consent to blanketly waive liability for his/her own health and safety.

In a conflict of confidentiality situation where doctor-patient confidentiality may be justifiably limited—where the physician owes a duty to both an organization (such as government agencies, astronaut corps (NASA), military, schools) and an individual—the individual-patient must be clearly informed of the limitations and scope of doctor-patient confidentiality [7]. This may also include scenarios involving remote medical assistance and telemedicine. In addition, where medical screening, profiling and the exchange of medical data are conducted for studying and augmenting medical knowledge, the situation requires delineated procedures and communications to the SFP. Ultimately, additional discussions between industry and government are still necessary to further clarify the ethical and legal parameters for conflicts of issues, as these scenarios will inevitably arise in the near future and the current personal liability waiver regime may not suffice in such instances.

2.4. Standardization

Another issue that has yet to be addressed is the lack of standardization in space medicine.

Standardization is a traditional tool in science and technology fields that seeks to reduce risk and enhance reliable quality results by implementing technical and operational control mechanisms across people, time and space [8]. Space agencies use medical standardization to some extent to determine astronaut flight readiness, establish baselines, guide countermeasures, assessments and any necessary post-flight rehabilitation to return the astronaut to preflight health status.

However, the global medical community acknowledges distinctions in medical practices across cultures. This is evident, for instance, in traditional Chinese medicine (TCM) approaches and Western medicine. Among national human spaceflight programs and agencies recognizable cultural differences exist in medical philosophies and approaches, diagnostic equipment and treatment in orbital human spaceflight [9]. Differences also exist internationally in the medical recognition of psychological disorders. In the case of the International Space Station (ISS), an Integrated Medical Group, comprised of representatives and experts from partner agencies, cooperate and compromise in leading the multilateral implementation plan for astronaut-patient health and space medical research [10]. As a treaty (agreement) based endeavor, differences in medical practices are limited to the cultures of partner nations. Commercial spaceflight, on the other hand, presents a broader challenge, especially as a prospective means of private and global public transportation.

Without standardized medicine and approaches for commercial spaceflight medical events there is potential opportunity for multilateral conflicts of priority, approach, and for SFPs to 'forum shop' physicians, spaceports or jurisdictions that suit their interests.

Some practical ethical questions and issues that remain to be clearly addressed for commercial space transportation include:

- Whether the *experimental* nature of spaceflight raises a higher duty of care or ethical concern among the private persons and entities involved (physicians, operators and SFPs)?
- What are the ethical implications for compiling, analyzing, and sharing medical data on commercial SFPs?
- Whether an appropriate medical consultation forum should be established (similar to aviation) to provide physicians, prospective and actual crew and SFPs etc. with a place to lodge substantive questions and concerns on issues of health and the physiological implications of spaceflight?
- What are the ethical parameters for contractual 'informed consent' and selection criteria when dealing with an open manifest spanning diverse health profiles and various physical differences (e.g. someone who is deaf versus a wheelchair bound individual), foreign language comprehension, and legal capacity?
- Whether insurance companies' interest in passenger screening, selection and clearance poses a potential medical and ethical conflict of interest?
- What protocols, safeguards and operations need to be established for telemedicine to guarantee medical standards, ethics and patient confidentiality are maintained where members of the public are engaged in spaceflight?

Obviously, this is a not an exclusive list. Additional professional and practical ethics questions for physicians and passengers alike will inevitably arise as spaceflight emerges from its current experimental phase and becomes a norm of transportation. When spaceflight qualifies as common carriage and flight access is granted to a wider sector of the public, known health and safety risks may require standard physician ethical responsibilities to patients as well as to society in determining and certifying fitness-to-fly. Ethical and medical concerns for conditions that may present safety concerns for spaceflight, may be analogous to cases of epilepsy and driving today. When does a physician, society and/or a private launch operator have a right to discriminate and restrict access to prospective spaceflight participants? These issues have yet to be discussed.

2.5. Compliance and enforcement

Professional ethical compliance is largely self-regulated within the medical profession and national ethics committees. While some conflicts of interest may appear straightforward other ethical issues may arise in the application of commercial human spaceflight—these scenarios will likely be fact, mission and personnel dependent. Legal consequences, such as malpractice suits, may also follow where a breach of duty occurs. For instance, where negligence occurs in assessing, screening or treating an SFP with a precarious medical condition, or failure to detect or divulge potential critical medical information to the SFP. In the event of an incident, contributory negligence on the part of the SFP or crewmember to honestly and promptly report medical events may also be raised in accordance with the appropriate jurisdiction and applicable law.

3. Commercial space stations

The FAA's regulatory authority over commercial spaceflight is limited to launch and reentry activities. The United States Code (USC), Title 51, Section 50905(c) further restricts the scope of regulatory authority over design and operations to where health and safety are concerned. Consequently, orbiting space stations (e.g. space habitats/ space hotels), long distance and long duration spacecraft (space cruiseliners), and extraterrestrial stations on celestial bodies are not regulated at the present time. The FAA voluntary guidance on human spaceflight requirements is likewise limited in scope to a brief human presence in space. Neither does the scope of the CSLAA extend to orbiting activities. Nonetheless, the FAA may recognize a vehicle as space capable and has related high altitude aerial tourism as a 'space activity' subject to national space law for purposes of determining and implementing an applicable regulatory regime.[1]

In the absence of a comprehensive space regulatory regime and the acknowledged fact that spaceflight is an inherently dangerous activity, the ethical obligations for commercial launch operations and space station operators are heightened. Ethical practices and public policy would suggest that operators adopt similar standards of care as that of other analogous public transportation providers—airlines and ocean cruise liners, for instance—even when not legally mandated. The higher the ethical standards adopted in practice the less likely a commercial space operator is to be found negligent in the event of an accident. In law, negligence constitutes a breach of duty (which implies an ethical and legal responsibility), so applying the

[1]For instance, the FAA acknowledged Paragon's World View high altitude balloon capsule for its space worthy technology and design, even if its function is limited to stratospheric flights. The US government, however, does not currently *certify* commercial space vehicles as safe for public transport.

highest standard of care towards SFPs as well as crewmembers from the outset, even if more costly, significantly serves the commercial operators best interests and greater good. For instance, a demonstration of a high standard of care may include: instituting clear informed consent procedures throughout the space flight and in-orbit residence; ensuring up-to-date emergency training, medical preparedness, technologies and accessibility; in-flight biometrics and health monitoring; engaging relevant expertise and experts on substantive questions and concerns; establishing reliable communications, event reporting practices; and customized personnel (passenger) assistance.

4. Ethics and public policy

4.1. Minors as SFPs

The US Commercial Space Launch Amendments Act of 2004 (CSLAA) and regulations are silent on the issue of minors as SFPs and a full public debate on ethics and public policy have yet to follow on this question. As a result the legal, public policy and ethical dimensions of this issue are convoluted. In 2006 the FAA responded to public commentators taking the position: "the FAA does not consider a person under the age of 18 someone who can provide informed consent" and "[g]iven the risks involved, parental consent may not substitute for the minor's inability to be informed" [11] The rationale for this age demarcation is that "[s]ocietally, the United States has acknowledged that it is reasonable to place restrictions on individuals under the age of 18, including restrictions on their ability to legally consent" and "[w]hile some states classify a person as a minor until the age of 21, in many states the age of majority is 18."

Nothing has been addressed in analogy, however, as to a minor's legal capacity to enter into enforceable contracts in daily society (e.g. ticket purchases) or to engage in other risky and extreme sports and surgeries, where minor consent is recognized or acceptably acquired from a consenting parent or guardian. The current regulatory exclusion of minors from spaceflight therefore suggest a distinct periphery of risk that society has not yet deliberately addressed.

4.2. Human test subjects

Research on humans in space may feature in several ways, as an active participant consenting to a medical study or test; or indirectly, as a SFP or crewmember whose medical screening and health data is collected and analyzed for studying the effects of spaceflight on healthy individuals, those with particular pathologies, or for comparative demographic purposes. Possible scenarios include, for instance:

- A commercial SFP may voluntarily consent to become a human test subject for pharmaceutical research or scientific knowledge on human physiology, drug metabolization, and spaceflight/ microgravity environments. In point of fact, NASA astronauts routinely serve as test subjects in space for medical and scientific purposes, and private astronauts to the International Space Station have been sponsored by national space agencies to conduct medical research and tests on themselves while in space. Consequently, it is foreseeable that these activities will occur with commercial SFPs.

- Given the novelty of the technology and scope of uncertainty in regard to spaceflight on human physiology, psychology and sociology, the pioneers in commercial spaceflight, whether deemed healthy crewmembers or SFPs, are in many ways participatory subjects of an ongoing experiment.
- To a greater extent, in the absence of relevant medical knowledge, and regulation, on more vulnerable health categories and age groups, the question on whether minors, geriatrics, physically impaired individuals, those with special conditions or overall deemed 'less fit-to-fly' individuals can engage in any human spaceflight activity and training remains open.

Collecting and analyzing space travelers' medical data is the only way to augment the human database of knowledge and lessen uncertainty. Moral and legal issues of collecting medical data from pre-flight and post-flight screenings and assessments from adult participants may be satisfied by applying ethically established protocols, procedures, and obtaining valid informed consent from each participant. Commercial companies and individuals interested in leading human studies and trials in space should follow the proper channels for conducting human research studies. This usually means obtaining authorization from the appropriate university or Institution Review Boards, independent ethics committees, national ethics committees etc. Following the example of governmental space agencies (e.g. NASA, ESA) commercial research investigators in space should also comply with the research principles of the World Medical Association incorporated in the Declaration of Helsinki, and relevant national regulations and guidelines (e.g. USFDA) as if the studies were conducted on Earth.

When the industry matures to the point of enabling safe routine flights and lawmakers allow for minors to participate in spaceflight activities as SFPs, these ethical concerns and duties will likely be heightened. As with adults, at some point healthy children will become space pioneers if the medical and space communities are to obtain any significant medical data on the physiological effects of spaceflight on this demographic. The ethical quandaries surrounding risk, especially concerning potentially irreversible damage to pediatric health, may be curtailed in the near future by technological advances in launch/reentry, safety, vehicle and mission design and architecture.

5. Ethics and culture

5.1. Cross-cultural ethics

In general, ethics is the study of what should to be done. What *should* be done does not equate with what *can* be done. In theory, philosophical arguments hold that ethics are universalizable—a valid ethical principle that applies to one should apply to all. However, in life value principles, rules and practices frequently diverge across cultural groups [12]. This is particularly true in the field of bioethics and where human initiatives carry risk and great uncertainty, such as exploration and invention. The more an action, implementation measure or enterprise deviates from the certainty of the status quo, the higher the unprecedented risk. National policies on the value of life, autonomy, community, risk and human space flight/exploration reflect contemporary community values on these issues. Some distinctions indicate comparative perspectives between western and non-western cultures and values. Yet distinctions can also be found within similar cultural regions. For instance, the United Kingdom's national space policy declined national participation in human space exploration programs until 2009, even though it is a founding member of the European Space Agency, which established and still operates a European Astronaut Corps since 1978.

Looking forward, decision-makers and international space forums will benefit from engaging in transcultural dialog and value reciprocity discussions on human risk and commercial ventures in space. Cross-cultural ethics, based on mid-level values, such as the framework recommended by the Institute of Medicine for NASA's long duration spaceflight in 2014, [13] can provide a starting point and balanced approach in establishing best industry practices, as well as future law, for space access and transportation. The fundamental ethics outlined in the IOM's ethical decision-making framework for long duration and exploration spaceflight may, for instance, be applied to private space ventures as follows: an ethical duty to

1. Avoid harm—by preventing harm, exercising caution, and removing or mitigating harms that occur.
2. Uphold beneficence—using spaceflight to benefit society, this includes transportation, entertainment, scientific research and exploration.
3. Seek and maintain a favorable and acceptable balance of risk of harm and potential for benefit in spaceflight operations.
4. Respect for individual crew and participants' autonomy – especially concerning voluntary decision-making.
5. Ensure fairness—in company procedures and operations; and
6. Fidelity—recognize individual contributions of crew and SFPs as appropriate, and honor societal obligations to employees.

5.2. Risk and culture

Risk is a social construction, a determination subjective to culture, context, perception and communication of an identifiable or potential hazard versus opportunity [14]. This understanding is particularly relevant to the nature, risks and perceived benefits inherent to human spaceflight. The United States and Russia, for instance, present great risk-taking cultures and histories, particularly in regard to rocket launches and spaceflight activities. On a practical level both these nations also have greater technological capabilities, government sponsorship, and the national resources to take on big risks, such as transporting humans into Earth orbit and beyond. On the cultural level individual independence and autonomy are deemed fundamental values in American culture, and this is particularly evident in the national space culture. Other countries may demonstrate a more risk adverse culture when it comes to

http://dx.doi.org/10.5772/intechopen.74716

evaluating the cost/benefit of spaceflight and may decline to participate in human spaceflight missions. While legally it may be permissible for individuals to engage voluntarily in extremely hazardous activities in countries and cultures that are neutral or favorable to voluntary risk, there is no clear comparative ethical evaluation on moral standards directly addressing the issue in the public debate on spaceflight.

One of the precepts of an ethical principle is its universalizability. If a principle should apply to one person, ethics generally dictates that it should apply to all. Fairness and equity for instance are principles applied in ethics, public policy and law, and yet even here societal value determinations can diverge between cultures. Meaning, a person's right to autonomous risk assessment, decision-making and action is not universally held. A particular gap can be seen between existing spacefaring nations and non-spacefaring nations. Which implies that voluntary risk is not purely a matter of individual autonomy but also coincides with a vested interest of society in maintaining a collective value. For instance, a person's right to autonomy or to engage in a hazardous activity—one that cannot be made safe no matter how much care is taken—can conflict with a society's right to not be harmed as a consequence of that activity, or to not have to expend state resources and efforts in rescue missions to assist those few voluntarily engaged in the hazardous activity. This is where public policy comes in and requires a practical balancing approach to leverage these two distinct ethical rights.

5.3. Indulging the wealthy?

With regard to ensuring public health and ethics in public policy, questions have been raised that distinguish the choices and consequences of the wealthy over the non-wealthy. For instance, distinguishing someone who can purchase the $250,000 plus spaceflight ticket, to one that wins a public competition to fly to space. But is there a moral distinction? Spaceflight advocates argue that people should be allowed to take risks that they voluntarily choose to participate in, and can pay for (this may include event tickets, mandatory insurance, and applicable fees). A prime example can be seen in extreme sports, such as undertaking to climb Mount Everest—a high-risk activity that does in fact claim lives every year, and that routinely calls on state resources for emergency response. Analogously, spaceflight is not only an inherently dangerous activity like climbing the world's highest mountain, but it is the least safe means of transportation. Risk mitigation at this point is limited – unlike an emergency row passenger on a commercial aircraft, who can decline the heightened risk/ duties and request reassigned seating, today's spaceflight technologies do not currently provide SFPs with any alternative safer options for participation in spaceflight.

From a legal perspective there is no significant distinction between one who voluntarily engages in an extreme sport or activity costing tens of thousands of dollars and the average person who wins or is gifted a ticket. Both are engaging in the activity and both persons must provide voluntary and informed consent to partake in the activity. From an ethical perspective, valid points of consideration call for an evaluation of the fundamental issue at play: should we indulge the whims of the wealthy because they can choose to pay for an experience? If so, how far does this autonomy extend? And how do principles of ethics and justice apply to commercial spaceflight? These are questions that merit acknowledgement and discussion looking forward.

5.4. Paternalism

Paternalism is the philosophical concept that a state can interfere with an individual's right of autonomy if it is in the individual's own best interest (this is a distinct notion from interference for the community or another's sake). There are varying degrees of paternalism that allow for less or more abrogation of personal autonomy (analogously seen, for example, with bioethical questions involving body modification, extreme surgeries, drug use etc.). The overall question that deserves to be acknowledged here for spaceflight is to what extent can the average reasonable person engage in an extreme and dangerous activity? And what are the ethical parameters for state interference in regard to this autonomy?

These ethics questions deserve acknowledgement because practically the conclusion may differ depending on the subjective country, culture, social norms and values, the type of legal system (e.g. common law vs. civil law system) and existing legal codes. Raising and evaluating these ethical and bioethical issues, serves to maintain the overarching human good—the fundamental reason for morality—through respect for human autonomy, dignity and life [12]. Consequently, the resulting issue presented here is how to identify, define and approach an optimal bioethical framework that can and should be applied to the commercial space transportation industry as a whole.

6. Astronaut ethics

The 1967 Outer Space Treaty applies only one requirement to individual spacefarers. Article V stipulates that astronauts in space "shall render all possible assistance" to other astronauts in space and on celestial bodies. This is the only personal duty required of astronauts under the international space law regime, and stems from traditional maritime principles and law of the sea. However, no uniform definition of 'astronaut' currently exists and US legislation governing SFPs is silent on this specific obligation. Thus it is unclear whether commercial launch operators and SFPs fall under this treaty provision.

The significance of distinguishing SFPs from astronauts under the treaty directly relates to implications of SFP health, safety and law. A legal duty to render assistance would exclude SFPs who are unwilling or unable to do so. For instance, Stephen Hawking would be unable to render assistance to another person on a suborbital flight even if cleared by a physician and launch operator. This also raises additional liability issues for the SFPs and the launch operator as the personal liability waiver is not generally concluded between passengers. Any commercial astronaut with limited fitness and related restrictions will likely fail to comply with this international obligation. What then?

The underlying ethical question raised here is whether a moral duty to render possible assistance to other persons in space exists, regardless of whether one is a SFP or crewmember. This is also a question of public policy. If the answer is yes, it follows whether the 'Good Samaritan' Principle, as applied in the US for instance, should also be extended to commercial human spaceflight and in-space activities to promote and protect prospective rescuers. The practical ethics and legal implications of this question have yet to be addressed by the greater space community.

7. Technology and ethics

The objectives of human health and safety are fundamental values where technology is concerned. Given the wide scope of commercial space activities proposed and human spaceflight experience gained to date, future health and medical events are a high possibility. This leads to two ethical implications for space technology and ethics pertaining to human spaceflight:

- *Duty to report safety concerns*—One of the acknowledged lessons of the 1986 Challenger accident is the reminder of professional responsibility and ethics of engineers and operation managers to voice concerns in regard to launch activities. Launch operators should also institute policy and procedures for reporting and evaluating concerns and issues, whether related to operational, personnel or technical issues, from employees, SFPs and any concerned third parties.
- *Engineering, mission design and ethics*—vehicle/station design should incorporate necessary structures to ensure human health, hygiene and safety, even when not mandated by law. This includes, medical equipment and appropriate facilities. Priority conflicts can arise, however, when space, weight, size, mission objective and fuel are limited. For instance, when determining which medical equipment and supplies should be included. Operators, will inevitably face competing interests, like NASA, in vehicle and mission design and will have to arrive at ethical determinations on these critical issues. A baseline approach is to maintain equal or equivalent health and safety requirements by analogous transportation systems, such as maritime and aviation. This may include: medical (trained) personnel, medications in various forms, and essential medical equipment.

8. Environment, health and ethics

Other inherent implications of spaceflight may affect environmental health and safety. The projected increase of routine commercial spaceflight activities alone will inevitably impact Earth's environment, atmosphere, and space. For instance, the type of vehicle, fuel, ejected debris, and biological contamination may all effect the Earth and space environment. Human space settlements and activities on celestial bodies are likely to raise additional and convoluted ethical and practical questions with regard to the environment and human health.

The Outer Space Treaty contains only one provision tangentially applicable to environmental health. Article IX provides for state measures to be taken to avoid forward and backward contamination of Earth from space and vice versa. This is the only provision in the foundational space law instrument that deals with the environment. The scientific advisory body to the United Nations, Committee on Space Research (COSPAR), has since issued international guidelines for controlling biological contamination of celestial biospheres. While these constitute voluntary guidelines, national space agencies like NASA have incorporated COSPAR guidelines in their planetary protection policies.

Other environmental principles may also be applicable to space through the specific application of international law to outer space. International environmental law is based on ethical principles such as the precautionary principle. The precautionary principle is a risk mitigation strategy that calls for decision makers or regulators to act preemptively to ensure that a harm does not occur rather than wait for scientific certainty on the actual or potential risks of harm from conducting a specific action or series of activities [15]. The precautionary principle can be applied to anything such as preventative exclusion of SFPs for a medical condition to restricting space mining activities in particular areas or on specific celestial bodies.

Another environmental health and ethics issue includes death and funerary rituals. Since 1997 private companies, such as Celestis, have been providing commercial space funeral rites, launching capsules with ashes of celebrities and customers into low orbital trajectory, to orbit the Earth a few times before burning up in the atmosphere [16]. Another company, Elyseum Space, is proposing a similar service to commence late this year, to include sending remains to the Moon and deep space, while providing value added services like Apple friendly app trackers [17]. The underlying idea is said to bring a poetic and celestial perspective to the human condition. These particular funerary activities in space are deemed to pose little to no risk and have not raised ethical concerns at the present time. Although, instituting extraterrestrial memorials on celestial terrains may trigger questions on planetary protection and environmental conservation.

Looking forward, however, it is not entirely clear what ethical implications and practical medical protocols will develop when a human (or perhaps even an animal companion or study subject) dies in space. Such events may result, for instance, from illness, accident, or SFPs planning the ultimate last adventure (e.g. terminally ill or elderly individuals who choose to end their days in space). Significantly, it may not always be practical or possible to return a deceased person to Earth. As commercial companies progressively seek to engage in long duration and distant missions these are inevitable questions that require societal forethought, moral respect and clear cross-cultural dialog.

9. Conclusion

Private and commercial human spaceflight present a myriad of bioethical, legal and policy implications for consideration. In many cases the ethical principles and legal/ policy positions on bioethical issues overlap. Understanding the fundamental ethical values at stake in the application of new technologies and societal opportunities therefore is a significant step in establishing a practical yet moral and sustainable framework for human expansion into space. Significantly, the inherent risks involved in spaceflight activities call for incorporating ethical risk management strategies and policies into industry standards and practices, even where not already instituted or mandated by law. As spaceflight progresses towards common carriage, spaceflight is likely to take on the legal and ethical vestiges of common carriers, with passenger rights and higher standards of care afforded to the launch operator as a common carrier. This chapter raises some of the complex issues and challenges that face the private spaceflight industry and that merit collaborative discussion across states, disciplines and the global space transportation community.

Author details

Sara M. Langston

Address all correspondence to: langstonsara@gmail.com

Senmurv Consulting LLC, Arlington, Virginia, USA

References

[1] Boston J, Bradstock A, Eng D, editors. Ethics and public policy. In: Public Policy: Why Ethics Matters. Canberra: ANU ePress; 2010

[2] Treaty on Principles Governing the Activities of States in the Exploration and Use of Outer Space, Including the Moon and Other Celestial Bodies, January 27, 1967, 18 UST 2410, 610 UNTS 205

[3] Van der Burg W. Law and bioethics. In: Kuhse H, Singer P, editors. A Companion to Bioethics. West Sussex: Blackwell Publishing Ltd; 2009. pp. 56-64

[4] FAA. Recommended Practices for Human Space Flight Occupant Safety, version 1.0, July 7, 2014

[5] Marsh M. Ethical and medical dilemmas of space tourism. Advances in Space Research. 2006;**37**:1823-1827

[6] FAA. Guide For Aviation Medical Examiners. https://www.faa.gov/about/office_org/headquarters_offices/avs/offices/aam/ame/guide/pharm/dni_dnf/ (last accessed March 30, 2016); FAA. Pilot Safety Brochures. http://www.faa.gov/pilots/safety/pilotsafetybrochures/media/meds_brochure.pdf (last visited March 30, 2016)

[7] Ball J, Evans C, editors. Safe Passage: Astronaut Care for Exploration Missions. Washington DC: National Academy of Sciences; 2001. p. 175

[8] Brunsson N, Jacobsson B. A World of Standards. New York: Oxford University Press; 2000. pp. 1-6

[9] Jones J. International Space Station Medical Operations, http://ntrs.nasa.gov/archive/nasa/casi.ntrs.nasa.gov/20080010877.pdf (last accessed March 30, 2016)

[10] International Space Station Medical Operations Requirement Documents (ISS MORD) (NASA SSP−50260, Revision B). Houston: NASA, 2003

[11] FAA. Human Space Flight requirements for crew and space flight participants. Final rule. Federal Register. 2006;**71**(241):75626

[12] Gbadesgesin S. Culture and Bioethics. In: Helga Kuhse H, Singer P, editors. A Companion to Bioethics. West Sussex: Blackwell Publishing Ltd; 2009. pp. 24-35

[13] IOM. Health Standards For Long Duration And Exploration Spaceflight: Ethics Principles, Responsibilities, and Decision Framework, Institute of Medicine. Washington, DC: The National Academies Press; 2014

[14] Dake K. Myths of nature: Culture and the social construction of risk. Journal of Social Issues. 1992;**48**(4):21-37

[15] Bodansky D. Scientific uncertainty and the precautionary principle. Environment. 1991; **33**(7):4-5

[16] Celestis. www.celestis.com (last accessed March 30, 2016)

[17] Elyseum Space. http://elysiumspace.com/ (last accessed March 30, 2016)

Chapter 2

Basic Methodology for Space Ethics

Tony Milligan

Additional information is available at the end of the chapter

http://dx.doi.org/10.5772/intechopen.75689

Abstract

The introduction sets out a standard concern that space ethics may be unduly constraining upon state and private sector activities in space. As a counter-picture, Section 2 sets up a distinction between 'standard space ethics' and 'special space ethics' which will allow us to explore ways in which space ethics enables as well as constrains. A case is then made in Section 3 for pragmatic constraints upon space ethics itself. Space ethics should be either 'policy apt' (able to directly shape space policy within a liberal democratic social context) or 'precursor apt' (able to contribute productively to broader, precursor discussions which may feed into policy apt deliberations). What makes any ethic satisfy either of these conditions will depend upon a range of factors. The ethic should have stability (dealt with in Section 3.1). It should not merely track transitory voting trends or the ebbs and flows of electoral politics. Secondly, it should have a high degree of political realizability (dealt with in Section 3.2). Finally, the ethic should be psychologically available. Section 4 then shows the usefulness of these basic constraints upon space ethics through a contrast between the emerging US and European agendas in astrobiology.

Keywords: astrobiology, ethics, liberal democracy, availability, realizability

1. Introduction

The past decade has seen a considerable shift in the volume and quality of work on space ethics [1–4]. Key standpoints have been articulated more clearly over arguments for and against the 'inherent value' of microbial life; over the just distribution of limited space resources; over the ethical issues of risk (and the distribution of risks and benefits); and over the connection between terrestrial environmental ethics and planetary protection elsewhere. Ethical considerations are, however, still sometimes regarded as constraints of an unhelpful, and possibly arbitrary if politically necessary, sort. Unhelpful, because they often tell us what we should not do, rather than what we have a duty to do, and arbitrary, because of suspicions that ethics in general does not answer to how the world is. Politically necessary, because a strict libertarian approach toward the world seems to fit poorly with a functioning liberal democratic society where there are requirements of 'solidarity' and some conception of a 'common good'.

Ethical constraining is, thus, part of any functioning society and, in the case of liberal democratic societies, obviously so. It is tempting, however, in some contexts (such as human activities in space), to include such constraints only as side considerations, pushed out to the margins, and linked perhaps to philosophical or metaphysical commitments of a peculiar sort. Here, we might think again about theories of inherent value or about ethics formed as an extension of some or other set of religious commitments.

To some extent, this is where ethics is placed in NASA's *Astrobiology Strategy* [5], i.e. on the margins, securing only 13 lines in a 256 page text. By contrast, at the time of writing, Working Group 5 of the COST Action TD1308 is finalizing a White Paper *Astrobiology and Society in Europe Today* (2018) where ethics is situated firmly within the humanities and the humanities firmly within astrobiology [6]. Of course, due allowances must be made here for NASA's statutory remit: it may be their business to support constructive reflection upon the societal aspects of Astrobiology, but it is not their business to tell us the difference between right and wrong. (Although, in fairness, the WG5 document also does not seek to do that.) Even so, the marginal positioning of ethics within the NASA document looks like more than statutory compliance. It looks like awkwardness and perhaps also caution about the constraints that ethics might introduce.

At the same time, NASA has helped to fund work on astrobiology and society (such as the recent project with the Princeton Centre for Theological Inquiry, 'Inquiry on the Societal Implications of Astrobiology' which ran from 2015 to 2017. Within such 'societal implications' work, ethics is more obviously central. Compliance with international space law (which *is* within NASA's remit) is also entangled with a range of ethical considerations, as are attempts to justify funding on space research. These reach beyond prudential considerations such as the generation of commercial spin-offs. My suggestion, then, is not that NASA or any other body might be seeking to evade the problems of ethics, but rather that the marginalization of ethics within the above document is both understandable and symptomatic. Ethics can be difficult to place, and particularly so at present, given the emergence of private sector space activity in combination with state-funded programs. It is an important and (given the context of liberal democratic societies) reasonable concern that ethical considerations should not *unduly* constrain either scientific exploration or the development of commerce. It is also a reasonable assumption that these two must find an accommodation with one another, and that ethics should not get in the way of their doing so unless the *way* in which it is done happens to be particularly objectionable (e.g. through the strict subordination of science to commerce—an unlikely scenario).

And so we have a picture of ethics as troubling and potentially constraining what is problematic about this picture is that it is a little one-sided. There are important respects in which ethics does constrain, but in the context of space it also enables. Indeed, it has a track record of doing so. This will be explored in Section 2 below. However, ethics can only do this if it itself is reasonably constrained or 'fit for purpose,' although what counts as 'reasonable' constraint, and what is 'fit for purpose' (hence *not* arbitrary) in relation to human activities in space, will need a little cashing out. Section 3 will attempt to make some headway on these matters. Overall, the chapter is about the shape that space ethics needs to have if it is to be appropriately situated. It is about basic methodology, 'basic' in the sense of 'rudimentary' rather than 'foundational.' There is no presumption that knowledge in this domain has a foundational structure. Rather, the chapter specifies requirements that any ethic should meet, irrespective of its preference for a terminology of rights, duties, consequences, integrity, or virtue.

The implications of the relevant, rudimentary, methodological considerations will be explored in Section 4. As an exemplar, in the light of these considerations, we may be better placed to make sense of the differences between the above, emerging, European and NASA agendas in astrobiology: as competing players in a zero-sum game (ethical constraint and regulation versus their marginalization), or as contributory responses, each playing a useful role within a much larger conversation. The primary concern of what follows is, however, basic methodology rather than any comprehensive untangling of the differences between these agendas.

2. Space ethics as enabling as well as constraining

Space ethics may be broken down into subdivisions in multiple ways. For convenience, the subdivisions appealed to here will be 'standard ethical considerations' and 'special ethical considerations' with the latter particularly (although not exclusively) associated with space. Issues of safety, risk and its disclosure belong within the standard ethical considerations. They apply equally to a broad range of other terrestrial human activities. By contrast, the ethics of SETI and at least some considerations of planetary protection (such as the avoidance of forward contamination) involve considerations which are peculiar to space, or at least to space together with a relatively small number of special terrestrial contexts. For the moment, I will simply assume that planetary protection, e.g. matters such as forward contamination, really is entangled with ethics, although this is an assumption which may seem more obvious to ethicists than to policy theorists or the developers of protocols. The entanglement of SETI with ethics is, by contrast, non-controversial: Do we have a duty to search for life? Do we have a responsibility to try to discover intelligent life before we signal our own presence? And so on.

My suggestion is not that this broad contrast between standard and special ethical considerations is a rigid dichotomy, but merely that it holds *up to a point* and is useful for present purposes. With regard to both sides of the division, I want to suggest that ethical considerations often play an enabling as well as constraining role. Even so, some caution is warranted. *Often*, is not *always*. Also, such an enabling role may seem suspiciously fortuitous. There is, however, a reason why ethics is often enabling. A large class of ethical problems in space involve issues of justice, and justice is readily recognizable as an enabler for agents. It entrenches both special entitlements and approximations to various kinds of equality of access and opportunity. Of course, not all matters of space ethics involve justice. And in those cases whether or not there is an enabling dimension to ethics may involve a more mixed story. For example, the more metaphysical arguments about inherent microbial value may offer little in the way of enabling to any agents. Or, if enabling, they may merely be so in the sense that they back up actions (e.g. forms of microbial protection) which enable science in contexts where we also have more straightforward prudential reasons for the same protective measures.

2.1. Standard space ethics as enabling

The enabling aspects of various standard ethical considerations are easily seen. As a point of standard ethics, space exploration should not reproduce familiar forms of gender bias, with rank upon rank of white, male (and for public purposes, heterosexual) astronauts traveling into space. When, for example, Senator Louie Gohment stood up in the US House of Representatives on May 26 2016, and insisted that there will be no gay space colonies, reasonable ethical concerns about gender bias were raised [7]. Nor should the safety of humans sent into space, or working on the ground with dangerous rocket fuels, be unreasonably compromised or put at risk without appropriate disclosure of these risks. Pressures to speed up the frustratingly slow pace of Virgin Galactic's space tourism program should not end up with lives lost and debris on the floor of the Mojave Desert. Reasonable precautions ought to be taken to avoid any repetition of their test crash in October 2014, and their rocket fuel accident of 2007 in which three people were killed on the ground. This is not, of course, a defense of risk aversion. Aviation kills, but so does road traffic, in appallingly larger numbers, in tens of thousands across the USA each year. Death by road accident in Tel Aviv is consistently higher than death by terrorism across the whole of Israel. In the UK, more than a quarter of a million people have died on the roads since the 1950s. In each case, the response of introducing constraints has not collapsed into a position of preventing movement, but rather has been geared to enable safety. Reasonable caution is not, therefore, the same as risk aversion. There are, admittedly, also prudential reasons for such caution in the case of space tourism, i.e. the publicity sensitivity of a high investment business. Yet the ethical side of these matters is also built in and accepted by Virgin Galactic.

More generally, research *ought* to be conducted and reported in line with appropriate sets of standards and researchers *ought* to meet the requirements of professional integrity. In both cases, these are ethical *oughts*. As an illustration of the enabling role of standard ethics, these are contexts where ethical failures tend to be entangled with missed opportunities and with technical failures. For example, we know more now about the bodily and psychological effects of prolonged stays in space than we would do if the all-male pattern of the Gemini and Apollo programs had been endlessly repeated. Opportunities would otherwise have been missed. This is also a point where standard and special considerations meet. To date, no woman has set foot upon the Moon. This is an ethical failing of a familiar terrestrial sort. In the UK, for example, at the time of writing, although there have been 10 elected or acting Labour Prime Ministers and First Ministers of Scotland, Wales and Northern Ireland since the 1990s, none have been women. Gender-based exclusions at the top are common even with respect to organizations that protest about the very same problem. However, the fact that no woman has ever set foot upon the Moon is also a more special consideration. It spans the standard/special divide. It is both about commonplace job discrimination (a narrow ethical consideration) and also about the broader ways in which we think of space exploration as a human project. (Where humanity is not monopolized by any single gender.)

More tragically, standard ethical failures concerning risk and its disclosure were bound into the tragedies of the Columbia Space Shuttle disaster in 2003 and the Challenger disaster of 1986. The 'normalization of deviance,' i.e. the undercritical acceptance of anomaly and underperformance in one case, and poor risk assessment and disclosure of a known problem with the o-rings in the other, will both qualify as ethical failures of a serious sort, on any plausible account of the proper concerns of the ethics of risk [8]. Ethical failures and technical failures are, in this way, joined. They can be different aspect of the same problems. Ethics of this standard sort may then be constraining, even inconvenient, but constraining in ways that can sometimes increase the overall likelihood of mission success and of mission participation from across groups who might otherwise be unfairly excluded.

Again, a qualification here is that technical problems of the sort mentioned above may also be dealt with in ways which are utterly unmotivated by ethics, and focused solely upon mission success. However, this fits poorly with actual NASA practice and with the behaviours of other reputable bodies for whom professional standards *are* treated routinely as matters of good ethical practice. If anything, an unusually high level of attention is given to the safety of astronauts by contrast with those involved in other, high risk but lower profile, lines of work (construction would be an obvious example, where fatality levels are high).

2.2. Special space ethics as enabling

Will a parallel argument apply also in relation to special ethical considerations in space? There are good reasons, related to planetary protection and fair access, to believe so. Planetary protection enables science. That is the primary goal of the safeguards adopted to avoid forward contamination. It is geared to the avoidance of contamination that might corrupt indications of life elsewhere, or make their detection unreliable. Yet we may still ask 'What does this have to do with ethics?' Is such protection genuinely an ethical issue, or merely a prudential safeguard for science? Although, here again, it is not obvious that the division between prudential and ethical is a hard distinction. After all, prudential considerations appeal ultimately to some sort of good or goods, or some sort of interest or interests. In the present case, this is the interests of science and, indirectly, the interest that *we* have in science as a social good. If science were not a social good then a great deal of taxation-based expenditure would face a major justification problem. Indeed, one of the more convincing defenses of space advocacy, in the work of James Schwartz, draws upon the value of science as the primary and most effective justification for human activities in space [9, 10].

How far we take this attitude toward planetary protection as, in some respects an ethical matter is a further, and difficult, question. The European White Paper, mentioned above, tends toward the view that it should be regarded as a form of environmental protection, thereby aligning it more closely with the broader range of ethical considerations that we now apply in relation to terrestrial sites. What ultimately justifies such terrestrial protection is, of course, also a matter of debate: human interests or something broader. Even so, clear extensions of terrestrial environmental ethics to space can be made. They include a safeguarding of the available range of human wilderness experiences in a solar system where few planets and moons are ever likely to be suitable for human habitation. Such an approach can also appeal to existing space law. The Outer Space Treaty does, after all, classify space as the 'common heritage of mankind' [11]. And even the requirement of avoiding forward contamination may be read in the light of the requirement that back contamination should also be avoided, a clear appeal to risk and the value of something: humans, or perhaps the terrestrial biosphere.

What this means at a practical level is, of course, a matter for debate. One proposed option, from Charles Cockell, which draws out the ethical dimensions of planetary protection, is that there is a case for something akin to national parks on the Moon and Mars, reserved areas protected from human *use* but not necessarily from all human activity [12]. Of course, none of this implies that the actual technical measures currently taken under the remit of planetary protection fall short. It is not to be confused with the claim that NASA and/or COSPAR are 'getting it wrong' when it comes to the protocols and the nuts and bolts of planetary landings. We are also a long way from the kind of presence that would make planetary parks at all feasible (or indeed worthwhile). This approach of treating planetary protection as a form of environmental protection is, as yet, simply about the best justifications and longer term scope of protective measures.

The view of planetary protection from NASA, and COSPAR, has admittedly tended to be more modest, with planetary protection regarded in a different light from environmental protection, as protection of a special sort, albeit with the thought of environmental protection sitting somewhere in the background [13]. In some ways, such an approach strengthens the separation between standard ethics (which includes environmental protection) and special ethics (which is distinctive). But if we soften this distinction, then a broader range of special ethical considerations can also be seen to play an enabling role, particularly in relation to the safeguarding of science, and of access to limited or scarce resources in space. There are, after all, only so many satellite niches [4]; only so many asteroids of the right type for mining that come within reasonable reach [14]; only so much ^{3}He, for energy production, in lunar and asteroid regolith [15]; only so many strategically valuable 'peaks of eternal light' on the Moon places where the chances of finding water ice and a continuous solar energy supply are unusually high [16]. There is only one Moon and one Mars. Our prospects of ever moving beyond the Solar System are unclear, and within the latter, there are serious difficulties in the face of establishing a stable presence on more than a few surface locations. Venus is too hot and the gas giants have extremely deep gravity wells, only their moons may be suitable.

http://dx.doi.org/10.5772/intechopen.75689

Given this, even if we hold that humans are all that ultimately matters, it will still be an ethical priority to safeguard resources *and opportunities* in ways that involve constraint. Otherwise, conditions of justice will not be realized within and between generations. With regard to future generations of humans, through such constraining of our actions other humans will be enabled. And, with respect to projects that are multi-generational (which surely includes certain aspects of the development of a human presence in space) and require opportunities to be available to future humans as well as ourselves, there will also be enabling as well as constraint.

With regard to the current generation, asteroid mining in particular poses issues of claims and entitlements. Identification of good candidate asteroids for mining and the setting up infrastructure will be major investments. It will be important to private sector players, and for the investment process, that their entitlements (however judged) are secured. The very last thing that an emerging asteroid mining industry needs is exposure to the risks of some form of claim jumping or free riding, where the benefits of preliminary investment by one company are reaped by others. Considerations of justice again enter into such matters. These will, of course, be areas for future space law, but guided by considerations of policy, legal principles and standards of *equity* and *fairness*, which involve an ethical dimension. Ethics can play an enabling role (or more precisely, a dual constraining and enabling role) for stable forms of commercial space activity.

3. Constraining ethics

As a reinforcer of the point that space ethics can work in this way, and will ordinarily tend do so when matters of justice are involved, there is a case for saying that ethics in general is ordinarily constrained so that the ways in which it limits action are themselves limited.

This is not, however, a case of arbitrarily rigging ethical deliberation in 'subjective' ways so that whatever projects we happen to prefer are then licensed. Rather, the social dimension of ethics draws it away from anything of this sort and calls instead upon an answer to the question: 'What is an ethic for?' And here, the issue is 'an answer' rather than 'the answer.' There are, after all, multiple kinds of ethics and it is far from clear that they all play exactly the same functional roles. The ethics of personal friendship is very different from the ethics of technology (even if there are areas of overlap). In the context of space, what we are dealing with is also a social ethic and not just an individual one. This means that we are in the territory of the kind of ethics that can feed into policy and legislation, rather than an ethic geared to the duties of love and friendship or to balance out the human need for solitude with the dangers of loneliness. I will also take it, as a background assumption already hinted at above, that the kind of social ethics we *ought* to be interested in presupposes something like liberal democracy. More strictly, it presupposes political organization that preserves the good features of the latter although, at present, the only political organization known to do so *is* liberal democracy.

In line with this, we can begin to set up a series of *adequacy conditions* that any plausible approach toward ethics in space should satisfy. This will not be guaranteed to yield a single outcome, a single correct ethical theory, with only one approach satisfying the conditions.

Rather, it is more likely to narrow down the field of plausible candidates. As a first pass, I will suggest that any such candidate ethic should satisfy at least one of the following two conditions:

1. It should be 'policy apt' in the sense of being able to directly shape space policy within a liberal democratic social context.
2. It should be 'precursor apt' in the sense of being able to contribute productively to broader, precursor, discussions which may then feed into 'policy apt' deliberations.

The second condition is less constraining than the first, and is geared toward some form of multiculturalism and the inclusion of religious perspectives and varying metaphysical perspectives within discussions. This will not violate any church/state division, or what liberals refer to as 'state neutrality over conceptions of the Good.' It will not require the privileging of one philosophical metaphysic over another because it is only at a precursor level. Although, in practice, the imperfections and pragmatism of liberal democracy as well as the periodic emergence of a broad consensus, will mean that the two sorts of discussions are likely to remain distinct only *up to a point*. Key religious commitments might, for example, turn out to have practical implications which are identical to those of the best secular discourses and may then be captured in a more neutral language that is also more 'policy apt' and does not give the impression that state policy is an endorsement of a particular denomination.

The point applies also to metaphysical disputes. (By which I mean disputes about moral ontology, the nature of truth in the domain of the moral, and disputes about what it is that ethical claims ultimately attempt to track.) Within this class of concerns, we may include disputes and theories about 'inherent value'. Disputes that might well run deep but still converge upon similar practical outcomes. This has been the case in the European discussions. Cockell [17, 18] and Milligan [3] have tended to be sympathetic toward extending notions of inherent value to microbial life; Persson [19] has tended to link such value more exclusively to sentience. Smith [20], a contributor form the US side, has considered structured complexity as a rival to sentience. Yet these positions converge over regarding microbial protection as an ethical requirement and, hence, planetary protection as, properly, a form of environmental protection. Strong pragmatists would no doubt say that 'there is no difference that makes no difference', and because of agreement about ways of proceeding, that these metaphysical disputes are empty. The claim here is a rather different one. It allows room for such disputes to be substantial and bring matters of importance to light, albeit at a precursor level. In fact, the presence of multiple ways of reaching similar practical conclusions will be a positive virtue from the standpoint of consensus building.

An upshot of the above is that the distinction between policy-apt discussions and precursor-apt discussions ensures that a place can be found for religious and metaphysical deliberation, but it does not try to run policy considerations directly from the latter. In this respect, it tackles a problem that has repeatedly emerged in other forms of ethical discussion about the non-human (e.g. within animal rights discussions and ecology) where the connection between metaphysical and practical has often been overly direct. Human moral

http://dx.doi.org/10.5772/intechopen.75689

psychology and any sense of political constraints have sometimes been set aside in favor of claims based upon special theories of inherent value. In other words, what is again being upheld here is a broadly liberal democratic standpoint and pragmatism about space ethics.

What makes any ethic satisfy one of the above conditions of policy aptness or precursor aptness will itself be a complex matter. However, the following three features can be set out.

3.1. Stability

While an ethic may be adapted to liberal democracy, it should not track the ebbs and flows of electoral fortunes, at least not in a close way. Any standpoint in space ethics, especially with regard to its special features (and the way that they often involve large questions about humanity) will be subverted by a preoccupation with 5 year terms of office. Developing a human presence in space is a multi-generational project and should be recognized as such. This is, of course, a familiar difficulty. Robert Zubrin (although I do not sympathize with his particular approach toward Mars settlement) has made the point repeatedly and well [21]. One of the main problems about getting a workable and ambitious US space program for a Mars landing is the fact that the government keeps on changing. Getting humans to Mars is a project that is likely to require at least two decades (the launch windows alone are a significant limitation). During that time, multiple changes of political priorities, personnel and party ascendancy, may occur. This may seem like, but is not, directed against the party that is currently ascendant. A space ethic that deliberately tracked President Obama's social attitudes, without special reasons to believe that the latter would endure, would have been as counterproductive as one which tracks President Trump's social attitudes. A viable space ethic has to accept the political constraints of a working liberal democracy, but must have greater stability than voting trends.

3.2. Political realizability

Even given its stability, a sense of realpolitik should incline us to reject a belief which is familiar within various protest movements, such as sections of environmentalism and the animal rights movement (movements with which I happen to sympathize): the belief that the great considerations of humanity's future, or the Earth's future, or the future of sentient life forms are capable of trumping political divisions because of their importance and moral depth. The latter do not license some kind of exceptional politics, set apart from our flawed democratic processes. Even given the assumption that global warming does threaten to be a disaster for humanity and for other creatures, this alone does not imply that there is a way to work around the party systems and regular divisions of liberal democracy, rather than working through them in the search for some stable form of consensus building or bipartisan co-operation. Similarly, the horrors of animal harm within the food chain may be every bit as terrible as suggested on the strongest animal rights critique, but the exceptional nature of the harms does not make an exceptional politics any more viable. Such causes will succeed within the bounds of what is politically realizable under liberal democracy, or else they will not succeed. This is a completely general, and pragmatic, consideration and applies as readily to space ethics as it does to ethics of other sorts. The fact that we are talking about humanity's future does not imply that we can set aside the regular constraints of politics.

3.3. Psychological availability

Finally, an ethic should not be so idiosyncratic or metaphysically laden that its practical implications could only be accepted by a small number of humans at any given time. Such an ethic would not only fail to meet the requirement of policy aptness, it would also fail to meet the requirements of precursor aptness because no one outside of the limited group could come to share the same vision of what is to be done. On its own, however, this constraint would only provide a minimal account of psychological availability. I will suggest a stronger version that involves two additional constraints although, under analysis, these may turn out to be aspects of a single consideration.

Firstly, the ethic should be of a sort that psychologically normal humans could live by as opposed to merely defending in public fora. I will take it that this will rule out any form of strict consequentialism, or any ethic that is built around the overriding importance of a single consideration even if that consideration happens to be an extension of humanity's survival. We are, naturally, pluralists about ethical norms and one of the strongest and best features of liberal democracy is that it draws this out: many things matter to us. So, for example, an overriding preoccupation with the manifest destiny of humanity should not lead us to regard tragic accidents as merely technical setbacks in a greater cause rather than regarding them as occasions for reflection upon issues of safety, responsibility and harm. In extremis, and contra Werner Von Braun, a single-minded ethical commitment to a human future in space cannot override the ethical issues of whether or not the labor force used in a space program is free or forced, protected by adequate safety measures or exposed to excessive risk, punishments and harms. The ethic should not be monomaniacal or otherwise fanatical. Neither of these is a good fit for liberal democracy.

Secondly, and perhaps more controversially, the ethic should be of a sort that does not conceal our real motivations behind forms of public justification. It should allow us to be honest about what actually drives our attitudes. For example, if we tend to support environmental protection because of a sense of the value of places in their own right, then we should not have to pretend that the real reasons for environmental protection always turn out to be a concern for humans. Public justifications and real motivations may never fully align, we are too dissonant for that, but they should not be forced apart in ways that make them radically distinct and lead to concealment. Our ethical theory for space should not force us into what has been called 'moral schizophrenia' where the real motivations are hidden [22]. The policy-apt/precursor-discussion apt distinction may be understood as geared toward this consideration, with the latter retaining ample opportunities to set out what it is that really motivates our concern.

4. Implications for ethical agendas in astrobiology

Given these adequacy conditions, how then do the contrasting NASA and WG5 astrobiology agendas fare? I will take it that both meet all three requirements, but doing so is not exactly a binary matter. One approach may meet any one of these conditions more strongly or more effectively than another. Neither the NASA approach nor the WG5 approach obviously wins out across the board. In terms of **political realizability**, the middling-constraint approach of NASA has the clear advantage of actually being linked to a robust program of activities in space. It has not only realizability but demonstrated realizability. More generally, given a certain degree of reluctance on the part of liberal democratic states to strongly constrain the actions of private agents, unless activities in space pose some clear *and immediate* threat to human interests, this is always likely to be an area where a more modest set of restrictions scores high.

By contrast, the WG5 approach is likely to score more strongly on both **stability** and, perhaps surprisingly, **psychological availability**. Indeed, there is a case for saying that if it scores higher on one, it is also likely to score higher on the other. In terms of stability, there are long-term socio-economic pressures in the USA toward a more robust inclusion of ethical considerations in policy formation. Ethical considerations have tended to spread through public sphere activity, and have also tended to spill beyond concern for the human alone. Other pressures are more bound up with the peculiar conditions of space. Firstly, middling regulation is all well and good for the USA while it is the dominant player. As that ceases to be true, or becomes heavily qualified, the pressures to secure position through regulation (through informal norms or international agreements) will tend to increase. Given the growing recognition that key, accessible, space resources (water, ^{3}He, orbital niches, accessible metallic asteroids) are scarce or at least limited, the commercial and scientific pressures toward regulation to ensure continuity/sustainability of access will increase over the course of time. Astrobiological research will then fit into a broader environment where issues of just distribution of opportunities and access claims are normal.

Secondly, there is an interesting and much-commented upon feature of domestic politics in the USA. At least notionally, protestant evangelism has political influence upon Republican politics and this denomination records an unusually high level of skepticism about the search for, and likelihood of finding, life elsewhere. This is partly because of a perceived connection with evolutionary theory. However, this skepticism need not imply that a crisis of faith would result from the discovery of life. Its presence varies enormously depending upon the pattern of local pastoral care, and it sits side by side with the emergence of a more robust environmental ethical commitment within the same denomination (on something close to classic Christian 'stewardship' grounds). If issues of planetary protection were reframed as matters of the latter sort, i.e. as matters of environmental protection, this would carry some obvious advantages in terms of winning over sympathies for a protection agenda. An ethical agenda based upon a strong conception of environmental duties (inclusive of a form of respect for life, and perhaps also for the integrity of places) might turn out to have greater stability than one based upon a lack of space-frontier constraint.

Finally, in terms of **psychological availability**, the WG5 approach may be slightly better placed to accommodate the fact that we do not (perhaps cannot) simply concern ourselves with human interests. Whatever our theories, what we actually value always tends to spill over these bounds. This does not, of course, require us to commit to peculiar doctrines, e.g. the quite different metaphysical claim that microbial life has *equal value* with human life, or with sentient life, or anything of that sort. Value egalitarianism would not itself be psychologically available in the relevant sense. It would be unlivable. Its practical implications would not then be either *policy apt* or *precursor apt*. Whether we speak about inherent value or about human practices of valuing, our natural pluralism should not be confused with an inability to recognize that we often have more reasons for protecting and valuing one thing rather than another. For many practical purposes, talk about 'inherent value' will translate into 'reasons for action,' and these can vary significantly from object to object in ways that can be difficult to connect to any such version of value egalitarianism. Nonetheless, we do value places, things and other creatures in ways that are more in line with thinking of planetary protection as a variety of environmental protection. The latter is a more readily available attitude than a stricter attempt to limit concern to the human, even if doing so currently remains close to a statutory requirement for NASA.

What this mixed picture also suggests is that the model of the 'one correct approach' toward space ethics within astrobiology will be a poor fit. Rather, even where two differing agendas both meet the basic adequacy conditions set out above, they are still likely to fare better by some criteria and less well by others. Given this, the appropriate model for ethical deliberation, within which we can situate the WG5 approach and the NASA Astrobiology Roadmap, is best thought of as a dialog rather than a zero-sum game. There is, of course, a great deal more to say about this dialog, and these particular contributions to it. Even so, the little that has been said illustrates the practical guiding role of the basic constraints set out above.

Acknowledgements

The importance of representing the differences of vision in the WG5 and NASA document approaches as a matter of contributions to a dialog/conversation rather than simply conflicting visions was originally suggested to me by Jacques Arnould.

Author details

Tony Milligan

Address all correspondence to: anthony.milligan@kcl.ac.uk

Department of Theology and Religious Studies, King's College London, UK

References

[1] Arnould J. Icarus' Second Chance: The Basis and Perspectives of Space Ethics. New York: Springer; 2011

[2] Impey C, Spitz AH, Stoeger W, editors. Encountering Life in the Universe: Ethical Foundations and Social Implications of Astrobiology. Tucson: University of Arizona Press; 2013

[3] Milligan T. Nobody Owns the Moon: The Ethics of Space Exploitation. North Carolina: McFarland; 2015

[4] Schwartz JSJ, Milligan T. Some ethical constraints on near-earth resource exploitation. In: Al-Ekabi C, Baranes B, Hulsroj P, Lahcen A, editors. Yearbook on Space Policy 2015, Access to Space and the Evolution of Space Activities. Heidelberg, New York and London: Springer; 2016. pp. 227-239

[5] NASA (2015). Astrobiology Strategy (NASA-NAI) https://nai.nasa.gov/media/medialibrary/2015/10/NASA_Astrobiology_Strategy_2015_151008.pdf

[6] Persson E, Capova K, editors. Astrobiology and Society in Europe Today (COST). 2018. Forthcoming

[7] Congress. "Issues of the Day," Congressional Record 162 (May 26, 2016), H3292. 2016. https://www.congress.gov/crec/2016/05/26/CREC-2016-05-26-pt1-PgH3290.pdf

[8] Pinkus R. Engineering Ethics: Balancing Cost, Schedule, and Risk – Lessons Learned from the Space Shuttle. Cambridge: Cambridge University Press; 2008

[9] Schwartz JSJ. Prioritizing scientific exploration: A comparison of the ethical justifications for space development and for space science. Space Policy. 2014;**30**(4):202-208

[10] Schwartz JSJ. Our moral obligation to support space exploration. Environmental Ethics. 2011;**22**:67-88

[11] United Nations. (1967). Treaty on principles governing the activities of states in the exploration and use of outer space, including the moon and other celestial bodies, January 27, 1967, article I, T.I.A.S. 6347, 30. pp. 209-214

[12] Cockell CS, Horneck G. Planetary parks – Formulating a wilderness policy for planetary bodies. Space Policy. 2006;**22**(4):256-261

[13] Rummel JD, Race MS, Horneck G. Ethical considerations for planetary protection in space exploration. Astrobiology. 2012;**12**(11):1017-1023

[14] Elvis M. How many ore-bearing asteroids? Planetary and Space Science. 2014;**91**:20-26

[15] Crawford I. Lunar resources: A review. Progress in Physical Geography. 2015;**39**:137-167

[16] Elvis M, Milligan T, Krolikowski A. The peaks of eternal light: A near-term property issue on the moon. Space Policy. 2016;**38**:30-38

[17] Cockell C. Environmental ethics and size. Ethics and the Environment. 2008;**13**(1):23-39

[18] Cockell C. The value of microorganisms. Environmental Ethics. 2005;**27**(4):375-390

[19] Persson E. Ethics and the potential conflicts between astrobiology, planetary protection, and commercial use of space. Challenges. 2017;**8**:12

[20] Smith KC. Manifest complexity: A foundational ethic for astrobiology? Space Policy. 2014;**30**(4):209-214

[21] Zubrin R. The Case for Mars: The Plan to Settle the Red Planet and why we must. New York: Free Press; 1996

[22] Milligan T. Valuing humans and valuing places: "Integrity" and the preferred terminology for Geoethics. Geosciences. 2018;**8**:25

Chapter 3

From the Individual to the Cultural Space Group

Carole Tafforin

Additional information is available at the end of the chapter

http://dx.doi.org/10.5772/intechopen.72357

Abstract

From a behavioral point of view, human crews into Space will have to both live and work in physical environment (microgravity, 1/3 g, 1/6 g), confined environment (spatial restriction, social constraints, and sensorial privation), and isolated environment (familiar privation, cultural background, and remote communication) that involve a multisystem adaptive model on a long-duration process. Physiological, medical, psychological, sociological, anthropological, and ethological impacts have been emphasized in a wide panel of investigations. The current results are presented with a focus on relevant methods in ethology based on the observation, description, and quantification of (i) the individual behavior from short-term orbital missions; (ii) the social behavior during inter-planetary missions simulated in terrestrial environments; and (iii) the cultural behavior in considering manned missions on Moon, on Mars, and beyond. Global analysis highlights that the crewmembers going into Space will be definitively interactive men and women with personal experiences, social rules, and new cultural habits. They will have their individual identities and they will be a group entity for extended periods of time.

Keywords: space exploration, human behavior, adaptation, orbital mission, isolated and confined environments

1. Introduction

When a human crew takes flight from Earth for the purpose of staying in Space, what could happen from an adaptive perspective? In the evolutionary history, the human being adopted the supine posture for walking. The possibility of movement of the individual then led to the evolution of kinds of socialities that resulted in cultural habits (**Figure 1**).

Firstly, just as gravity plays an essential role in terrestrial locomotion [1], microgravity is the new environmental factor that will lead to changes in sensory-motor activity. Secondly, the crew will be exposed to unusual environment factors as the space habitat will be isolated

Figure 1. Evolutionary steps of human being on Earth.

and confined. Coping with monotony and facing up to autonomy thus becomes essential for adaptability. Furthermore, when moving out of space habitats during extra-vehicular activities (EVAs), such hostile environment is another factor to consider for survivability. Thirdly, all of these factors have to be taken into account in synergy with the temporal aspect of going far from Earth and for extended durations.

In sum, the human crew has to adapt to a new, unusual, and hostile environment for living and working in Space, and for extended periods of time. Many studies in ethology, i.e., the science of behavior, were performed in these extreme settings (**Figure 2**), in real situations (orbital missions), in simulations (parabolic flights and bed rests), during confinement experiments, and in analogue environments (polar stations). On the one hand, missions in orbital station or in space shuttle are new conditions of weightlessness; parabolic flights create the physical characteristics; bed rests simulate the physiological consequences. On the other hand, confined habitats are unusual conditions and South Pole stations or Artic expeditions are hostile environments analogous to space exploration missions.

Generation and applications of extraterrestrial environments on Earth [2] contribute to an exhaustive and global knowledge of what constitutes Space, from its physical characteristics to its behavioral aspects within the relationship between the individual and the environment.

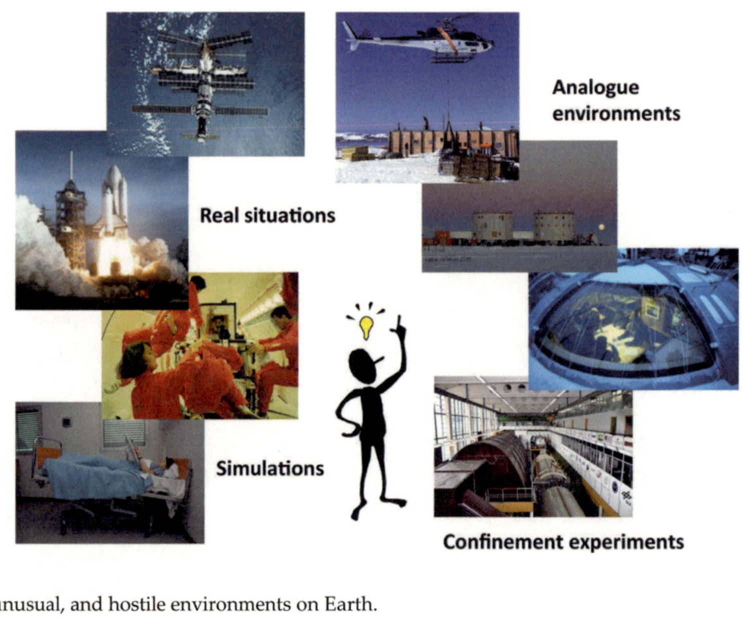

Figure 2. New, unusual, and hostile environments on Earth.

Interplanetary missions are regarded as prime opportunities to highlight the psycho-physiological issues of man and woman under both microgravity and isolation and confinement conditions over very long durations. During these space travels, the crewmembers will have to adapt to a wide range of environmental factors such as weightlessness, social constraint, closed module, monotonic panorama, familial privation, and cultural diversity. They will also have to cope with the restrictions imposed by life-support system. The ethological works were particularly concerned with these interfaces [3, 4].

In this complex network, the adaptation model to space environment (**Figure 3**) considers the individual as a whole with all the facets, concerning maladaptive reactions and adaptive strategies. It puts into action a "hard" system (*left facet of the model*), in terms of conservative regulation, which tends to recover the initial states of the sensorial mechanisms and the physiological mechanisms. For instance, deficits or variations in the environment induced by microgravity will generate, at the physiological level, a redistribution of fluids and electrolytes followed by a cardiovascular reaction leading to endocrinal and metabolic changes. At the sensorial level, the weightlessness will transfer the information from the inhibited vestibular function to the visual function, which will be reinforced. Consequently, a "soft" system (*middle facet*) will be activated, in terms of innovative regulation, to express new behavioral strategies by the adjustments or modifications of body orientations, postures, and movements. New motor learning will conduct to a new representation of the space environment (*upper facet*) involving psychological functions, cognitive skills, sociability, and cultural ability.

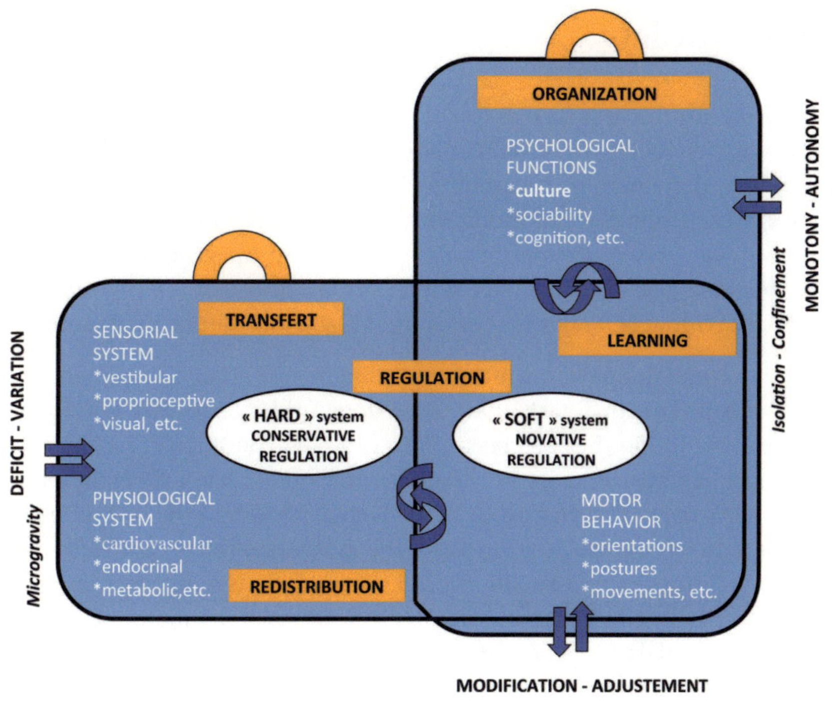

Figure 3. Multisystem adaptive model in Space.

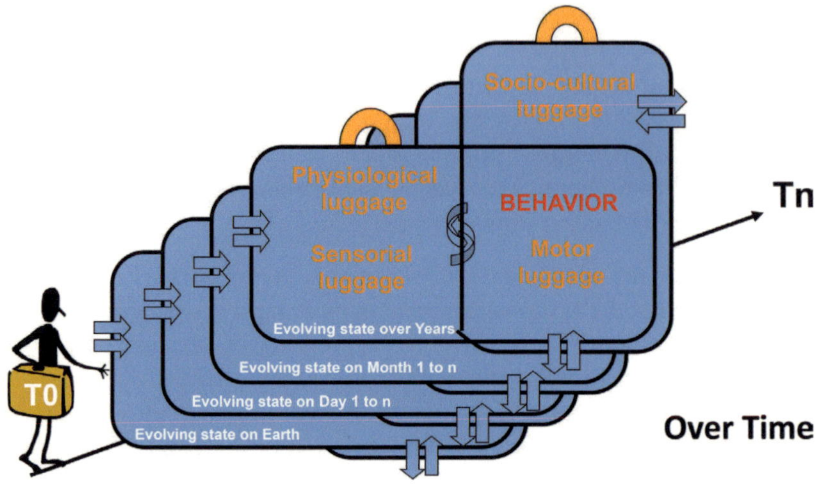

Figure 4. Dynamics states of adaptation to Space.

Overall, the space traveler, fitted with his physiological luggage and sensorial luggage, equipped with his motor luggage and impacted by his sociocultural luggage, will follow an adaptive dynamics, day after day, month after month, year after year as evolving states of the individual behavior, the social behavior and the cultural behavior (**Figure 4**).

2. Individual behavior

When experiencing for the first time, reduced gravity in parabolic flights, neophyte subjects were in confusing referential cues with inadequate body moving [5]. Their performance, described by the optimization of orientations in a three-dimensional space, improved over time. As they gradually develop harmonious coordination between sensory-motor and cognitive experience, spontaneous, preliminary, followed by integrative stages of adaptation underlie new relations between the body references and those of the physical environment. As a result, the individual behavior is characterized by a multidirectional motor activity.

Once in short-term orbital flights, the man or woman in space, i.e., the spationaut, has to perform domestic and professional tasks like those done under terrestrial gravity, whereas the "body tool" available to him/her has been placed in new conditions that require significant behavioral changes [6]. Microgravity has the most obvious effect of diversifying movements, postures, and orientations. The vertical position is thus no longer the only one possible. This means that in order to efficiently perform tasks, the individual has to invent new motor strategies that transform the quality of locomotion and manipulations with new orientation possibilities in weightlessness. In ethological studies, modifications of the motor behavior are treated as observable evidence of the human adaptation to space. In addition to the conservative physiological homeostasis, the quantitative description of what the moving individual is doing in microgravity is postulated as innovative regulation in the multisystem adaptive model.

Whether onboard the orbital Mir station or in the space shuttle where we analyzed video recordings, the results showed how humans in space elaborated a new world of perceptions and actions through the changes of body orientations. As an example of spationaut adapting to this new environment, after three days of space sickness induced by vestibular-ocular conflicts, we observed the occurrence of head-down orientations as main changes in the motor activity (**Figure 5**).

Holistically, the physiological challenges facing man and woman going into Space have been well documented in books and chapters, with examples such as fundamentals of space physiology [7, 8] and space medicine [9, 10]. Exposure to microgravity and radiation has important physiological implications for the maintenance of medical health. Also, behavioral health is of prime importance to keep the missions operational. Psychology of space exploration is a new challenge for long-duration interplanetary flights and stays on planets far from Earth [11]. Humans in Space might answer issues regarding psychological hurdles [12].

On the one hand, individuals' health status has to be optimum, with medical and physical requirements. The spationaut, with his/her physiology, has to regulate loss of weight, loss of bone, loss of muscle, loss of vision, and loss of proprioception. Microgravity effects are minimized with medical training before going into Space, special countermeasures on physical activity when staying in Space, and physical reconditioning when returning to Earth. Specific emphasis is placed on cardio-vascular responses by developing individual countermeasures [13] and on musculo-skeletal responses because there is an individual variability in spite of exercise regimes [14]. Emphasis is also placed on neuro-cognitive responses while maintaining a high level of performance [15] from body-disorientation to related stress in microgravity. That has an impact on the individual behavior in terms of behavioral health.

On the other hand, the individual's behavioral health should be positive with congruence and assertiveness requirements [16]. The spationaut, with his/her psychology, has to cope with loss of mental ability, loss of wide sociability, loss of close family contacts, loss of privacy, loss of large landscape, and a multitude of odors. Isolation and confinement effects are prevented with

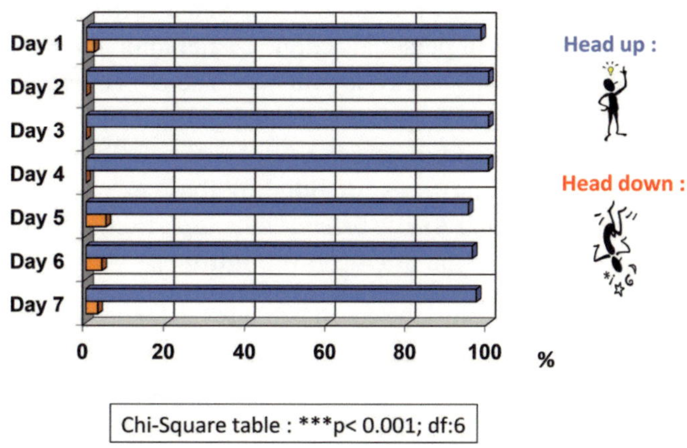

Figure 5. Head/body orientations in microgravity during a 7-day orbital flight.

ground support [17] as far as communication with Earth is possible. Relevant researches are conducted on personal value [18], personality trait [19], emotional state [20], communicative profile [21], and mental health, with the need of defining individual characteristics upon strong motivation [22]; in other words, all that has an impact on the individual behavior in ethological terms.

Good health, both medico-physiological and etho-psychological, is the key to quality of life and to successful work. The individual behavior in its optimal and positive meaning is the first link of the well-being and good-spirit of the crewmember within social contexts constrained by isolation and confinement once he/she has settled into Space.

3. Social behavior

The isolated and the confined crewmember observed using the ethological approach has demonstrated that confinement generates stress manifestations versus isolation; that isolation enhances social relationships versus confinement, and that the crew adapted positively to both environments [23]. From the individual in orbital flight to a small group inside confinement chambers or inside polar stations, research works need continued sharpening on adaptation of the human being on Earth to the human being in Space, period by period [24].

The main results from the research showed three adaptive periods: initial, mid and final periods over a 28-day Isolation Study of the European Manned Space Infrastructure (ISEMSI) campaign; over a 60-day Experimental Campaign for European Manned Space Infrastructure (EXEMSI); and over a 135-day Human Behavior in Extended Spaceflight (HUBES) campaign. In these limited habitats, the personal distances decreased and the public distances increased among the crewmembers. We observed high values of social distances and body mobility from the initial period to the final period. An increasing spatial dispersion with decreasing social orientations was also noted among the crew at the midpoint of the medium-term simulations [25]. Over a 520-day Mars-500 experiment, the crew simulated how to live and work together like a real mission with a 250-day Earth-Mars travel, a 30-day Mars landing, and a 240-day Mars-Earth travel. The results showed that time had a major impact on the individual and interindividual behaviors in terms of personal actions, visual interactions, object interactions, body interactions, facial expressions, and collateral acts. The crewmembers followed phasic, periodic, and punctual behavioral changes in extended periods of time, as it was observed at the Antarctic Concordia station and during the Artic Tara expedition, to avoid monotony. Be it space simulators or analogue settings provided by these extraterrestrial environments on Earth for going into Space (**Figure 6**), the behavioral adaptation is mainly related to social interactions with an emphasis on visual interactions (**Figure 7**).

Along with physical and mental health, the social behavior in terms of nonverbal interactions and verbal communications has become of new interest to investigate interplanetary missions from the perspective of multidisciplinary approaches. For instance, correlation of etho-social and psycho-social data during the simulated Mars-500 interplanetary mission aimed at identifying crewmembers' behavioral profiles for better understanding Space groups of future explorations. We found significant negative correlations between anxiety and interpersonal

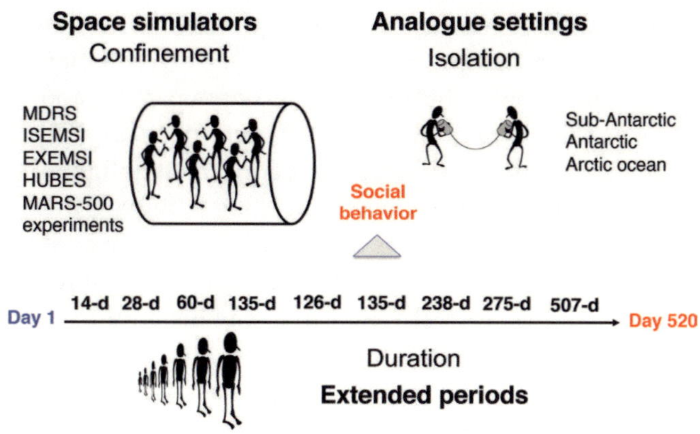

Figure 6. Environmental and temporal factors simulated on Earth.

communications, and between the sociometric parameter "popularity in leisure environment" and anxiety level. We also found significant positive correlations between the sociometric parameter "popularity in working environment" and interpersonal communications, and facial expressions; and between the sociometric parameter "popularity in leisure environment" and interpersonal communications, and facial expressions. This highlighted complementary viewpoints in the field of life sciences and social sciences: objective versus subjective, active versus discursive, exhaustive versus restrictive, and descriptive versus introspective [26].

Crew relations also play important roles in the success of missions where crewmembers stay one year and beyond in isolation and confinement. Wireless monitoring of interactive behaviors correlated with individual questionnaires, and video analyses of collective activities showed that the amount of time spent together during free time is highly associated with the intensity of relationships [27].

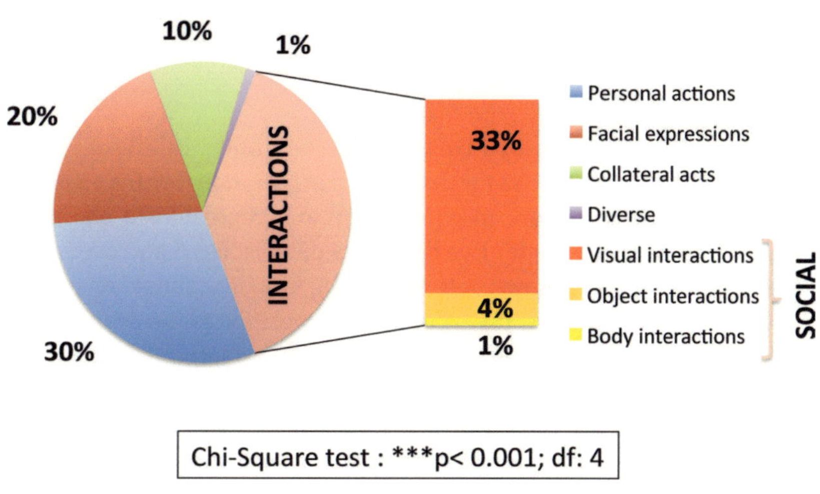

Figure 7. Social interactions in isolation during a 506-day Tara-Artic expedition.

Interpersonal and organizational issues were already raised within the first long-duration stays onboard the orbital Mir station and onboard the International Space Station (ISS). Negative effects included low crew cohesion, poor leadership skills, and crew tension resulting from environmental stress and related to crew heterogeneity [22]. However, positive effects of group experience inside space habitats need to be highlighted. For instance, from the human point of view, being in Space along with being in microgravity are unique opportunities for evidence of Earth observation as a whole. It is actually a salutogenic experience, as reported by the ISS inhabitants [28]. Despite constraints of social monotony, social isolation, and social confinement, the crewmember can create distance from the crew by widening the vision of the surrounding to a faraway environment. According to the classification of Hall [29], there is a shift from a social space to a public space (120–360 cm ad infinitum). Salutogenis means that individuals who adapt positively to an inhospitable or extreme environment can derive benefit from their experience [30]. One example of positive coping strategy in dealing with the stress of being in Space is seeking social support, which is defined as effort to obtain sympathy, help, information, or emotional support from other persons [31]. It is a component of social behavior.

For interplanetary missions, selection criteria should include social compatibility. Then training should enhance interpersonal skills, leading to the best way of psychosocial adaptation in Space [32]. The crew's effectiveness and safety will be thus enhanced. Interpersonal interactions have gained attention in studies conducted on different groups in analogue space stations referred to as Lunar Palace 1 [33]. Significant individual differences were identified, as well as crew structure was a determinant, even with mononational crewmembers. Results showed that group climate was a good state for a successful mission. However, intercultural interactions are new aims to emphasize. Culturally related differences in values and behavioral norms could influence cohesive group formation [34]. Multinational crews become cultural space groups on a long-duration process by sharing values from their individual experiences and perceptions of different terrestrial regions.

4. Cultural behavior

Small groups of three to six members, mixed-gender and multicultural compositions would be core features of terrestrials gathered for deep space exploration. Such a group is a dynamic organization where all the forces are in equilibrium and regulated to obtain optimal efficiency structures [35]. The rules of adaptive dynamics of an isolated and confined crew could thus be compared to the laws governing self-organizing systems. These laws are based on the heterogeneity of their own elements. International composition is the first characteristic that may have an impact on the cultural behavior for building adaptive strategies. Also, gender composition is an element of heterogeneous organization and helps in mitigating the interpersonal conflicts for developing cooperative strategies [36]. This leads to new goals of investigations into intercultural relations with an emphasis on use of space, use of language, and use of time. The analysis of mean durations of collective activity, while free grouping at meal times during the Mars-500 experiment (**Table 1**), showed significant differences regarding

Subject	RU1	RU2	RU3	FR1	IT1	CH1
Mean (minutes)	23	22	14	25	16	16
Standard deviation	±6	±7	±5	±6	±8	±6

Table 1. Collective activity duration in confinement during a 520-day Mars-500 experiment.

cultural background and mission goal. On one side, the French group-member (FR1) spent the longest time at meal as it is customary in daily life activities in a given country or region. On the other side, the Russian-Italian-Chinese group (RUS, IT1, and CH1) spent the shortest time at meal as a specific fact attributed to Martian crew versus Orbital crew (RU1, RU2, and FR1). The latter stayed on the Mars-orbit, while the new space group simulated planet landing and staying. These results suggest that if individual differences could generate conflicts within the group members, cultural differences could enhance cohesion of the group, with cultural behavior viewed as positive way to live and work together very far from mother Earth and for a very long time away from family links.

Constructing micro-society models for cultural space groups is thus relevant from an anthropological perspective. The "notion of space" holds a major place in field studies and has an obvious relation with the notion of culture [37]. This approach takes into account spatial relations as a central variable that influences the cultural behavior and the underlying cognitive process. In the history of humanity, there has been a revolution in cognitive capabilities and in learning skills that brought *Homo sapiens* from real to virtual integration [38]. In the future of space exploration, there will be an evolution of the human adaptability to autonomy that will bring *Homo spatius* to virtual integration of the surrounding world, thus avoiding etiological factors of the environment (isolation, confinement, monotony, etc.). We know that cooperation was a behavioral response of survival in the ancient civilizations and ethnicities [36]; equivalent responses might occur in future micro-societies and on remote planets, from surviving to adapting and then evolving. Hence, it is of prime importance to consider the cultural values of the space group [4].

Consider a manned mission on Moon, under hypogravity (1/6 g). Life-support system, rover exploration, navigation innovation, and basic technologies have resulted in the highest readiness for the development of planetary habitats in the conception of a Lunar village [39]. After having walked on the lunar surface, living, and working on the Moon need to be considered. With that aim, test beds were performed in analogue environments with an emphasis on field investigations at the Mars Desert Research Station (MDRS) that examined, for instance, communications in multilingual crews. The international and mixed-gender composition of small Euro-Moon-Mars crews properly simulated the very next isolated and confined groups who would land on the planet. Ethological studies on language skills showed that verbal communications and nonverbal interactions were influenced by cultural background such as the mother tongue [40]. Crew-members using nonnative languages compensate with interaction abilities. With the evolving of daily life habits in Space and over time, some will actively interact, others will actively communicate, and the whole will progress in a common cultural behavior.

Consider a manned mission to Mars, under reduced gravity (1/3 g). In addition to key technologies and habitat designs, communication time to Earth is delayed by 20 min. Autonomy and auto-organization of the cultural space group become crucial. In case of lack of contact with mother Earth, in actual isolation, it is also crucial to break the monotony and to find new centers of interest within the group. A personal account of the Mars-500 experiment said that multiculture was seen as an advantage rather than a disadvantage because the crew attempted to understand each other and looked for new knowledge accumulated by every crewmember from their own living and working experiences [41]. It is important to take into account the need of rituals. In African villages, for instance, far from any civilization, there are small ethnic groups defined by their customary dwellings in small living places and in grouped huts to promote exchanges. Cultural behavior is associated with a region as well as a nation. In a Mars village, rituals of the inhabitants should be invented.

Considering the early humans, the evolutionary context began with the genus *Homo*. Originating in Africa, *Homo* dispersed widely across the terrestrial globe and was exposed to biological changes but more importantly influenced by culture, upon development of complex behaviors including advances in technology [42].

The discussion is open. Human nature would develop further with interplanetary humanization and Space colonization, beginning with physical evolution and extending over cultural evolution.

5. Conclusion

In future research, the adaptation strategy of each spationaut and of the whole space group will be shown as a multisystem integration, from survivability factors (cardio-vascular deconditioning, hormone regulation, immune response, radiation reaction) to adaptability factors (motor behavior, cognitive demand, social interactions, verbal communications, cultural profiles, living habits). The main interest of transdisciplinary approaches is to investigate synergetic effects of the multiple determinants involved in human well-being, as positive facts, from the medical, physiological, psychological, sociological, anthropological, and ethological viewpoints. This would contribute in preventing negative environmental impacts on future interplanetary missions.

Once the space traveler reaches the Moon, Mars, or beyond, who will he or she be? They will definitively be an interactive man or woman, with their own social rules and cultural habits. He or she will have their individual identity and they will be as a group entity, by building the same language code and communication rules based on multinationalities.

From an individual behavior with new body orientations in microgravity, a social behavior with specific interactions and interindividual distances in confinement, to a cultural behavior with proper rules of communications within the space group after separation for a long time (**Figure 8**), we may draw an adaptive scenario of human crew in Space.

http://dx.doi.org/10.5772/intechopen.72357

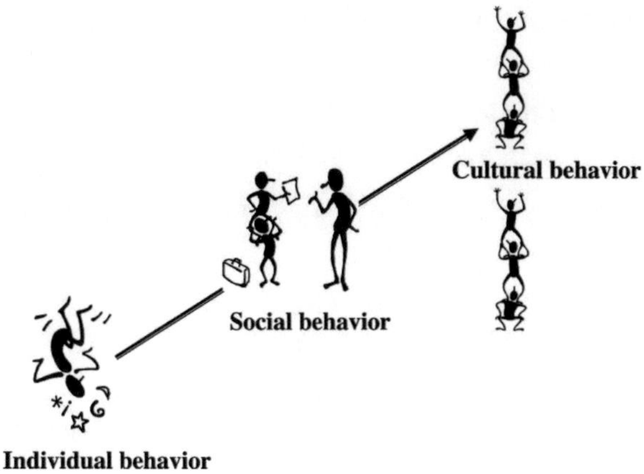

Figure 8. Adaptation scenario of the crews' behavior in Space.

Acknowledgements

The author thanks the French Space Agency for its scientific and financial support in the entire ethological studies.

Author details

Carole Tafforin

Address all correspondence to: ethospace@orange.fr

Ethospace, Toulouse, France

References

[1] Schlacht IL, Rittweger J, Foing B, Masali M, Daumer M, Cremasco MM. How measurements from hypogravity locomotion studies can inform the architectural design of planetery habitats. In: Proceedings of the 67th International Astronautical Conference; October 2016; Guadalajara, Mexico. IAC-16.A1.2.6

[2] Beysens DA, Van Loon JWA. editors. Generation and Applications of Extra-Terrestrial Environments on Earth. Aalborg, Denmark: River Publishers; 2015. 281 p

[3] Tafforin C, Thon B, Guell A, Campan R. Astronaut behavior in orbital flight situation: Preliminary ethological analysis. Aviation, Space and Environmental Medicine. 1989;**60**: 949-956

[4] Tafforin C, Giner Abati F. Cultural ethology as a new approach of interplanetary crew's behavior. Acta Astronautica. 2017;**139**:102-110

[5] Tafforin C. Initial moments of adaptation to microgravity of human orientation behavior, in parabolic flight conditions. Acta Astronautica. 1996;**38**:963-971

[6] Tafforin C, Campan R. Ethological experiments on human orientation behavior within a three-dimensional space-in microgravity. Advances in Space Research. 1994;**14**:415-418

[7] CNES. In: Proceedings of Space Physiology. Toulouse, France: Cepadues; December 1983. 493 p

[8] Grigoriev A, Agadjanyan AA, Baranov VM, Polyakov VV. On the contribution of space medicine in the public health care. Acta Astronautica. 1997;**41**:531-536

[9] Clément G, editor. Fundamentals of Space Medicine. The Space Technology Library. El Segundo, California: Microsom Press. Dordrecht, Boston, London: Kluwer Academic Publishers; 2003. 361 p

[10] Nicogossian AE, Williams RS, Huntoon CL, Doam CR, Polk JD, Schneider VS, editors. Space Physiology and Medicine: From evidence to Practice. New York, United States: Springer; 2016. 508 p

[11] Vakoch DA, editor. Psychology of Space Exploration. Washington, United States: The NASA History Series; 2011. NASA SP-2011-4411. 254 p

[12] Kanas N, editor. Humans in Space. The Psychological Hurdles. New York, United States: Springer. Chichester, United Kingdom: Praxis Publishing; 2015. 149 p

[13] Hughson RL, Shoemaker JK, Arbeille P. CCISS, vascular and BP Reg: Canadian space life science research on ISS. Acta Astronautica. 2014;**104**:444-448

[14] Convertino VA. Exercise as a countermeasure for physiological adaptation to prolonged spaceflight. Medicine and Science in Sports Exercise. 1996;**28**:999-1014

[15] De la Torre GG. Cognitive neuroscience in Space. Life. 2014;**4**:281-294. DOI: 10.3390/life4030281

[16] Tognini M. Open Discussion. 30th Planetary Congress of Astronauts. Toulouse, France: October 2017

[17] Feichtinger E, Sundblad P, Fuglesang C, Zell M, Urbina D, Charles R, Kuipers A. The role of communications for psychological crew support during human exploration mission simulation Mars-500. In: Proceedings of the 63rd International Astronautical Conference; October 2012; Naple, Italy. IAC-12.A1.1.24

[18] Sandal GM, Bye HH, Van de Vijver FJR. Personal values and crew compatibility: Results from a 105 days simulated space mission. Acta Astronautica. 2011;**69**:141-149

[19] Mittelstädt JM, Pecena Y, Outbaid V, Maschke P. Psychometric personality differences between candidates in astronaut selection. Aerospace Medicine and Human Performance. 2016;**87**:933-939

[20] Polackova Solcova I, Lacev A, Solcova I. Study of individual and group affective processes in the crew of a simulated mission to Mars: Positive affectivity as a valuable indicator of changes in the crew affectivity. Acta Astronautica. 2014;**100**:57-67

[21] Gushin V, Shved D, Vinokhodova A, Vasylieva G, Nitchiporuk I, Ehmann B, Balazs L. Some psychophysiological and behavioral aspects of adaptation to simulated autonomous Mission to Mars. Acta Astronautica. 2012;**70**:52-57

[22] Kanas N, Sandal G, Boyd JE, Gushin VI, Manzey D, North R, Leon GR, Suedfeld P, Bishop S, Fiedler ER, Inoue N, Johannes B, Kealey DJ, Kraft N, Matsuzaki I, Musso D, Palinkas LA, Salnitskiy VP, Sipes W, Stuster J, Wang J. Psychology and culture during long-duration space missions. Acta Astronautica. 2009;**64**:659-677

[23] Tafforin C. Confinement vs. isolation as analogue environments for Mars missions from a human ethology viewpoint. Aerospace Medicine and Human Performance. 2015;**86**:131-135

[24] Tafforin C. From the individual in orbital flight to a small group in a polar base. Human Ethology Bulletin. 1999;**14**:5-7

[25] Tafforin C. Global ethological analysis of crew's adaptive dynamics in extreme settings for future human planet exploration. In: Proceedings of the IAA Symposium on the Future of Space Exploration; July 2015; Torino, Italy

[26] Tafforin C, Vinokhodova A, Chekalina A, Gushin V. Correlation of etho-social and psycho-social data from "Mars-500" interplanetary simulation. Acta Astronautica. 2015;**111**:19-28

[27] Johannes B, Sitev AS, Vinokhodova AG, Salnitski VP, Savchenko EG, Artyukhova AE, Bubeev YA, Morukov BV, Tafforin C, Basner M, Dinges DF, Rittweger J. Wireless monitoring of changes in crew relations during long-duration mission simulation. PLoS-One. 2015;**10**:e0134814. DOI: 10.1371/journal.pone.0134814

[28] Robinson J, Slack K, Olson V, Trenchard MH, Willis KJ, Baskin PJ, Boyd JE. Patterns in Crew-Initiated Photography of Earth from ISS—Is Earth Observation a Salutogenic Experience ? On Orbit and Beyond. New York, United States: Springer; 2013. pp. 51-68

[29] Hall ET. La dimension cachée. Paris, France: Seuil; 1971. 254 p

[30] Ritsher JB, Kanas NA, Ihle EC, Saylor SA. Psychological adaptation and salutogenesis in space: Lessons from a series of studies. Acta Astronautica. 2007;**60**:336-340

[31] Suedfeld P, Brcic J, Johnson PJ, Gushin V. Coping strategies during and after spaceflight: Data from retired cosmonauts. Acta Astronautica. 2015;**110**:43-49

[32] Fiedler ER, Harrison AA. Psychosocial adaptation to a Mars mission. Journal of Cosmology. 2010;**12**:3694-3710

[33] Wu R, Wang Y. Psychological interaction during a 105-day isolated mission in Lunar Palace 1. Acta Astronautica. 2015;**113**:1-17

[34] Vinokhodova AG, Gushin V. Study of values and interpersonal perception in cosmonauts on board of International Space Station. Acta Astronautica. 2014;**93**:359-365

[35] Anzieu D, Martin JY. La dynamique des groupes restraints. Paris, France: Presse Universitaire de France PUF; 2013. 397 p

[36] Giner AF. Raices Biológicas del Comportamiento Agresivo en el Hombre. Salamanca, Spain: Universidad de Salamanca; 2015

[37] Kokot W. Culture and space: Anthropological approaches. Ethnoscripts. 2007;**9**:10-23

[38] Harari YN. Sapiens. Une brève histoire d'humanité. Paris, France: Alban Michel; 2015. 499 p

[39] Foing B. Towards a Moon village: Enabling technology and precursor missions. In: Proceedings of the 67th International Astronautical Conference; October 2016; Guadalajara, Mexico. IAC-16.A3.3.2

[40] Tafforin C, Giner Abati F. Interaction and communication abilities in a multicultural crew simulating living and working habits at Mars Desert Research Station. Antrocom Journal of Anthropology. 2016;**12**:97-109

[41] Urbina D, Charles R. Symposium keynote: Enduring the isolation of interplanetary travel. A personal account of the Mars500 mission. Acta Astronautica. 2014;**93**:374-383

[42] Smith C. An adaptive paradigm for human space settlement. Acta Astronautica. 2016;**119**:207-217

Chapter 4

Acute and Chronic Effects of Hypobaric Exposure upon the Brain

Paul Sherman and John Sladky

Additional information is available at the end of the chapter

http://dx.doi.org/10.5772/intechopen.74231

Abstract

Exposure to the hypobaric environment presents numerous physiological challenges to both aviators/pilots, mountain climbers and astronauts. Decompression sickness (DCS) is one of the most commonly experienced maladies and may present variably in protean fashion from mild symptoms such as the bends to severe neurological or pulmonary (i.e. chokes) symptomatology. Furthermore, exposure to extreme non-hypoxic hypobaric environments such as those experienced by our U-2 pilots, irrespective of clinical history of decompression sickness, incites development of white matter hyperintensity lesions that are diffuse in nature. Additionally, non-hypoxic hypobaric exposure also impacts white matter integrity independent of presence of white matter hyperintensities as measured by fractional anisotropy. Functionally, this translated into subtle but significantly lower neurocognitive test performance in U-2 pilots exposed to extreme non-hypoxic hypobaric conditions when compared to pilots without repeated exposure and correlated with degree of white matter lesion burden. In this chapter, we discuss results of our U-2 pilot studies along with published research on high-altitude climbers. We also review ongoing and future directional research and discuss operational implications due to our findings of non-hypoxic hypobaric exposure. Lastly, we examine the incidence of DCS in our astronaut population as well as the risks of performing extravehicular activity (EVA).

Keywords: decompression sickness, hypobaria, MRI, U-2 pilots, astronauts, extravehicular activity, white matter hyperintensities, fractional anisotropy, acute mountain sickness, military free fall operations

1. Introduction

Normoxic hypobaric (low atmospheric pressure) exposure, such as experienced by U.S. Air Force (USAF) U-2 pilots, and inside safety personnel operating altitude chambers (low pressure chambers), is associated with increased subcortical white matter hyperintensity (WMH) burden [1, 2]. Astronauts conducting extravehicular activity (EVA), also known as "space walks," are exposed to a hypobaric environment similar to U-2 pilots. WMHs are regions of accumulation of extra-cellular water due to focal degradation of the myelin sheath [3], and the volume of WMHs is a non-specific marker of cerebral integrity sensitive to multiple etiologies [4]. Repetitive normoxic hypobaric exposure is associated with a decrease in axonal integrity as quantified by global decrease in Fractional Anisotropy (FA) using diffusion tensor imaging (DTI) technique in magnetic resonance imaging (MRI) [5]. Further, this decrease in axonal integrity and increased subcortical WMH burden are associated with a decrement in neurocognitive function [6]. A small convenience sample of astronaut brain MRI data suggests similar WMH change to U2 pilots [7], although the astronaut group was an average of 9 years older. The neuropathophysiological mechanism for decreased axonal integrity and formation of WMHs related to decompressive stress is poorly understood and ongoing human and animal studies are addressing this operational concern. The long-term ramifications of repeated hypobaric exposure are uncertain, but are relevant to current United States Air Force (USAF) military operations and deep space mission plans with frequent EVAs.

WMHs have also been demonstrated in high-altitude mountain climbers [8, 9]. Extreme mountain climbers have demonstrated transient white matter volume change and diffusion tensor imaging (DTI) changes [10, 11]. WMHs have also been demonstrated in other dysbaric environments such as occupational and recreational diving [12, 13]. Results of a meta-analysis of experienced, healthy divers, without a history of neurological DCS, suggest that repeated hyperbaric exposure increases the prevalence of WMHs [14]. Divers included military, commercial and recreational divers, caisson workers, and hyperbaric chamber attendants. It is unknown if WMHs associated with diving or altitude exposure behave in a similar fashion.

2. Background

The U-2S Dragon Lady is a high-altitude military reconnaissance aircraft capable of flying at altitudes over 70,000 feet (21,336 m) for up to 15 h. The U-2 has performed high altitude reconnaissance operations for nearly 60 years and remains heavily utilized by the USAF today. The cabin pressurization system exposes U-2 pilots to cabin pressures equivalent to 29,500 feet (8992 m), approximately the altitude on the summit of Mount Everest. The aircraft was designed with a partially pressurized cabin to save weight and thereby increase attainable altitude. Pilots are required to wear a full pressure suit in case of unexpected cabin decompression. The current pressure suit is designed to automatically inflate in the event of any cabin decompression or bailout to maintain a physiological pressure equivalent of 35,000 feet (10,668 m) or less. U-2 pilots routinely fly operational sorties lasting 8–11 h every 3–4 days. This prolonged hypobaric exposure subjects U-2 pilots to high risk for decompression sickness (DCS). Altitude DCS generally occurs in individuals exposed to a cabin pressure equivalent to 18,000 feet (5486 m) or higher [15]. Pilots can voluntarily inflate the suit during flight to lower DCS risk or to reduce any DCS symptoms they are experiencing. To mitigate risk of DCS, U-2 pilots also undergo a standard denitrogenation ("prebreathing") procedure by breathing 100% oxygen for 60 min before flight. Prebreathing establishes an oxygen gradient to offload nitrogen from tissues to the blood, thereby decreasing nitrogen stored in the body. This prebreathe has been proven highly beneficial in reducing the incidence and delaying the onset of DCS [16]. In 2010 a 10-min exercise-enhanced pre-breathing (EEP) period was added to the standard 60-min resting prebreathe prior to any high-altitude sortie. This further enhances the denitrogenation process and, therefore, may further lower the risk of DCS for the U-2 pilot population [17]. During flight, the pilots breathe 100% oxygen provided by a separate dedicated life support system.

Low-pressure chamber inside safety monitors undergo routine exposure to 25,000 feet (7620 m) for approximately 30 min per training flight, with total time above 18,000 feet (5246 m) not exceeding 60 min. Exposure frequency is variable but generally not more often than every third day, although occasionally mission demands require every other day exposure. The flight profile includes a 30-min denitrogenation period on 100% oxygen and the monitor remains on oxygen (never experiences hypoxia) for the duration of the flight (see **Figure 1** below, Air Force Instruction [AFI] 11-403).

Altitude chamber inside observers experience a similar hypobaric environment to U-2 pilots at a much shorter duration and without increased radiation exposure risk.

2.1. DCS physiology

Following the creation of the first vacuum pump circa 1670, Robert Boyle first noted the formation of numerous gas bubbles in his animals exposed to the reduced atmospheric pressures of the pump [18]. However, it was not until 1862 when the first human episode of DCS was recorded by Paul Bert in his hot air balloon. Bert documented his experience of transient left arm flaccid paralysis that occurred following rapid ascent (305 m/min) to 29,000 feet (8838 m) which resolved following his return to surface [19]. Both hypobaric and hyperbaric exposures can result in decompression sickness (DCS) and neurologic DCS (NDCS). Each of these dysbaric environmental exposures share similar clinical and pathophysiological features, but there are inherent differences to both. One of the fundamental differences is the time of onset; in a hypobaric environment (aviators and astronauts), the symptoms occur during the exposure to a low atmospheric pressure. Conversely, following hyperbaric exposure (divers), the symptoms typically occur after the exposure with majority occurring within 24 h. Furthermore, the pathophysiology is felt to be different as arterial gas embolism occurs primarily during hyperbaric exposure and rarely during hypobaric exposure. Furthermore, the spinal cord is more

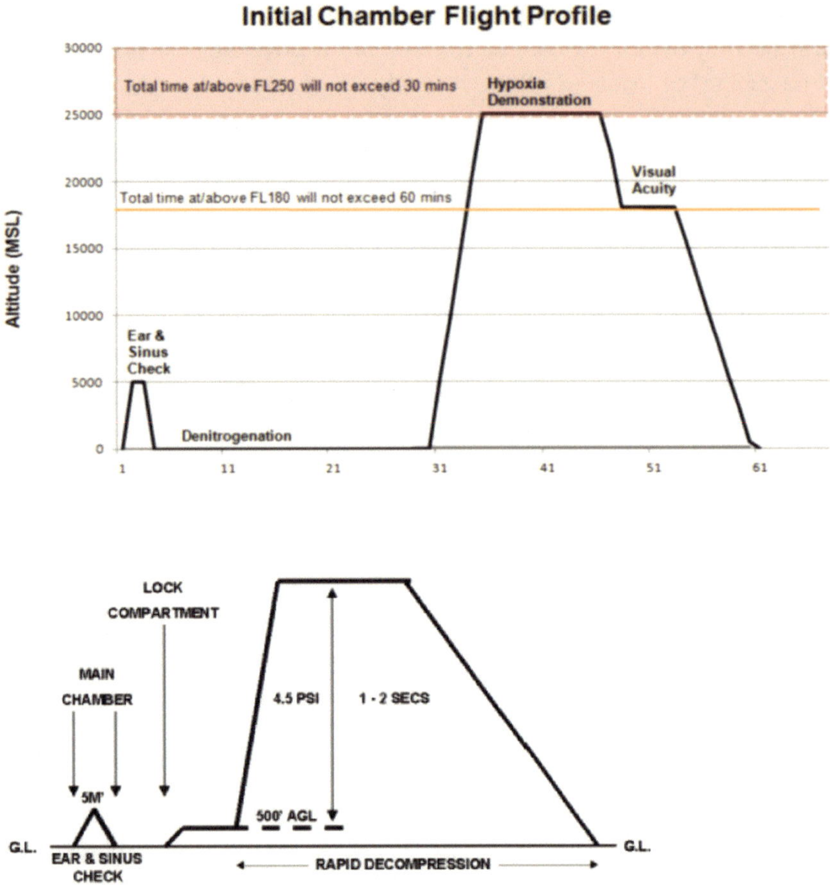

Figure 1. AFI 11-403; Nov 30, 2012. Initial altitude chamber flight profile with rapid decompression.

preferentially affected and vulnerable during hyperbaric exposure [20]. Arterial gas embolism (AGE) arises as expanding gas ruptures alveolar capillaries allowing the entry of alveolar gas into the arterial circulation. Venous gas emboli (VGE) in small quantities are common in diving but are typically asymptomatic as they are effectively filtered by the lung. However, VGE is not a desired condition and in large quantities can cause cough, dyspnea, substernal chest pain, pulmonary edema (referred to as the "chokes") and further cardiorespiratory distress. The presence of a patent foramen ovale (PFO) or other right-to-left shunt can cause the VGE to enter the arterial circulation. The incidence of PFO is approximately 25% in the general population [21].

There are two conditions requisite for development of DCS. The first requires the supersaturation of an inert gas in the surrounding tissue. Supersaturation is defined simply as the partial pressure of the inert gas is greater than the surrounding ambient pressure. The second condition is the development of a gas bubble from the presence of bubble nuclei from the supersaturated tissue. This typically occurs when the decompression rate of the ambient pressure exceeds the rate of inert gas wash-out from the tissues. During diving, caisson work or operation in a compressed air tunnel, supersaturation results due to increase in inert gas partial pressure in the tissues as a direct result of inspiring the air at high pressures. Hypobaric conditions such as aviation or extravehicular activity (EVA) in space predispose to supersaturation of pre-existing dissolved nitrogen at sea level (~570 mm Hg) which can then form bubbles when exposed to reduced barometric pressure. Gas supersaturation in the tissue can be mitigated with phase transition. The issue with DCS is when a gas space arises due to partial or complete desaturation of a pocket of supersaturated tissue. This sets up a pressure difference or "deformation pressure" within the tissue [22]. It is the pressure difference and not necessarily the volume of gas involved which causes the pain observed in pain-only DCS.

Formation of bubbles can result in direct mechanical, embolic and even biochemical effects and the results can range from trivial to fatal. The bubbles can result in mechanical distortion of tissues resulting in pain or may occlude vascular structures resulting in stroke-like signs and symptoms. There are three potential sources of microemboli: (1) micro-bubbles of gas; presumably nitrogen; (2) small thrombi secondary to platelet activation and deposition; (3) microparticles. Other effects also include endothelial injury resulting in leakage of plasma and increased leukocyte endothelial adhesion. The classic symptoms of DCS including joint pain, paresthesias and skin changes are thought to be secondary to either direct pressure of the gas bubble on the tissue itself, blockage of small arteriolar vessels, and/or interaction with serological proteins [20]. While echo imaging has shown the presence of venous gas emboli in tissues [23], the presence of arterial gas emboli is quite uncommon, reported in only 6 of more than 1500 altitude chamber exposure cases [24]. Occlusion of small cerebral vessels by activated platelets due to accelerated coagulation in the presence of venous nitrogen gas bubbles was demonstrated in both medium and large-sized arteries in mice after DCS and therefore remains another possibility for the development of WMH [25].

2.2. Signs and symptoms of DCS

Signs and symptoms of decompression (DCS) are protean and range from mild to severe including death. Historically DCS has classically been divided into arterial embolism, Type I, Type II and skin bends. However, due to inconsistencies in applying this classification system, it has largely been replaced by the all-inclusive term decompression illness (DCI) and based on system involvement. For consistency, we will keep with the prior classification system. Type 1 is typically referred to as "pain only" DCS symptoms or the "bends" with localized pain in the joints (lower extremities; particularly knee involvement) and may be accompanied by cutaneous manifestations (pruritus and mottling) and constitutional symptoms. Type II symptoms are systemic and more severe and generally involve both the central nervous system and cardiopulmonary systems (see **Table 1** below). Skin bends refer to the marbled appearance of the skin whereas the characteristic rash of livedo reticularis (cutis marmorata) is a more severe form of skin bends and is nearly pathognomonic for decompression sickness in the appropriate clinical context [26]. In one prospective study that looked at 447 cases of DCS over an 11 year period at the Armstrong Laboratory, the most collective symptom was musculoskeletal in 83% of the cases of which knee pain was the most common. This was followed

DCS classification	Signs and symptoms	Location of bubbles
Type I: "Pain-only"		
"Bends"	• Localized deep joint pain • Pain may often occur at altitude but may occur during descent or even hours later	Large joints: • Elbows • Hips • Wrists • Shoulders
Mild skin changes	• Pruritus • Mottling (mild) • Formication (feeling of ants crawling on skin)	Skin
Type II: More severe systemic involvement		
Neurological	• Confusion/memory loss • Visual changes: diplopia, scotomas • Headache • Seizures, vertigo, unconsciousness	Brain
	• Dysthesias and paresthesias around lower chest • Constriction pain/pressure around chest or abdomen • Ascending paralysis • Bowel/bladder incontinence	Spinal cord
	• Fasciculations or muscle twitching • Paresthesias/numbness	Peripheral nerves
Cardiopulmonary: aka "Chokes"	• Dyspnea (shortness of breath) • Dry cough • Pain worsened with breathing • Deep burning chest pain	Lungs
Skin bends: cutis marmorata	• Livedo reticularis rash • Pitting edema	Skin

Table 1. DCS signs and symptoms.

by paresthesias in 10.8%, chokes in 2.7%, cutaneous lesions in 2.2% and neurological deficits in 0.5% [27]. Neurological decompression sickness (NCDS) resulting from hypobaric exposure typically involves the brain more than spinal cord and may range from mild symptoms such as slowed thought processes to severe including confusion, aphasia, unresponsiveness and even permanent cognitive decline [28].

2.3. Incidence of DCS in U-2 pilots

The USAF U-2 pilots are at considerable risk of development of DCS due to the extreme altitudes and long duration sorties. The risk of DCS is dependent on both the rapidity of ascent as well as the duration of exposure to altitude (typically defined as >18,000 feet) in a hypobaric environment. In a 1996 survey of 416 active/retired U-2 pilots (60% response rate), more than 75% of the pilots attested to DCS symptoms such as joint pain or skin manifestations. 12% of those surveyed cited at least one episode that was severe enough to abort or alter the profile of their mission [29]. The risk of DCS per flight increased from 0.076% pre-2006 to 0.23% during the 2006–2010 operation years [30]. Furthermore, 44% of episodes were diagnosed as NDCS including 5 life-threatening cases with symptoms ranging from mild, such as complaints of slowed thought processes to severe, including anomia, confusion, unresponsiveness, and cognitive decline. Neuropsychiatric symptoms persisted in 6 pilots which may represent permanent injury. This upsurge in NDCS was felt to be a consequence of more frequent and longer periods of flight/hypobaric exposure for the pilots [28].

2.4. Incidence of DCS in astronauts

Astronauts are also at risk for the development of DCS. Before the very first EVA (extravehicular activity) occurred, NASA realized that DCS was a risk to be mitigated. Earth-normal atmospheric pressure at sea level is 760 mm Hg (14.7 psia or 1ATA). The current NASA space-suit referred to as the EMU (extravehicular mobility unit) operates at 4.3 psia or 222 mm Hg above the vacuum of space whereas the Russian Orlan space-suit operates at 5.8 psia. Increasing the space-suit pressure or reducing cabin inert pressure are the two ways to reduce the pressure gradient differential between environments to help minimize risk of DCS. However, increasing suit pressure typically results in reduced operational capacity by the astronaut due to increased fatigue, reduced dexterity and mobility [26]. DCS is a known risk during EVA but complete elimination of DCS is practically impossible. Therefore, mitigation plans between USAF, United States Navy (USN), NASA and the academic research community were undertaken to define "acceptable risk." The current definition implemented by the International Space Station (ISS) protocols is the following: (1) DCS < 15%; (2) Grade IV VGE < 20%; (3) No type II DCS.

However, despite the above concerns of DCS, there have been no recorded cases of DCS among astronauts and cosmonauts during EVA's working in pressurized space-suits between 3.7 and 5.8 psia. This is in stark contrast to both Russian and American altitude chamber technicians who report symptoms or signs of DCS ~ 20–40% of the time [26, 31]. There are three possible explanations for this disparity: (1) Potential bias not to report symptoms; (2)

Masked DCS symptoms; (3) Potential operational and gravitational benefits of the spaceflight environment.

Regarding the bias not to report, an EVA is considered the pinnacle of any astronaut's career and the willingness to divulge mild DCS symptoms such as pain that was not operationally limiting would be nominal. It is important to note that NASA's current policy states that any DCS symptom incurred by a crewmember or test subject who participate in hypobaric or hyperbaric operations needs to be reported [26]. It is known that under-reporting of DCS symptoms occurred in the U-2 pilot population as reporting of DCS symptoms during hypobaric operational training could lead to disqualification. This was discussed earlier in Bendrick's article on 275 U-2 pilots of whom 75% reported DCS symptoms via an anonymous questionnaire at least once in their career but rarely reported it to their Flight Surgeon [29]. Interestingly, Webb et al. in 1996 published an article citing an incidence of DCS in 77% of test-subjects undergoing the 60 min U-2 pre-breathing protocol [32]. This again highlights the disparity between operational vs. research reports of DCS and underscores that for numerous reasons, astronauts and pilots are not inclined to report every slight discomfort they experience.

However, in addition to under-reporting bias, there are valid reasons why mild symptoms of DCS may be masked during an EVA. Astronauts frequently take aspirin prior to any EVA to pre-emptively mitigate any aches or pain. In addition, the actual operation of the EMU spacesuit that the astronaut dons can be painful. It would be near difficult for an astronaut to discern pain from "pain-only" DCS vs. the natural discomfort incurred from working within the confines of the EMU. Furthermore, as most "pain-only" DCS symptoms resolve following re-pressurization after completion of the EVA, there is no driving force for astronauts to report [26].

Furthermore, it is distinctly possible that DCS has not occurred during an EVA. There is a stark contrast between a test subject wearing an O_2 mask in a shirt-sleeve training environment at 1-G and an astronaut maneuvering in an uncomfortable spacesuit in micro-G environment, surrounded by 100% $O_{2.}$ One aspect to reduced incidence of DCS is simply due to limited motion in the both the Orlan (Russian) and EMU (American) spacesuits. Another possible explanation is the longer pre-breathing exposure during EVA's compared to those tested in a chamber along with exposure to a micro-gravity (μG) environment. The latter situation is unique in that during adaptation to a μG condition, there are substantial fluid shifts from the legs to the torso and head with a net reduction in total body water. As a response to these fluid shifts, denitrogenation may be more efficient and accelerate nitrogen wash-out from the tissues [33]. Furthermore, astronauts are physically active during their prebreathe protocols and it is well documented that exercise during prebreathe enhances N_2 washout from the tissues [34].

2.5. Acute mountain sickness

A review of clinical and MRI findings in acute mountain sickness (AMS) and high altitude cerebral edema (HACE) demonstrates parallels to recent findings in our U-2 pilots and low-pressure chamber inside safety monitors, with the hypobaric environment as the common element. The primary difference is the presence of hypoxia in AMS and HACE. AMS generally occurs above 2500 m and has been defined by the Lake Louise Consensus Group as the presence of a headache with one or more of the following: gastrointestinal symptoms (nausea, vomiting, anorexia), insomnia, dizziness and lassitude or fatigue [35]. Determining factors include the rate of ascent, altitude reached, altitude at which a person sleeps and individual physiology. Most consider HACE to be a clinical and pathophysiologic extension of AMS. HACE is an encephalopathy, characterized by disturbances of consciousness that may progress to coma, ataxic gait, increased intracranial pressure and retinal hemorrhages.

A growing body of evidence suggests that is not only hypoxia, but hypobaria that contributes to the development of AMS [36–40]. The underlying pathophysiology of AMS remains poorly understood. Hypoxia-induced cerebral vasodilatation or its effectors, such as nitric oxide, may produce the headache, possibly via the trigeminovascular system or by causing mild cerebral edema [41–43]. Whether this edema is cytotoxic (intracellular), vasogenic (extracellular), or both remains controversial. However, MRI has demonstrated reversible abnormalities in HACE, such as areas of increased T2 and fluid-attenuated inversion recovery signal intensity within the splenium of the corpus callosum (white matter structure), with associated increased apparent diffusion coefficient (ADC) values consistent with increased water diffusivity. These findings are indicative of vasogenic edema. Hemodynamic factors such as sustained vasodilation, impaired cerebral autoregulation and elevated capillary pressure may contribute to vasogenic edema [44–47]. Hypoxia-induced biochemical alteration of the blood brain barrier may also be important. Current high altitude human research demonstrates increased cerebral blood flow after a single hypoxic hypobaric exposure to 7620 m for occupational training which persists at 72 h. These findings will be described in detail later in this chapter.

Central nervous system MRI changes demonstrated in AMS have similarity with recently published astronaut data. These include intracranial fluid redistribution, increased intracranial pressure in microgravity, and brain structural plasticity changes from pre-to-post spaceflight [48–50].

2.6. Brain magnetic resonance imaging (MRI) techniques

Our original U-2 pilot brain MRI evaluations were performed on a Siemens (Siemens AG, Erlangen, Germany) Magnetom Tim Trio 3-Tesla scanner at the Research Imaging Institute (RII), University of Texas Health Science Center San Antonio (UTHSCSA), with a 12-channel phased array coil. All subsequent human and animal MRI data has been acquired on the same Siemens Magnetom Verio 3-Tesla scanner at Wilford Hall Ambulatory Surgical Center (WHASC), Joint Base San Antonio, Texas using a 32-channel phased array coil. This includes all imaging on control/normal subjects and low-pressure chamber inside safety monitors. Both scanners are operated under quality control and assurance guidelines in accordance with recommendations by the American College of Radiology.

Three-dimensional imaging parameters were for T1 magnetization prepared rapid acquisition gradient echo (MPRAGE), repetition time (TR) = 2200 ms, echo time (TE) = 2.85 ms, isotropic resolution = 0.80 mm, and for fluid-attenuated inversion recovery (FLAIR), TR = 4500 ms,

TE = 11 ms, isotropic resolution = 1.00 mm. FLAIR image processing was previously reported [2, 51–54]. WMH regions were coded as ependymal regions, contiguous with CSF structures, and as subcortical regions as previously described [57]. WMHs were quantified in number (count) and total volume.

Calibration MRI data was performed in 46 patients (CAL) on both the RII and WHASC scanners for the cross comparison and analysis of advanced imaging sequences such as diffusion imaging [5]. The calibration for the average FA values in subjects imaged on both scanners showed excellent correlation ($r = 0.85$), with coefficients of variation were similar to what has been previously reported [56, 57].

High angular resolution diffusion imaging (HARDI) was utilized for diffusion tensor imaging (DTI) and fractional anisotropy (FA) assessment as previously reported [51, 58]. Briefly, DTI data were collected using a single-shot echo-planar, single refocusing spin-echo, T2-weighted sequence with a spatial resolution of 1.7 × 1.7 × 3.0 mm with sequence parameters of TE/TR = 87/8000 ms, We chose the ENIGMA-DTI analysis protocol [59] because it can effectively overcome the impact of the punctate WMH lesions on FA values compared to simple averaging of FA values within a region of interest, effectively limiting analysis of FA values to that of the normal-appearing WM. DTI is a quantitative MRI technique that has an advantage over T2-weighted fluid attenuated inversion recovery (FLAIR) imaging because it can ascertain subtle WM damage in normal-appearing WM prior to development of WMH lesions [60]. FA is a widely used quantitative measure of WM microstructure, extracted from DTI [61]. FA is an important biomarker in clinical studies as it can sensitively track WM changes in neurological and psychiatric diseases [57, 62, 63] and in normal development and aging [64].

Two additional advanced sequence techniques were used in our current hypobaric research: pseudo-continuous arterial spin labeling (pCASL) and proton magnetic resonance spectroscopy (MRS). pCASL technique is a noninvasive method for calculation of estimate cerebral blood flow which does not require intravenous contrast injection. MRS demonstrates quantifiable neurometabolite concentrations regions of interest, both gray and white matter. pCASL imaging data for gray and white matter were collected using gradient-echo echo-planar imaging with TE/TR = 16/4000 ms as previously reported [51]. Further, pCASL data were processed using the pipeline described elsewhere [65]. Perfusion-weighted images were calculated based on the methods described by others [67, 68].

Proton magnetic resonance spectroscopy (MRS) data were acquired from voxels placed in frontal white matter (FWM) and the anterior cingulate cortex (ACC). For the frontal white matter region, short TE and long TE data were acquired using point resolved spectroscopy localization (TR = 1500 ms, short TE = 30, long TE = 135 ms, number of signals averaged (NEX) = 256, volume of interest (VOI) ~ 3.4 cm^3). Data were acquired in both hemispheres and averaged together. For the anterior cingulate, the same short TE point resolved spectroscopy localization parameters were used with a voxel size of 6 cm^3. Standard neurometabolites were evaluated using available software and methods as previously reported [69, 70] We have demonstrated a high degree of consistency across structural and physiological measurements with brain MRI [51].

2.7. White matter integrity in high altitude pilots exposed to hypobaria

The number and volume of WMH regions are sensitive markers of cerebral health, commonly used to study the extent of the cerebral injury [71]. Healthy cerebral white matter tracts are myelinated with compounds containing long-chain fatty acids with very short T2-relaxation time and thus appear dark on T2-weighted images. Local edema, often associated with degradation of the myelin sheath, results in localized accumulation of extracellular water, which leads to an increased signal intensity on a T2-weighted image. HWM lesions also form in normal aging, where they begin to occur during mid adulthood (fourth-fifth decade of life). In both normal subjects and patients who suffered brain injury, the number and volume of HWM lesions are correlated with a decline in cerebral integrity [72], reduction in cerebral white matter and gray matter volumes [73, 74], cerebral blood flow [75], and cerebral glucose metabolism [76]. Increasing numbers and volumes of HWM regions have also been linked to cognitive declines, particularly in executive functioning [77], processing speed [78], and general cognitive status [79], and were correlated with the severity of neurocognitive deficits in neuropsychiatric and neurological disorders [80].

The etiology of HWM is nonspecific and is commonly associated with cerebral ischemia and disruptions of cerebral circulation [81]. Histopathological findings indicate there are two distinct types of HWM lesions: subcortical and ependymal. Subcortical HWM regions are more closely associated with ischemic factors [3]. In contrast, periventricular ependymal HWM lesions are thought to be of non-ischemic origin and potentially produced by pulse-wave encephalopathy [55, 82, 83]. This condition refers to the microtears in the ependymal lining caused by the pulsatile movements of ventricular cerebrospinal fluid (CSF) [83–85].

Our initial study evaluated 50 U-2 pilots (avg. age 37.4 ± 5.2 year), 12 (avg. age 38.9 ± 6.1 year) of whom had suffered neurological decompression sickness (NDCS) [86]. The NDCS pilots demonstrated a significantly higher total WMH lesion volume ($p = 0.026$) compared to the non-NDCS pilots, but not a significant increase in total lesion count ($p = 0.120$). Analysis of the lesion by type (subcortical vs. ependymal) did not demonstrate a significant difference between NDCS pilots and non-NDCS pilots ($p = 0.059$). Examination of regional measurements revealed pilots who experienced NDCS had significantly higher number and volume of insular subcortical lesions ($p = 0.020$ and $p = 0.018$, respectively). No difference was noted in the presence of mild hypertension or mild hyperlipidemia. No difference was noted with total flight hours or average high-flight hours per month between the two groups. No pilot had a history of significant head injury, significant scuba diving history, episode of decompression illness associated with diving, or high-altitude exposure other than that associated with USAF flight duties. The initial hypothesis for the elevation of WMH volume in pilots that suffered NDCS was hypobaric-related gas microemboli (<30 μm) which may have led to loss of permeability or occlusion of small cerebral vessels and subsequent immune mediated gliosis (**Figure 2**).

What was noteworthy was the prevalence of WMHs in high altitude pilots that had not suffered NDCS, mandating comparison of high altitude pilots to a normal control group.

Subsequent MRI evaluation was performed on 105 total high-altitude U-2 pilots (U2P; mean age 37.7), 83 low pressure chamber aerospace physiology inside observers (AOP; mean age

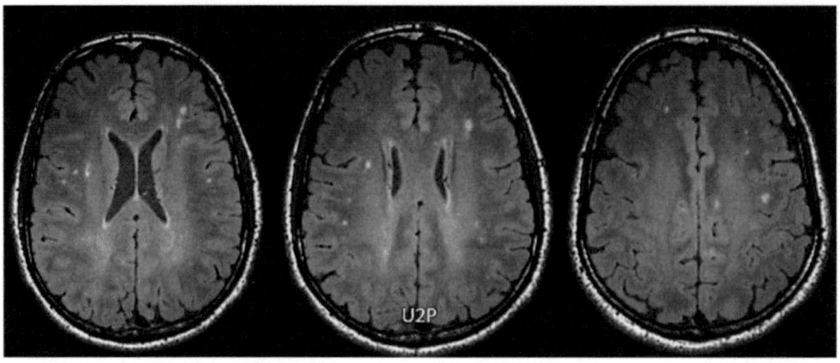

Figure 2. Axial FLAIR images demonstrating multiple subcortical WMHs in a U-2 pilot, without NDCS.

36.5) and 148 age and health matched advanced/doctorate degree control subjects (CTRL; mean age 34.6) [1, 2, 5, 6]. All study subjects were active duty members of the US Armed Forces. All participants were between the ages of 26 and 50 years, were healthy at time of study without any history of central neurologic or psychiatric disease, and had undergone a routine annual medical examination within 12 months prior to study. All participants at the time of testing met USAF Flying Class II neurological standards [87]. Briefly, exclusionary criteria for Flying Class II include a history of any of the following: head trauma with any loss of consciousness or amnesia; migraine headache; psychiatric or psychological disease requiring any medication or hospitalization; hypertension (HTN) requiring more than a single angiotensin-converting enzyme inhibitor (ACE-I) for control; hyperlipidemia (HLD) requiring more than a single statin for control; diabetes or glucose intolerance; ischemic cardiac disease; any neurological disease including infection, seizure, or stroke; any medical condition associated with neurological injury; or substance or drug abuse or dependence. All AOP had experienced >50 occupational exposures to >25,000 feet altitude. Two (2.4%) of AOP and 16 (15%) of U2P reported mission-related symptoms of NDCS. All U2P and AOP undergo standardized hypoxic hypobaric chamber exposure as part of routine aircrew qualification training every 5 years; these exposures are of 30- to 60-min duration with hypoxia relieved via 100% oxygen aviator mask with the onset of physiological symptoms. Fourteen CTRL had experienced a single episode of aircrew hypoxic hypobaric chamber exposure as part of initial flight surgeon qualification training. No subject experienced NDCS related to this periodic aircrew chamber training.

Group-wise analysis demonstrated that both AOP and U2P, two groups occupationally exposed to nonhypoxic hypobaria, had significantly elevated WMH volume/count compared to CTRL. Although the WMH volume/count were higher in U2P than in AOP, neither was statistically significant. Comparable results were obtained in group-wise analysis after excluding any subject with HTN or HLD (14 CTRL, 10 AOP and 20 U2P) and after excluding AOP or U2P who had experienced NDCS. Equivalency of U2P to AOP WMH volume was noted on the Kolmogorov–Smirnov test (p50.388). The Jonckheere–Terpstra test demonstrated CTRL < PHY_U2P on WMH volume (p50.024) and count (p50.012); AOP < U2P was not significant ($p > 0.10$). The Spearman correlation coefficients between WMH volume/count and age and hours of hypobaric exposure were positive but not significant. Linear regression of combined

AOP and U2P total hours of hypobaric exposure versus WMH volume/count was not significant (WMH volume/count, r250.002/r250.009, respectively). The total hours of exposure were not significantly associated with WMH presence in either group. Likewise, the Spearman correlation coefficient between 2 measures of WMH burden and age were positive but not significant (all $r^2 < 0.03$; all $p > 0.10$).

Regional analysis revealed that frontal lobe lesions constituted the largest fraction of both volume and number of WMH loci in both U-2 pilots (50 and 56% for volume and number, respectively) and normative controls (69 and 70% for volume and number, respectively). This is presumably because higher metabolic demand and cerebral blood flow. Pilots had a higher volume ($p < 0.03$) of WMH in the frontal, insula, limbic, sublobar, and temporal regions and a higher number ($p < 0.01$) of WMH in the insula, limbic, temporal, and sublobar regions. WMH were normally uniformly distributed throughout the brain in U2P than in controls and did not increase with age.

The relationship between hypobaric exposure and WMH is complex. We observed no significant correlations between WMH measurements and the total number or hours of hypobaric exposure. This suggests that other factors may modulate the hypobaria-related WMH change, including hyperoxemic pre-exposure nitrogen degassing, exposure duration, level of physical and mental activity during exposure, frequency of exposure episodes, and amount of rest between exposures, as well as other yet unknown environmental and genetic susceptibility risk factors. Injury secondary to microemboli, of nitrogen gas, platelet-based thrombi or microparticles, remain a potential source of this injury.

Diffusion tensor imaging (DTI) and FA findings in U2P were noteworthy for demonstrating effects suggesting a global process, affecting normal appearing white matter, not just subcortical white matter damage, presumably secondary to repetitive hypobaric exposure. Whole-brain average FA values for all pilots were significantly lower than in controls (KS $p < 0.001$; GLM $p < 0.001$). After Bonferroni correction of p-values, we observed two regional findings: pilots had significantly decreased FA values for the sagittal striatum ($p < 0.001$), while pilots had significantly higher FA values for fronto-occipital fibers ($p = 0.003$). Functionally, the striatum coordinates multiple aspects of cognition, including motor and action-planning, decision-making, motivation, reinforcement, and reward perception. The fronto-occipital tract integrates auditory and visual association cortices with the prefrontal cortex. Other FA tracts were not significantly different.

We separated the pilots into lower two-thirds (U2PL)/upper one-third (U2P-U) based on WMH burden (U2P-L/U2P-H). There was no significant difference in WMH burden between U2P-L and controls (CTRL) (WMH volume/count $p = 0.17/0.52$, respectively), while there was a significant difference between U2P-H/U2P-L ($p < 0.001/$, 0.001) and U2P-H/CTRL ($p < 0.001/<0.001$). Comparing FA values of U2P-H and U2P-L to CTRL demonstrated significantly lower FA values in both pilot groups for whole brain average FA ($p < 0.001/p < 0.001$, respectively, U2P-H/U2P-L) and sagittal stratum ($p = 0.005/p = 0.01$). Comparing mean values of U2P-H to U2P-L demonstrated a nonsignificant trend toward lower FA values in U2P-H than U2P-L for whole-brain average FA and all tracts except fronto-occipital where U2P-H = U2P-L).

Lower average FA findings are consistent with a diffuse disruption in white matter integrity. This finding trended with higher WMH burden previously described. Reduced sagittal stratum FA has been shown to be genetically associated with processing speed deficits in two independent cohorts [88]. We observed a decrease in processing speed in U2P compared to a USAF pilot cohort control [11] and this may suggest the reduced sagittal stratum FA in U-2 pilots may explain this decrease in processing speed. Additionally, USAF pilots are uniquely high-functioning individuals with exceptional visual-spatial abilities [89], which may account for the higher FA values in the fronto-occipital fibers in U-2 pilots, reflecting this associative cognitive ability, and provide an anatomical basis for the superior spatial performance noted in all USAF pilots. Historically, the pathophysiological theory of hypobaric related brain damage has been arterial gas emboli, but there are other recent studies which also suggest a more diffuse process [23, 90]. It is improbable that gas emboli alone could produce the diffuse disruption of axonal integrity demonstrated by our MRI findings. Our studies provide support for other potential pathophysiological explanations, including neuroinflammation and microparticle damage [91, 92].

2.8. Neurocognitive changes

WMH are also relevant surrogates for cerebral activity in neurological disorders and also normal aging. As stated earlier, these WMH have also been linked to cortical and subcortical functions particularly executive function, processing speed, overall cognition along with motor/gait function [76–78]. Therefore, we compared neurocognitive performance in U-2 pilots with repeated hypobaric exposure to pilots without repeated hypobaric exposure and also assessed whether cognitive performance correlated with severity of WMH burden. All participants were between the ages of 26–47 years old and had to meet Flying Class II standards and could not have any prior history of neurological or psychiatric disease. 106 U-2 pilots were compared against 83 active duty (AD) pilots who were also matched for age at time of cognitive testing. Computer-based Multi-Dimension Aptitude Battery-II (MAB-II) and Assessment of Cognitive Function (MicroCog) assessments were utilized. MAB-II yields an overall evaluation of neurocognitive ability based on the Wechsler Adult Intelligence Scale and generates three intelligence quotient (IQ) scores: full-scale IQ, verbal IQ and performance IQ. The MicroCog is a separate computer-based cognitive assessment that comprises18 subsets resulting in 9 index scores. The MicroCog was specifically chosen to provide more accurate information regarding reaction time and processing speed, both critical functions to any active aviator. While there were no significant differences between U-2 and AF pilots on the MAB-II testing, there were subtle but significant differences on the Micro-Cog assessment. Specifically, U-2 pilots scored significantly in the following domains (see **Table 2**): reasoning/calculation ($p < 0.001$), memory ($p = 0.007$), information processing accuracy ($p = 0.016$), and general cognitive functioning ($p = 0.002$). Furthermore, within the U-2 pilot population, significantly lower scores on reasoning/calculation, memory, general cognitive functioning and proficiency were observed in those pilots with higher WMH burden [6]. However, it is relevant to note that despite the differences in the U-2 pilots, their overall neurocognitive performance continues to remain commensurate with age and cohort-specific normative data tempering concerns for any immediate clinical significance. The long-term sequela is unknown.

MicroCog category	U-2 pilots (n = 93)	Air Force pilots (n = 80)	*p*-Value, t-test (2-tailed)
Attention/mental control	104.4 ± 9.3	103.8 ± 10.8	0.696
Reasoning Calculation	99.4 ± 12.5	106.5 ± 10.9	<0.001
Memory	105.5 ± 12.5	110.9 ± 13.7	0.007
Spatial Processing	109.1 ± 9.4	109.1 ± 9.4	0.989
Reaction Time	107.3 ± 6.7	104.8 ± 9.2	0.047
IPS	103.6 ± 12.5	106.5 ± 10.5	0.100
IPA	102.1 ± 9.8	105.8 ± 10.0	0.016
GCF	103.5 ± 10.0	108.5 ± 10.6	0.002
GCP	105.4 ± 9.4	108.6 ± 10.2	0.037

IPS: information processing speed; IPA: information processing accuracy; GCF: general cognitive function; GCA: general cognitive proficiency.

Table 2. Microcognitive testing results in U-2 pilots compared to Air Force pilot controls.

2.9. Current research and future directions

Current human research is focused upon occupational exposure in military environments, specifically the serologic, neurometabolite and brain MRI changes after a single exposure to low pressure chamber altitude training (training profile as per AFI 11–403 above). Study volunteers are active duty members who have recently completed basic military training and are completing an aircrew fundamentals course prior to additional training for aircrew duties (AFC). AFC students experience hypoxia symptoms for 3–5 min to meet the training objective. An additional active duty age-matched control group (CTRL) group was also recruited. Brain MRI technique is similar to technique previously described; all performed on the same WHASC Siemens 3-Tesla scanner, with discussion of pCASL and MRS.

Preliminary evaluation of pCASL and MRS techniques has been performed on 96 AFC trainees and 68 healthy CTRL subjects. MRI evaluation was obtained 24 h before, 24 h after, and 72 h after low pressure chamber exposure and at the same time intervals for CTRL without the hypobaric exposure. A GAM which controlled for age and gender differences was used to the compare the two groups. There is a statistically significant increase in cerebral blood flow (CBF) in white matter in the AFC group ($p < 0.001$). The difference is dependent upon age as a covariable, although there is no significant difference in age between the two groups ($p > 0.10$). It is possible that this might reflect a difference in central nervous system maturation. Increased CBF persists on the 72-h post exposure MRI and it is unknown how long CBF remains elevated. Findings reflect an increased metabolic demand upon the brain and suggest a transient injury from a single exposure to hypobaria. There was a significant difference in most neurometabolites within the ACC and in GSH within the FWM in aircrew personnel with hypobaric exposure as compared to controls. These differences may be representative of changes at a cellular level in response to, or preceding, changes in blood flow to these regions versus age-related differences or differing WMH between the two groups. This remains a subject of ongoing evaluation.

2.10. Military free fall operations

U-2 pilots are not the only group at risk of DCS during operational movements or exercises. This would include our high-altitude high-opening (HAHO) and high-altitude low opening (HALO) parachutists. Due to improvements in both parachutes and life-support systems, military parachutists are now able to drop from altitudes in excess of 25,000 feet. These higher altitudes carry an increased risk of DCS. Furthermore, slow descent (HAHO operations), colder temperatures (ambient temperature at 35,000 feet is approximately −56°C), and even moderate exercise at altitude [93] increase the risk of VGE. Another practical issue is that the presence of any facial hair can impair the seal of the oxygen mask over the parachutist's face rendering prebreathing ineffective. All parachutists engaged in military free fall (MFF) must undergo strict prebreathing protocol as outlined by Air Force Instruction (AFI) 11-409 (See **Table 3** below) [94].

Furthermore, in an effort to reduce the physical demands and risk of DCS on MFF parachutists, the following military protocol was issued: (1) MFF parachutists may not conduct more than two jumps between 13,000 and 17,999 feet in a 24-h period; (2) Conduct no more than one oxygen jump above 18,000 feet in a 24-h period; (3) not conduct MFF operations within 24 h of making a non-oxygen dive and (4) not wear dark goggles on MFF operations that require prebreathing to facilitate viewing of the eyes of the jumpers by the jumpmaster to ensure they are not experiencing physiological difficulties [95]. Despite such extensive and potentially mitigating protocols there is little information in the literature regarding the actual incidence of DCS affecting parachutists engaged in HAHO or HALO operations. One small study of 10 experienced parachutists underwent blinded exposure in a hypobaric chamber to both 17,500 and 35,000 feet respectively separated by 48 h. Participants underwent 60 min 100% O_2 prebreathe, and then decompressed to respective altitude over 7 min where they remained for 15 min followed by slow descent over 35 min. They were suspended to reduplicate the effects of the harnesses on blood flow in the lower limbs. VGE detection was accomplished by precordial

Altitude	Oxygen requirement	Pre-breathe time	Maximum
10,000–12,999 feet	Aircrew: supplemental jumpers: see below[a]	N/A	Unlimited
13,000–19,999 feet	Supplemental	N/A	Unlimited
20,000–24,999 feet	100% O_2	30 min	110 min
25,000–29,999 feet	100% O_2	30 min	60 min
30,000–34,999 feet	100% O_2	45 min	30 min
35,000 feet or greater	100% O_2	75 min	30 min

[a]Supplemental oxygen: parachutists/jumpers may perform unpressurized operations between 10,000 and 13,000 feet without supplemental oxygen not exceed 30 min. For unpressurized flight above 13,000 feet or exceeding the 30-min envelope between 10,000 and 13,000 feet, a continuous supply of supplemental oxygen will be used.

Table 3. AFI 11-109.

2D and Doppler echocardiography. Following exposure, the 10 parachutists then engaged in ground level moderate exercise consisting of a 4 km/h. march on a treadmill while carrying a Bergen weighing 40 lbs. While the study sample was small, there was no evidence of VGE or DCS during the altitude profile of the study and no evidence for resurgence of VGE or exercise intolerance during the ground profile of the study [96]. These findings corroborated Webb's work in 2002 that indicated that exercise at ground level would not trigger a resurgence of VGE or DCS symptoms following a 2-h exposure at an altitude of 35,000 feet [97].

2.11. Extravehicular activity (EVA)

As discussed earlier in the chapter under DCS subsection, there have been no reported cases of DCS during EVA. This concern has been mitigated using various strategies such as implementation of a lower pressure high oxygen environment utilized in the Gemini, Apollo space missions and Skylab space station coupled with single 4-h pre-launch oxygen prebreathe. This resting 4 h-in-suit prebreathe protocol [98] has been utilized six times during space-flight without reported incidents of DCS. Other protocols included the "Camp-Out" protocol (last used on May 6, 2011) which involved exposure to a mildly hypoxic environment requiring a single 40–75 min in-suit prebreathe, along with several exercise-enhanced protocols. The two most common of the exercise prebreathe protocols include the "cycle ergometer with vibration isolation and stabilization" (CEVIS) and "in-suit light exercise (ISLE) protocols. These were developed to help minimize scheduling constraints of EVA's following delivery of the International Space Station (ISS) *Quest* airlock in 2001. The theory behind these protocols is that since denitrogenation is a perfusion-limited process, the implementation of exercise into the prebreathe protocol may facilitate denitrogenation. The CEVIS protocol uses a short but intense prebreathe exercise protocol (10 min duration) utilizing cycle ergometry with escalating workload peaking at 75% $VO_{2\ max.}$ After completion of exercise, the astronaut then prebreathes 100% oxygen for the next 50 min followed by depressurization to 10.2 psia in the ISS airlock over 30 min. It is during this depressurization that the spacesuit is donned. As of May 6, 2016, the CEVIS protocol has been utilized 52 times with no reported signs or symptoms of DCS. In contrast to the CEVIS protocol, ISLE prebreathe protocol replaces the bouts of short, intense exercise with longer period of mild exercise in the EMU (spacesuit). While it shares many steps with the CEVIS exercise protocol it does differ in that only 40 min are spent prebreathing followed by 20 min depressurization to 10.2 psia. Once the suit is donned, mild exercise consisting of arm and leg circular motions are performed over 4 min followed by 1-min rest period. This cycle continues for total duration of 50 min achieving a $VO_{2\ max}$ of 6.8% (compared to 75% in the CEVIS protocol). This is followed by an additional 50 min prebreathe of 100% oxygen culminating in a final depressurization of the airlock to vacuum. The ISLE protocol has been used over 40 times since May 6, 2016 without any occurrences of DCS is currently the prime protocol used by the ISS [26].

These complex prebreathing protocols were designed to meet operational demands but in doing so have left knowledge gaps regarding DCS risk factors particularly in space. These include risk of bubble formation in space, micronuclei generation, implications of tissue saturation across different gas and pressure environments, and nitrogen elimination in space. Furthermore, there are numerous physiological factors to consider such as age, body habitus, aerobic conditioning, presence of PFO, gender, hydration status and even timing of menstrual cycle that can influence the development of DCS in a hypobaric environment or vacuum such as space. The relative risk and importance of these physiologic risk factors in the genesis of DCS is unknown until a multivariate analysis such as logistic regression or survival analysis is undertaken. Lastly, these DCS risk-mitigation protocols will not likely be sufficient or applicable to future space exploration missions that utilize suitports, variable pressure suits, and require the ability to rapidly deorbit for medical therapy. Historically EVA has been a single event in a flight day. However, the standard operational concept for future exploration missions is the possibility of multiple EVA's in 1 day or performing a single EVA several days in a row. Development of the Exploration Atmosphere coupled with use of suitports is going to push the boundaries of EVA operations and the subsequent potential risk of DCS is unknown [26].

3. Conclusion

Over the past 50–60 years, we have seen rapid developments in our aeronautics and space capabilities with planes such as the U-2S operating in the stratosphere along with the launch of the International Space Station in 1998. As we push the technical boundaries to attain various strategic and tactical advantages, we need to remain wary of the physiological effects that extreme and austere environments can impose on our military personnel and astronauts. While the effects of hypobaria and decompression sickness have been known for decades (brought to the forefront by Fulton's seminal work in 1951 [22]), the increase in NCDS experienced by our U-2 pilots from 2006 to 2010 has brought increased scrutiny. Furthermore, the findings of increased WMH coupled with subtle but significantly lower neurocognitive profiles (even among those that did not experience a clinical event of NCDS) heightens the concerns regarding the short and long-term effects that recurrent exposure to hypobaria carries. While the physiology has classically been thought to be secondary to VGE, other inflammatory factors may likely play a role which are being actively investigated. If elucidated, this allows the development for other therapeutic interventions in addition to the numerous prebreathing protocols to mitigate the risk of DCS. This has major significance for both military operations and further space exploration as we continue to press both the technological boundaries as well as our equipment and physiological limits of our personnel.

Acknowledgements

Opinions, interpretations, conclusion and recommendations are those of the authors and are not necessarily endorsed by the United States Air Force. The authors thank Dr. Stephen McGuire for his critical evaluation of our chapter content, to Ms. Elaine Kawano for scientific editing of the chapter and to Ms. Debbie Middleton for assisting in procuring the numerous articles in preparation for this chapter.

Conflict of interest

Neither Dr. Sherman nor Dr. Sladky report any conflicts of interest.

Author details

Paul Sherman* and John Sladky

*Address all correspondence to: paul.m.sherman3.civ@mail.mil

711th Human Performance Wing, United States Air Force School of Aerospace Medicine, Dayton, Ohio, United States

References

[1] McGuire S, Sherman P, Profenna L, Grogan P, Sladky J, et al. White matter hyperintensities on MRI in high-altitude U-2 pilots. Neurology. 2013;**81**(8):729-735

[2] McGuire SA, Sherman PM, Wijtenburg SA, Rowland LM, Grogan PM, et al. White matter hyperintensities and hypobaric exposure. Annals of Neurology. 2014;**76**(5):719-726

[3] Fazekas F, Kleinert R, Offenbacher H, Schmidt R, Kleinert G, et al. Pathologic correlates of incidental MRI white matter signal hyperintensities. Neurology. 1993;**43**(9):1683-1689

[4] Kochunov P, Williamson DE, Lancaster J, Fox P, Cornell J, et al. Fractional anisotropy of water diffusion in cerebral white matter across the lifespan. Neurobiology of Aging. 2012;**33**(1):9-20

[5] McGuire SA, Boone GR, Sherman PM, Tate DF, Wood JD, et al. White matter integrity in high-altitude pilots exposed to hypobaria. Aerospace Medicine and Human Performance. 2016;**87**(12):983-988

[6] McGuire SA, Tate DF, Wood J, Sladky JH, McDonald K, et al. Lower neurocognitive function in U-2 pilots: Relationship to white matter hyperintensities. Neurology. 2014;**83**(7):638-645

[7] Sherman PM, McGuire S. Subcortical white matter hyperintensity burden in astronauts [Abstract]. Aerospace Medicine and Human Performance. 2015;**86**(3):309

[8] Fayed N, Modrego PJ, Morales H. Evidence of brain damage after high-altitude climbing by means of magnetic resonance imaging. American Journal of Medicine. 2006;**119**:168.e1-168.e6

[9] Anooshiravani M, Dumont L, Mardirosoff C, Soto-Debeuf G, Delavelle J. Brain magnetic resonance imaging (MRI) and neurological changes after a single high altitude climb. Medicine and Science in Sports and Exercise. 1999;**31**(7):969-972

[10] Kottke R, Pichler Hefti J, Rummel C, Hauf M, Hefti U, Merz TM. Morphological brain changes after climbing to extreme altitudes – a prospective cohort study. PLoS One. 2015;**10**(10):e0141097

[11] Wong SH, Turner N, Birchall D, Walls TJ, English P, Schmid ML. Reversible abnormalities of DWI in high-altitude cerebral edema. Neurology. 2004;**62**(2):335-336

[12] Erdem I, Yildiz S, Uzun G, Sonmez G, Senol MG, et al. Cerebral white-matter lesions in asymptomatic military divers. Aviation, Space, and Environmental Medicine. 2009;**80**(1):2-4

[13] Gempp E, Sbardella F, Stephant E, Constantin P, De Maistre S, et al. Brain MRI signal abnormalities and right-to-left shunting in asymptomatic military divers. Aviation, Space, and Environmental Medicine. 2010;**81**(11):1008-1012

[14] Connolly DM, Lee VM. Odds ratio meta-analysis and increased prevalence of white matter injury in healthy divers. Aerospace Medicine and Human Performance. 2015;**86**(11):928-935

[15] Webb JT, Pilmanis AA, O'Connor RB. An abrupt zero-preoxygenation altitude threshold for decompression sickness symptoms. Aviation, Space, and Environmental Medicine. 1998;**69**(4):335-340

[16] Webb JT, Pilmanis AA. Altitude decompression sickness between 6858 and 9144 m following a 1-h prebreathe. Aviation, Space, and Environmental Medicine. 2005; **76**(1):34-38

[17] Webb, JT, Fischer MD, Heaps CL, Pilmanis AA. Exercise-enhanced preoxygenation increases protection from decompression sickness. Aviation, Space, and Environmental Medicine. 1996;**67**(7):618-624

[18] Boyle R. New pneumatical experiments about respiration. Philosophical Transactions. 1670;**5**(62):2011-2031

[19] Bert P. La pression barométrique: rechercehes de physiologie expérimentale. Paris: Libraire de L'Académie de Médecine; 1878

[20] Vann RD, Butler, FK, Mitchell SJ, Moon RE. Decompression illness. Lancet. 2011; **377**(9760):153-164

[21] Hagen PT, Scholz DG, Edwards WD. Incidence and size of patent foramen ovale during the first 10 decades of life: An autopsy study of 965 normal hearts. Mayo Clinic Proceedings. 1984;**59**(1):17-20

[22] Nims LF. Environmental factors affecting decompression sickness. Part I: A physical theory of decompression sickness. In: Fulton JF. Decompression Sickness. Philadelphia (PA): WB Saunders; 1951. pp. 192-222

[23] Balldin UI, Pilmanis AA, Webb JT. Central nervous system decompression sickness and venous gas emboli in hypobaric conditions. Aviation, Space, and Environmental Medicine. 2004;**75**(11):969-972

[24] Pilmanis AA, Meissner FW, Olson RM. Left ventricular gas emboli in six cases of altitude-induced decompression sickness. Aviation, Space, and Environmental Medicine. 1996;**67**(11):1092-1096

[25] Tanoue K, Mano Y, Kuroiwa K, Suzuki H, Shibayama M, Yamazaki H. Consumption of platelets in decompression sickness of rabbits. Journal of Applied Physiology (1985). 1987;**62**(5):1772-1779

[26] Conkin J, Norcross JR, Abercromby AF. Evidence Report: Risk of Decompression Sickness (DCS). Houston (TX): National Aeronautics and Space Administration; 2016

[27] Ryles MT, Pilmanis AA. The initial signs and symptoms of altitude decompression sickness. Aviation, Space, and Environmental Medicine. 1996;**67**(10):983-989

[28] Jersey SL, Hundemer GL, Stuart RP, West KN, Michaelson RS, Pilmanis AA. Neurological altitude decompression sickness among U-2 pilots: 2002-2009. Aviation, Space, and Environmental Medicine. 2011;**82**(7):673-682

[29] Bendrick GA, Ainscough MJ, Pilmanis AA, Bisson RU. Prevalence of decompression sickness among U-2 pilots. Aviation, Space, and Environmental Medicine. 1996;**67**(3):199-206

[30] Hundemer GL, Jersey SL, Stuart RP, Butler WP, Pilmanis AA. Altitude decompression sickness incidence among U-2 pilots: 1994-2010. Aviation, Space, and Environmental Medicine. 2012;**83**(10):968-974

[31] Balldin UI, Pilmanis AA, Webb JT. The effect of simulated weightlessness on hypobaric decompression sickness. Aviation, Space, and Environmental Medicine. 2002; **73**(8):773-778

[32] Webb JT, Fischer MD, Heaps CL, Pilmanis AA. Exercise-induced preoxygenation increases protection from decompression sickness. Aviation, Space, and Environmental Medicine. 1996;**67**(7):618-624

[33] Balldin UI. Effects of ambient temperature and body position on tissue nitrogen elimination in man. Aerospace Medicine. 1973;**44**(4):365-370

[34] Pendergrast DP, Senf C, Lundgren CE. Is the rate of whole-body nitrogen elimination influenced by exercise? Undersea and Hyperbaric Medicine. 2012;**39**(1):595-604

[35] Roach RC, Bärtsch P, Hackett PH, Oelz O, Lake Louise AMS Scoring Consensus Committee. The Lake Louise acute mountain sickness scoring system. In: Sutton JR, Houston CS, Coates G, editors. Hypoxia and Molecular Medicine. Proceedings of the 8th International Hypoxia Symposium. Burlington (VT): Queen City Printers; 1993. pp. 272-274

[36] DiPasquale DM, Muza SR, Gunn AM, Li Z, Zhang Q, et al. Evidence for cerebral edema, cerebral perfusion, and intracranial pressure elevations in acute mountain sickness. Brain and Behavior. 2016;**6**(3):e00437

[37] DiPasquale DM, Strangman GE, Harris NS, Muza SR. Hypoxia, hypobaria, and exercise duration affect acute mountain sickness. Aerospace Medicine and Human Performance. 2015;**86**(7):614-619

[38] Fulco CS, Muza SR, Beidleman BA, Demes, R, Staab JE, et al. Effect of repeated normobaric hypoxia exposures during sleep on acute mountain sickness, exercise performance, and sleep during exposure to terrestrial altitude. American Journal of Physiology. Regulatory, Integrative and Comparative Physiology. 2011;**300**(2):R248-R436

[39] Schommer K, Wiesegart N, Menold E, Haas U, Lahr K, et al. Training in normobaric hupoxia and its effects on acute mountain sickness after rapid ascent to 4559 m. High Altitude Medicine & Biology. 2010;**11**(1):19-25

[40] Roach RC, Loeppky JA, Incenogle MV. Acute mountain sickness: increased severity during simulated altitude compared with normobaric hypoxia. Journal of Applied Physiology (1985). 1996;**81**(5):1908-1910

[41] Hackett PH, Roach RC. High altitude illness. The New England Journal of Medicine. 2001;**345**(2):107-114

[42] Hackett PH, Roach RC. High altitude cerebral edema. High Altitude Medicine & Biology. 2004;**5**(2):136-146

[43] Carod-Artal FJ. High-altitude headache and acute mountain sickness [Article in English, Spanish]. Neurologia. 2014;**29**(9):533-540

[44] Matsuzawa Y, Kobayashi T, Fujimoto K, Shinozaki S, Yoshikawa S, Sekiguchi M, editors. High Altitude Medicine. Matsumoto (Japan): Shinshu University; 1992. pp. 300-304

[45] Jensen JB, Sperlin B, Severinghaus JW, Lassen NA. Augmented hypoxic cerebral vasodilatation in men during 5 days at 3,810 m altitude. Journal of Applied Physiology (1985). 1996;**80**(4):1214-1218

[46] Devine BD, Zhang R, Roach RC. Dynamic cerebral autoregulation at high altitude. In: Roach RC, Wagner PD, Hackett PH, editors. Hypoxia: Into the Next Millennium. New York (NY): Kluwer Academic/Plenum Publishers; 1999. pp. 319-322

[47] Krasney JA. A neurogenic basis for acute altitude illness. Medicine and Science in Sports and Exercise. 1994;**26**(2):195-208

[48] Koppelmans V, Bloomberg JJ, Mulavara AP, Seidler RD. Brain structural plasticity with spaceflight. NPJ Microgravity. Dec 19, 2016;**2**:2. DOI: 10.1038/s41526-016-0001-9

[49] Koppelmans V, Pasternak O, Bloomberg JJ, Dios YE, Wood SJ, et al. Intracranial fluid redistribution but no white matter structural changes during a spaceflight analog. Scientific Reports. Jun 9, 2017;**7**(10):3154. DOI: 10.1038/s41598-017-03311-w

[50] Roberts DR, Albrecht MH, Collins HR, Asemani D, Chatterjee AR, et al. Effects of spaceflight on astronaut brain structure as indicated on MRI. The New England Journal of Medicine. 2017;**377**:1746-1753

[51] McGuire SA, Wijtenburg SA, Sherman PM, Rowland LM, Ryan M, Sladky JH, Kochunov PV. Reproducibility of quantitative structural and physiological MRI measurements. Brain and Behavior: A Cognitive Neuroscience Perspective. Aug 2, 2017;**7**(9):e00759

[52] Smith SM. Fast robust automated brain extraction. Human Brain Mapping. 2002;**17**(3): 143-155

[53] Smith SM, Jenkinson M, Woolrich MW, Beckmann CF, Behrens TE, et al. Advances in functional and structural MR image analysis and implementation as FSL. Neuroimage. 2004;**23**(Suppl 1):S208-S219

[54] Mazziotta JC, Toga AW, Evans AC, Fox PT, Lancaster JL. Digital brain atlases. Trends in Neurosciences. 1995;**18**(5):210-211

[55] Henry Feugeas MC, De Marco G, Peretti II, Godon-Hardy S, Fredy D, Claeys ES. Age-related cerebral white matter changes and pulse-wave encephalopathy: Observations with three-dimensional MRI. Magnetic Resonance Imaging. 2005;**23**(9):929-937

[56] Grech-Sollars M, Hales PW, Miyazaki K, Raschke F, Rodriguez D, et al. Multi-centre reproducibility of diffusion MRI parameters for clinical sequences in the brain. NMR in Biomedicine. 2015;**28**(4):468-485

[57] Teipel SJ, Wegrzyn M, Meindl T, Frisoni G, Bokde AL, et al. Anatomical MRI and DTI in the diagnosis of Alzheimer's disease: a European multicenter study. Journal of Alzheimer's Disease. 2012;**31**(Suppl 3):S33-S47

[58] Kochunov P, Chiappelli J, Wright SN, Rowland LM, Patel B, et al. Multimodal white matter imaging to investigate reduced fractional anisotropy and its age-related decline in schizophrenia. Psychiatry Research. 2014;**223**(2):148-156

[59] Jahanshad N, Kochunov PV, Sprooten E, Mandl RC, Nichols ET, et al. Multi-site genetic analysis of diffusion images and voxelwise heritability analysis: A pilot project of the ENIGMA-DTI working group. Neuroimage. 2013;81:455-469 (Erratum in: Jahanshad N, Kochunov PV, Sprooten E, Mandl RC, Nichols TE, et al. Corrigendum to "Multi-site genetic analysis of diffusion images and voxelwise heritability analysis: a pilot project of the ENIGMA-DTI working group." Neuroimage. 2014;**90**:470-471)

[60] Maclullich AM, Ferguson KJ, Reid LM, Deary IJ, Starr JM, et al. Higher systolic blood pressure is associated with increased water diffusivity in normal-appearing white matter. Stroke. 2009;**40**(12):3869-3871

[61] Basser PJ, Pierpaoli C. Microstructural and physiological features of tissues elucidated by quantitative-diffusion-tensor MRI. Journal of Magnetic Resonance. Series B. 1996;**111**(3):209-219

[62] Kochunov P, Coyle T, Lancaster J, Robin DA, Hardies J, et al. Processing speed is correlated with cerebral health markers in the frontal lobes as quantified by neuroimaging. Neuroimage. 2010;**49**(2):1190-1199

[63] Tremblay S, Henry LC, Bedetti C, Larson-Dupuis C, Gagnon JF, et al. Diffuse white matter tract abnormalities in clinically normal ageing retired athletes with a history of sports-related concussions. Brain. 2014;**137**(Pt 11):2997-3011

[64] Kochunov P, Williamson DE, Lancaster J, Fox P, Cornell J, et al. Fractional anisotropy of water diffusion in cerebral white matter across the lifespan. Neurobiology of Aging. 2012;**33**(1):9-20

[65] University of South Carolina, Chris Rorden's Neuropsychology Lab. ASL Analysis. 2012. Available from: http://www.mccauslandcenter.sc.edu/CRNL/tools/asl [Accessed: Mar 1, 2015]

[66] Wang J, Alsop DC, Li L, Listerud J, Gonzalez-At JB, et al. Comparison of quantitative perfusion imaging using arterial spin labeling at 1.5 and 4.0 Tesla. Magnetic Resonance Imaging. 2002;**48**(2):242-254

[67] van Gelderen P, de Zwart JA, Duyn JH. Pittfalls of MRI measurement of white matter perfusion based on arterial spin labeling. Magnetic Resonance Imaging. 2008;**59**(4):788-795

[68] Soher BJ, Young K, Bernstein A, Aygula Z, Maudsley AA. GAVA: Spectral simulation for in vivo MRS applications. Journal of Magnetic Resonance. 2007;**185**(2):291-299

[69] Provencher S. LCModel. 2016. Available from: http://s-provencher.com/lcmodel.shtml [Accessed: Oct 23, 2017]

[70] Gasparovic C, Yeo R, Mannell M, Ling J, Elgie R, et al. Neurometabolite concentrations in gray and white matter in mild traumatic brain injury: An 1H-magnetic resonance spectroscopy study. Journal of Neurotrauma. 2009;**26**(10):1635-1643

[71] Ding K, Marquez de la Plata C, Wang JY, Mumphrey M, Moore C, et al. Cerebral atrophy after traumatic white matter injury: Correlation with acute neuroimaging and outcome. Journal of Neurotrauma. 2008;**25**(12):1433-1440

[72] Kochunov P, Thompson PM, Coyle TR, Lancaster JL, Kochunov V, et al. Relationship among neuroimaging indices of cerebral health during normal aging. Human Brain Mapping. 2008;**29**(1):36-45

[73] Du AT, Schuff N, Chao LL, Kornak J, Ezekiel F, et al. White matter lesions are associated with cortical atrophy more than entorhinal and hippocampal atrophy. Neurobiology of Aging. 2005;**26**(4):553-559

[74] Wen W, Sachdev PS, Chen X, Anstey K. Gray matter reduction is correlated with white matter hyperintensity volume: a voxel-based morphometric study in a large epidemiological sample. Neuroimage. 2006;**29**(4):1031-1039

[75] Kraut MA, Beason-Held LL, Elkins WD, Resnick SM. The impact of magnetic resonance imaging-detected white matter hyperintensities on longitudinal changes in regional cerebral blood flow. Journal of Cerebral Blood Flow and Metabolism. 2008;**28**(1):190-197

[76] Kochunov P, Ramage AE, Lancaster JL, Robin DA, Narayana S, et al. Loss of cerebral white matter structural integrity tracks the gray matter metabolic decline in normal aging. Neuroimage. 2009;**45**(1):17-28

[77] Kochunov P, Robin DA, Royall DR, Coyle T, Lancaster J, et al. Can structural MRI indices of cerebral integrity track cognitive trends in executive control function during normal maturation and adulthood? Human Brain Mapping. 2009;**30**(8):2581-2594

[78] Tate DF, Jefferson AL, Brickman AM, Hoth KF, Gunstad J, et al. Regional white matter signal abnormalities and cognitive correlates among geriatric patients with treated cardiovascular disease. Brain Imaging and Behavior. 2008;**2**(3):200-206

[79] Galluzi S, Lanni C, Pantoni L, Filippi M, Frisoni GB. White matter lesions in the elderly: Pathophysiological hypothesis on the effect on brain plasticity and reserve. Journal of the Neurological Sciences. 2008;**273**(1-2):3-9

[80] Sachdev P, Brodaty H. Quantitative study of signal hyperintensities on T2-weighted magnetic resonance imaging in late-onset schizophrenia. The American Journal of Psychiatry. 1999;**156**(12):1958-1967

[81] Pantoni L, Garcia JH. The significance of cerebral white matter abnormalities 100 years after Binswanger's report. A review. Stroke. 1995;**26**(7):1293-1301

[82] Bateman GA. Pulse-wave encephalopathy: A comparative study of the hydrodynamics of leukoaraiosis and normal-pressure hydrocephalus. Neuroradiology. 2002;**44**(9):740-748

[83] Bateman GA. Pulse wave encephalopathy: a spectrum hypothesis incorporating Alzheimer's disease, vascular dementia and normal pressure hydrocephalus. Medical Hypotheses. 2004;**62**(2):182-187

[84] Miura K, Soyama Y, Morikawa Y, Nishijo M, Nakanishi Y, et al. Comparison of four blood pressure indexes for the prediction of 10-year stroke risk in middle-aged and older Asians. Hypertension. 2004;**44**(5):715-720

[85] Nair GV, Chaput LA, Vittinghoff E, Herrington DM, Heart and Estrogen/Progestin Replacement Study Investigators. Pulse pressure and cardiovascular events in postmenopausal women with coronary heart disease. Chest. 2005;**127**(5):1498-1506

[86] McGuire SA, Sherman PM, Brown AC, Robinson AY, Tate DF, et al. Hyperintense white matter lesions in 50 high-altitude pilots with neurologic decompression sickness. Aviation, Space, and Environmental Medicine. 2012;**83**(12):1117-1122

[87] Air Force US. Medical Examinations and Standards. Air Force Instruction 48-123. Washington, DC: Department of the Air Force; 2009

[88] Kochunov P, Fu M, Nugent K, Wright SN, Du X, et al. Heritability of complex white matter diffusion traits assessed in a population isolate. Human Brain Mapping. 2016;**37**(2):525-535

[89] Chappelle W, Ree MJ, Barto EL, Teachout MS, Thompson WT. Joint use of the MAB-II and MicroCog for improvements in the clinical and neuropsychological screening and aeromedical waiver process of rated USAF pilots. Brooks City-Base (TX): Air Force Research Laboratory, 711th Human Performance Wing, Aerospace Medicine Department; 2010. Technical Report AFRL-SA-BR-TR-2010-0002

[90] Foster PP, Butler BD. Decompression to altitude: assumptions, experimental evidence, and future directions. Journal of Applied Physiology (1985). 2009;**106**(2):678-690

[91] Thom SR, Yang M, Bhopale VM, Huang S, Milovanova TN. Microparticles initiate decompression-induced neutrophil activation and subsequent vascular injuries. Journal of Applied Physiology (1985). 2011;**110**(2):340-351

[92] Vann RD, Butler FK, Mitchell SJ, Moon RE. Decompression illness. Lancet. 2011; **377**(9760):153-164

[93] Krutz Jr RW, Dixon GA. The effects of exercise on bubble formation and bends susceptibility at 9100 m (30,000 ft; 4.3 psia). Aviation, Space, and Environmental Medicine. 1987;**58**(9 Pt 2):A97-A99

[94] U.S. Air Force. High Altitude Airdrop Mission Support Program. Washington (DC): Department of the Air Force; 2015 (Air Force Instruction 11-409)

[95] U.S. Army. FM 3-05.211. Special Forces Military Free-fall Operations. Washington (DC): Department of the Army; 2005 (FM 3-05.211/MCWP 3-15.6/NAVSEA SS400-AG-MMO-010; AFMAN 11-411(I))

[96] Lupa H, D'Oly T. Exercise following simulated parachuting from 35,000 feet: Is tolerance reduced and is the probability of post descent decompression sickness increased? In: Strategies to maintain combat readiness during extended deployments – A human systems approach. In: Proceedings of the RTO Human Factors and Medicine Panel Symposium; 2005 Oct 3-5; Prague, Czech Republic. Meeting Proceedings RTO-MP-HFM-124, Paper 16. Neuilly-sur-Seine (France): NATO RTO; 2005. pp. 16-1-16-6

[97] Webb JT, Pilmanis AA, Fischer MD. Moderate exercise after altitude exposure fails to induce decompression sickness. Aviation, Space, and Environmental Medicine. **73**(9): 872-875

[98] Waligoria JM, Horrigan DJ, Conkin J, Hadley III AT. Verification of an Altitude Decompression Sickness Protocol for Shuttle Operations Utilizing a 10.2 Psi Pressure Stage. NASA Technical Memorandum 58259. Houston: Johnson Space Center; Jun 1984

Chapter 5

Spaceflight Induced Changes in the Central Nervous System

Alex P. Michael

Additional information is available at the end of the chapter

http://dx.doi.org/10.5772/intechopen.74232

Abstract

Although once a widely speculated about and largely theoretical topic, spaceflightinduced intracranial hypertension is more accepted as a distinct clinical phenomenon; yet, the underlying physiological mechanisms are still poorly understood. In the past, many terms were used to describe the symptoms of malaise, nausea, vomiting, and vertigo though longer duration spaceflights have increased the prevalence of overlapping symptoms of headache and visual disturbance. Spaceflight-induced visual pathology is thought to be a manifestation of increased intracranial pressure (ICP) because of its similar presentation to cases of known intracranial hypertension on Earth as well as the documentation of increased ICP by lumbar puncture in symptomatic astronauts upon return to gravity. The most likely mechanisms of spaceflight-induced increased ICP include a cephalad shift of body fluids, venous outflow obstruction, blood-brain barrier breakdown, and disruption to CSF flow. The relative contribution of increased ICP to the symptoms experienced during spaceflight is currently unknown though as other factors recently posited to contribute include local effects on ocular structures, individual differences in metabolism, and the vasodilator effects of carbon dioxide. Spaceflight-induced intracranial hypertension must be distinguished from other pathologies with similar symptomatology. The following chapter discusses the proposed physiologic causes and the pathological manifestations of increased ICP in the spaceflight environment and provides considerations for future long-term space travel.

Keywords: increased intracranial pressure, intracranial hypertension, spaceflight, space adaptation syndrome, VIIP, visual impairment, space flight-associated neuro-ocular syndrome, SANS

1. Introduction

Since the first manned spaceflight, many terms have been used to describe a collective group of seemingly related neurologic, ophthalmologic, and neurovestibular symptoms. Up to one half of astronauts are incapacitated by malaise, nausea, vomiting, and vertigo within the first few hours or days spent in space [1]. This constellation of symptoms, first described by Titov [2], was originally referred to as "space motion sickness" (SMS) [3] because of its similarity to motion sickness in the terrestrial environment. It is hypothesized that two physiologically distinct mechanisms converge to produce the symptoms of SMS [4, 5]: Cephalad fluid shifts are thought to alter the response properties of vestibular receptors while loss of tilt-related otolith signals in microgravity create a conflict between the actual and the anticipated signals collected from the external environment. The breadth of symptoms that astronauts report is likely due to a complex interaction between the neurovestibular system and autonomic nervous system [6]. A separate yet related term, "space adaptation syndrome," had similarly been used to include not only motion sickness, but also symptoms of head congestion and headaches brought on by a cephalad fluid shift into facial structures [7].

Most astronauts require only 2–3 days to acclimate to motion sickness in space and few continue to have residual symptoms during short term spaceflight [4]. As more time is spent in space, physiologically distinct yet overlapping symptoms seem to arise including headache and visual disturbance. These findings were noted to be similar to the cases of intracranial hypertension in the terrestrial environment which are caused by an elevation in intracranial pressure (ICP) [8]. Since the launch of the International Space Station (ISS) in 2000, the time that astronauts spend in space has dramatically increased. In addition, a 2015 partnership between the United States and Russia established the concept of 1-year mission onboard the ISS. Long-duration exposure to spaceflight has brought forth concern from the aerospace medicine community because its effects on the central nervous system are unknown.

2. Spaceflight increased intracranial pressure

Post-flight surveys of 300 astronauts from 1989 to 2011 found that approximately 29% reported deficits in distant and near visual acuity following short-duration missions (less than 2 weeks) while 60% reported deficits with long-duration space travel (typically 6 months) [9]. Starting in 2008, more detailed clinical data were collected for seven astronauts following 6 months of continuous orbital flight onboard the ISS. Complete visual and structural eye examinations before and after spaceflight revealed pathologic changes in the eye including optic disk edema, nerve fiber layer thickening, choroidal folds, posterior globe flattening, cotton wool spots, and decreased near vision accompanied by hyperopic shift. Of these seven crewmembers, lumbar punctures performed in the four cases with disk edema revealing opening pressures of 220, 210, 280, and 285 mm H_2O at 60, 19, 12, and 57 days post-mission, respectively. It should be noted that no in-flight opening pressure have ever been attempted. With this clinical evidence, the authors hypothesized that the observed findings were due to cerebral venous congestion due to CSF flow disturbance from spaceflight-induced cephalad fluid shifts. A retrospective review of data has since identified eight additional cases of pathologic visual changes [10] since the original publication.

In a follow-up study, Kramer, Sargsyan [11] evaluated 27 post-flight crewmembers using T2-weighted orbital and conventional brain sequences. They found various combinations of optic nerve sheath distention, posterior globe flattening, optic disk protrusion, increased optic nerve diameter, and greater concavity of the pituitary gland with posterior stalk displacement. Optic disk protrusion was only found with longer mission duration, indicating that clinical severity is associated with increasing spaceflight exposure. Repeat scans showed that some crewmembers continued to have posterior globe flattening 100 days after spaceflight suggesting that this condition may have a prolonged course or may not entirely return to normal. The authors determined that these findings were representative of intracranial hypertension due to elevated ICP.

Clinically, visual pathology is considered a sensitive measure of increased ICP, as the perineural subarachnoid space of the optic nerve is contiguous with the intracranial subarachnoid space and therefore vulnerable to ICP fluctuations. This contiguity has been verified in a cadaver study which found that the subarachnoid pressures of the optic nerve sheath have a linear relationship to ICP [8, 12]. The National Aeronautics and Space Administration (NASA) has since referred to this spaceflight pathological phenomenon as vision impairment and intracranial pressure (VIIP) [10, 13] and recognized it as a serious threat to long duration spaceflight.

The Space Life Sciences at Johnson Space Center convened a summit in February of 2011 to address the topic of VIIP. At that meeting, a research and clinical advisory panel was created to provide guidance for the future clinical and fundamental research. After further investigation, the visual pathology seen in astronauts seemed to differ from those with intracranial hypertension in the terrestrial environment. Choroidal folds and hyperoptic shifts are sometimes seen in terrestrial intracranial hypertension but seemed to occur disproportionately in astronauts. Retinal cotton-wool spots are not typically seen in terrestrial intracranial hypertension but are prominent features in the visual pathology seen after space flight. Also, astronauts may experience unilateral pathology, which is again uncommon with global increases in intracranial pressure [14, 15]. Because of these discrepancies, visual pathology in astronauts has now been referred to as space flight-associated neuro-ocular syndrome (SANS) [14]. After considering all evidences, the panel concluded that the increase in ICP may not be the sole cause of visual disturbances following spaceflight and chose to examine other possible influences on visual pathology [16].

2.1. Intraocular pressure

Space flight-induced compartmentalization of cerebrospinal fluid in the subarachnoid space with locally elevated cerebrospinal fluid sheath pressures has been proposed as an additional alternative hypothesis. Local orbital effects may explain ophthalmic structural and functional changes following spaceflight without an accompanying rise in ICP. This hypothesis purports that a local disruption of CSF dynamics surrounding the optic nerve sheath results in an orbital compartment syndrome [17]. A microgravity-induced cephalad fluid shift may lead to choroidal engorgement and subsequent expansion of the choroid against the rigid scleral tissue leading to a sudden increase in intraocular pressure (IOP) [18, 19]. The initial spike in IOP is followed by a decrease over a period of days likely due to a compensatory decrease in aqueous volume [19]. Thus, in-flight, post-flight, and HDT studies suggest the possibility that a lowering of IOP may occur during extended microgravity exposure. Ocular hypotony, generally defined as an IOP of <6.5 mmHg, is well-documented to cause disk edema, posterior globe flattening, choroidal folds, and a hyperopic shift very similar to some of our observed changes [17].

The lamina cribrosa is a mesh-like structure that acts as a pressure barrier between the intraocular space and cerebrospinal fluid space of the optic nerve sheath [20]. The difference in IOP and CSF pressure across the lamina cribrosa is known as the translaminar pressure difference. Small yet chronically elevated CSF pressure in combination with ocular hypotony would lead to a significant pressure gradient toward the intraocular space and could thereby be responsible for the ophthalmic structural and functional changes seen in astronauts secondary to spaceflight exposure.

3. Factors contributing neuro-ocular symptoms

3.1. Microgravity-induced fluid shifts

Early studies found that exposure to both microgravity and simulated microgravity led to a cephalad shift of plasma fluid into the interstitial spaces of the head and neck [21, 22]. This led researchers to believe that microgravity-induced cephalad fluid shifts caused increased ICP and were a prominent contributor to both space adaptation syndrome [1] and space motion sickness [23]. The initial support for this mechanism was sought through the use of the head-down tilt (HDT) method which simulates the fluid shifts that occur in the spaceflight environment. In an early study by Murthy et al. [24], 10 min of 6° HDT was found to significantly increase the ICP of six healthy males as indicated by tympanic membrane displacement. Increasing the angle to 15°, HDT generated a further increase in ICP. Although no long-term monitoring of the HDT method has been attempted in humans, ICP was evaluated for 7-days of 45° HDT using a subarachnoid catheter in rabbits [25]. An immediate increase in ICP was observed which peaked at 12 h of HDT and then decreased gradually toward the pre-HDT baseline value. These findings suggest that rabbits begin to adapt to HDT within the first few days.

Since cephalad fluid shift has been found to increase fluid in the interstitial soft tissue space of the head, it may seem intuitive that the increased filtration of plasma into the intracranial interstitium would lead to increased ICP. However, when autoregulatory mechanisms are intact, they prevent a sustained increase in cerebral blood flow (CBF) in the presence of an elevated cerebral perfusion pressure (CPP) [26, 27]. Kawai et al. [28] used transcranial Doppler to examine CBF in the middle cerebral artery of humans following 6° HDT. CBF velocity was found to increase immediately upon initiation of HDT, reach a peak at 3 h and then begin to decrease toward baseline after 9 h of HDT. Similarly, no significant differences in CBF velocity were found on transcranial Doppler measurements after up to 2 weeks of spaceflight when compared to pre-flight baseline values [7, 28, 29]. These findings suggested preserved or possibly improved cerebrovascular autoregulation during short-duration spaceflight. When time spent in microgravity is extended, though, there is evidence that autoregulation may become altered.

Ex vivo examinations of mice following HDT technique in the terrestrial environment revealed increased intrinsic vasoconstrictor responsiveness of cerebral arteries [30–32], thickening of the medial smooth muscle cell layer in some cerebral arteries [33, 34], and decreased cerebral blood flow [31, 34]. These findings provide histological evidence for appropriate autoregulatory increases in sympathetic tone of cerebral vessels. However, similar examination of post-spaceflight mice, following 13 days on-board the STS-135 shuttle mission [27], differed from terrestrial HDT technique by exhibiting less vasoconstriction, more vascular distensibility, and lower effective elastic modulus and stiffness. These findings suggest a decrease in cerebral vascular resistance (CVR) and thus an increase in CBF [CBF = (P_a – ICP)/CVR, where CBF is proportional to arterial pressure (P_a) and ICP and inversely proportional to CVR]. This finding supports the fact that increased arterial perfusion pressure alone, as in the HDT, does not lead to increased CBF but that CBF may still be elevated in microgravity and may further contribute to an increase in ICP [27].

3.2. Endothelial breakdown

Although autoregulatory mechanisms in the cerebrovasculature have evolved to provide a steady CBF in the face of wide fluctuations of cerebral perfusion pressure, endothelial dysfunction may lower the threshold pressure required to increased deposition of fluid into the intracranial interstitial. Using an *in silico* model for intracranial pressure dynamics, Stevens et al. [35] originally determined that increased interstitial fluid volume in the brain lead to a decrease in ICP in microgravity. After modifying the model to account for reduction in the integrity of the blood-brain barrier, they found a much more significant increase in intracranial interstitial fluid as well as elevation of ICP high enough for symptoms to manifest [36].

Endothelial cell gap junctions are held closed by the combined pressure of the interstitial fluid in the brain and the intracranial capillary pressure. Lakin et al. [36] proposed that in a 1-G environment, hydrostatic pressure is transmitted from the brain to the capillaries, thus increasing the pressure needed to close endothelial cell junctions. In spaceflight, the brain is unable to contribute its weight to maintaining the pressure balance, thereby allowing fluid to leak from the intracranial capillaries into the interstitial fluid.

3.3. Venous outflow obstruction and CSF hydrodynamics

Aside from increased CBF, it has also been proposed that cephalad fluid shifts contribute to elevations in ICP by increasing the post-capillary venous pressure [1] through downstream venous congestion [31, 37]. While supine, a majority of the cerebral outflow occurs through the internal jugular veins. However, internal jugular veins collapse in the standing position and blood is shunted through secondary venous channels (e.g., vertebral plexus and deep cervical veins). This has been confirmed using time of flight MRI techniques in the sitting versus supine positions [38, 39]. Termed "Space obstructive syndrome," Wiener [40] proposed that internal jugular vein compression along with loss of gravitational-induced cranial outflow of venous blood in the vertebral veins may lead to venous hypertension. Cerebral outflow may divert through the internal jugular veins when standing if there is a significant increase in CVP (e.g., with a Valsalva maneuver) [41]. This is relevant to spaceflight as the gravitational unloading of the thoracic space causes CVP to paradoxically decrease [42–44]. Decreased venous flow may lead to a rise in pressure high enough to disturb the gradient between the CSF and cerebral venous sinuses. CSF normally circulates through the subarachnoid space and is absorbed through arachnoid granulations into the cerebral venous sinuses. Similarly, cine phase-contrast MRI examining CSF flow in the upright posture found that a considerably smaller amount of CSF oscillated between the cranium and the spinal canal than in the supine position [38, 39, 41].

3.4. Carbon dioxide

Carbon dioxide (CO_2), a natural byproduct of cellular respiration, is known to be a potent vasodilator in the cerebral vasculature. This normal physiologic event occurs to increased CBF to the brain in times of respiratory compromise [45]. Nominal CO_2 levels on the ISS are between 2.3 and 5.3 mmHg [46] and the astronauts presenting with VIIP symptoms were exposed to levels less than 5 mmHg [46]. Although these levels are 20× higher than the normal 0.23 mmHg CO_2 on Earth, this CO_2 level is still relatively low and not thought to have detrimental physiological effects. However, as there is no natural convection in microgravity, astronauts may be exposed to localized areas of high CO_2 when working in a small space, during exercise [46] and possibly during sleep [47]. In a computational fluid dynamics analysis, Son et al. (2012) determined that without natural convection of gases and ventilation, pCO_2 could rise above 9 mmHg around a sleeping astronaut's mouth within just 10 min. These pockets of CO_2 would not be detected by the major constituent analyzers onboard the ISS, and therefore would go unreported. Regular exposure to slightly increased ambient CO_2 as well as potential exposure to pockets of high concentrations of CO_2 may compromise the integrity of the blood-brain barrier impairing cerebrovascular resistance thus leading to increased CBF and ICP [45, 48]. The response CBF and CVR to CO_2 was found to be reduced after long-duration missions on the ISS indicating impaired autoregulation and reduced cerebrovascular CO_2 reactivity [27, 49].

3.5. One-carbon metabolism

It has also been shown that variation in an important metabolic pathway, the one-carbon metabolism cycle, is associated with the occurrence of the VIIP syndrome in astronauts [50]. Zwart et al. (2012) found significantly higher serum levels of several one-carbon metabolites in astronauts affected by the VIIP syndrome compared to unaffected astronauts, including serum homocysteine, cystathionine, 2-methylcitric acid, and methylmalonic acid. These findings suggest that polymorphisms in enzymes of the one-carbon pathway may interact with microgravity to cause ophthalmic changes.

3.6. Radiation

It has also been proposed that radiation exposure outside of Earth's atmosphere may disrupt the integrity of the blood-brain barrier [36]. The two cosmic sources of radiation that are considered to impact mission success are solar particle events and galactic cosmic rays. Sanzari et al. [51] found that exposure to doses of ionizing radiation similar to that experienced by astronauts during a solar particle event led to significant long-term elevation in ICP in a porcine model. Experiments involving cell phone radiation found that small amounts of radiation may activate endothelial cell proteins causing the endothelial cells to shrink and widen the gap junction [52–54]. Increased vessel permeability in turn leads to extravasation of albumin into brain parenchyma leading to cerebral edema [53]. There is little evidence, though, that the radiation generated by solar particle events or galactic cosmic rays produce effects similar to that of radiofrequency waves.

3.7. Exercise

There have been several studies showing that resistive exercise during spaceflight may lead to a significant increase in IOP [55, 56]. The effect it has on increased ICP though remains controversial. Heavy loading and resistance exercise are important to prevent musculoskeletal losses, especially bone density [57]. For that reason, resistance exercises have been encouraged aboard the ISS. Inducing a Valsalva maneuver during weight lifting has been shown to increase intrathoracic pressure which may in turn elevate ICP [58]. Aerobic exercise though has not been found to increase ICP likely because it is accomplished without a Valsalva [59].

3.8. Sodium intake

Prepackaged foods for the International Space Station were originally high in sodium at up to 5300 g per day [60]. High sodium levels create an osmotic shift of body fluid from the interstitial to the intravascular space contributing to increased venous volume, congestion and ultimate jugular venous outflow obstruction. In 1974, a prospective trial of sodium restriction reportedly lead to remission of papilledema in all 9 patients with idiopathic intracranial hypertension that were involved [61]. It is likely that improvement occurs due to concomitant weight loss and not entirely due to sodium and water distribution. However, it is suggested that astronauts consume a lower sodium diet in attempt to prevent long term visual damage. NASA has since reformulated to substantially reduce the intake of sodium in the daily diet of astronauts to less than 3 g per day [60].

4. Cognitive and structural changes in the brain

Spaceflight imposes a short-term risk to mission operational success by contributing to headaches, malaise, and visual impairment and further may lead to long-term risks that have not yet fully been elicited. The long-term risks of spaceflight-induced intracranial hypertension may be best estimated through observations of chronically increased ICP on earth. Individuals with idiopathic intracranial hypertension (IIH) are plagued with well-documented symptoms of severe headache and vision loss but may also experience pulsatile tinnitus, ataxia, memory disturbances, and cognitive dysfunction [62–64]. Several small population studies have revealed significant cognitive deficits in patients with IIH especially within verbal and memory tests [65, 66]. In a study by Yri et al. [62], 31 patients with IIH performed significantly worse on tests of reaction time, processing speed, visuospatial memory, and attention compared to a demographically matched healthy control group. Individuals with IIH continued to exhibit cognitive dysfunction after 3 months of pharmacologic therapy despite improvement in ICP and headache. Further, quality of life measures have been found to be lower compared with population norms [67].

At this time, there is no evidence for gross structural damage as a cause of cognitive dysfunction in IIH, as brain morphometric and volumetric analysis have also been insignificant compared to healthy controls [68]. Subtle disturbances to white or gray matter substance due to mechanical compression similar to that in normal pressure hydrocephalus has also been proposed [63].

The evidence to suggest impaired cognition in astronauts related to spaceflight is sparse, but terrestrial data could potentially predict long duration sequela and may influence how we monitor astronauts in the future. In 2017, scientists released the results of a study using MR imaging to compare the brain morphology of astronauts after long and short duration space flight. Astronauts who participated in long-duration flights had significantly more narrowing of the central sulcus, upward shift of the brain, and narrowing of CSF spaces [69]. Another study compared MRIs of the brain before and after spaceflight from 27 astronauts and found decreased volume of the frontotemporal gray matter and an increase in the volume of the medial primary sensorimotor cortexes. This finding was attributed to neuroplasticity during adaptation to microgravity [70]. Long duration spaceflight has also been associated with an increase in periventricular white matter hyperintensities seen on MRI. These hyperdensities are linked to an increase in ventricular CSF volume leading to transependymal CSF flow from the ventricles into the brain parenchyma. It appears it is at least partially reversible on return to normal gravity [71]. Similarly, a significantly increased number of white matter hyperintensities were found in high-altitude U-2 pilots compared to age-matched healthy controls [72]. The presence of these white matter changes were associated with cognitive impairments ranging from slowed though processes to confusion, unresponsiveness, and even permanent cognitive decline [73, 74].

5. Future considerations

Under NASA's Human Research Roadmap and its Path to Risk Reduction, VIIP/SANS continues to be identified as a top risk that may affect astronauts on long duration missions and remain under intensive investigation by space agencies. Projects are currently planned to characterize fluid distribution and compartmentalization during long-term space travel to determine systemic and ocular factors of individual susceptibility to the development of ICP elevation and to evaluate noninvasive ICP monitoring devices for the clinical evaluation of ICP preflight, in-flight, and post-flight [75].

On land and in orbit, astronauts are subjected to a multitude of visual examinations including visual acuity tests, amsler grids, tonometry, fundoscopy, and optical coherence tomography. Following long duration space travel, researchers are applying MR imaging, visual field perimetry and cycloplegic refraction. Noninvasive techniques for in-flight ICP, intraocular pressure, and cerebral blood flow measurements are also being investigated including ophthalmodynamometry, tympanic membrane displacement and optic nerve ultrasound. A linear correlation has been found between central retinal vein pressure and ICP due to pressure gradients across the optic nerve sheath [76]. Ophthalmodynamometry is a useful method for determining the central retinal artery pressure and is therefore a useful indirect measure on ICP. Tympanic membrane displacement has been used to detect elevated ICP in hydrocephalus children in the terrestrial environment [77]. Because cerebrospinal fluid and perilymph communicate through the cochlear aqueduct, an increase in ICP is directly transmitted to the footplate of the stapes and resulting in inward displacement. The optic nerve ultrasound also seems to be a reliable non-invasive measure as optic nerve sheath diameter has been found to be highly sensitive and specific for the detection of elevated ICP using [78]. Noninvasive approaches though are correlation based and must be calibrated to each patient based on known ICP baseline measurements. This may lead to a high margin for error. Researchers are currently investigating how to correlate pre-flight to in-flight data across multiple modalities.

As space tourism increases, there will be spaceflight participants that are not as physical fit and have not undergone the rigorous training as that of NASA astronauts. Intracranial hypertension may also pose a risk to future commercial spaceflight. The incidence of intracranial hypertension and visual pathology may rise with the increase in civilian space travelers who are not as physiological adept as their astronaut counterparts. Further, increased ICP in the spaceflight environment may become more concerning in someone who has a predilection, or underlying disease process that, combined with increased ICP, could cause in-flight or post-flight problems [40].

The ultimate prevention of neuro-ocular dysfunction due to spaceflight would be reproduction of the normal 1G environment. This could theoretically be introduced by the Coriolis force through rotation of the entire space vehicle, part of the vehicle, or using an on-board centrifuge. Reintroduction of gravity is the only single measure than can protect all physiological systems in all individuals against the effects of weightlessness. Until that concept comes to fruition, other countermeasures are actively being researched.

Pharmacological agents are capable of lowering ICP in the terrestrial environment and are being studied as a means of reducing the risk of visual impairment. Acetazolamide acts as a carbonic anhydrase inhibitor leading to decreased production of CSF at the choroid plexus. However, it also increases the risk for renal calculi and would lower intraocular pressure which could worsen choroidal swelling and potentially optic disk swelling. Other diuretics (e.g., furosemide and hydrochlorothiazide) are more potent diuretics and although may theoretically aid in decreasing CSF production would produce undesirable metabolic side effects. Topiramate has been used in the treatment of migraine headaches and has a weak carbonic anhydrase effect which may lower ICP. It too has undesirable side effects such as cognitive slowing [79].

6. Conclusion

Many terms have been used to describe the symptoms of head congestion, nausea and vomiting, and visual disturbance in the space-flight environment. Over the years, attempts have been made to connect these seemingly related symptoms to a number of diverse pathophysiological origins. At this time, the contribution of increased ICP to the symptoms experienced during spaceflight is unknown. Although direct measurements of CSF pressure have not been performed in actual spaceflight conditions, the best evidence comes from the presentation of symptoms shared with cases of known intracranial hypertension on Earth as well as the documentation of increase ICP in symptomatic astronauts upon return to gravity. Documentation of CSF opening pressure via a lumbar puncture during spaceflight would provide definitive proof of elevated ICP during spaceflight but carries with it inherent procedural risks of post-lumbar puncture headache, hemorrhage, infection and spinal cord injury [80]. For that reason, noninvasive techniques are being studies though they too have inherent drawbacks.

Spaceflight-induced visual disturbance, first termed by NASA as VIIP, has been identified as a serious risk to astronauts during future long-duration space travel, having already affected over 40% of ISS inhabitants [81]. Although VIIP was originally attributed to spaceflight-induced elevated ICP, further factors now seem to contribute. For that reason, it has more recently been referred to as space flight-associated neuro-ocular syndrome [14].

Although prior research has provided better insight into the mechanisms of increased ICP in space, the exact pathophysiology is still unclear. It is likely that no entity discussed previously is the sole contributor to the neurological phenomena experienced in long-term spaceflight but a combination of many. Cephalad fluid shift plays a large role along with major contributions from venous outflow obstruction, blood-brain barrier breakdown, alterations in cerebrovascular tone, and disruption of CSF flow. Since not all individuals manifest with symptoms, it is likely that a combination of genetic, anatomical, and lifestyle related factors make some astronauts more susceptive to spaceflight-induced visual pathology as well as intracranial hypertension [10].

Little is known as to how the spaceflight environment setting will alter the anatomical and physiological integrity of our nervous systems and related structures, but aerospace physicians and astronauts should be educated in the current understanding of how human physiology reacts to this extreme environment. The goal of extending the duration of missions and sending individuals further into space than ever before will challenge the current capabilities of aerospace medicine. It will be critical to develop countermeasures to these known obstacles so that astronauts can participate at their peak in these missions and return safely to earth.

Author details

Alex P. Michael

Address all correspondence to: alexmichael87@gmail.com

Division of Neurosurgery, Southern Illinois University School of Medicine, Springfield, IL, USA

References

[1] Jennings T. Space adaptation syndrome is caused by elevated intracranial pressure. Medical Hypotheses. 1990;**32**(4):289-291

[2] Titov GCM. I Am Eagle. Indianapolis: Bobbs-Merrill; 1962

[3] Schneider RC, Crosby EC. Motion sickness: Part I—a theory. Aviation, Space, and Environmental Medicine. 1980;**51**(1):61-64

[4] Heer M, Paloski WH. Space motion sickness: Incidence, etiology, and countermeasures. Autonomic Neuroscience: Basic & Clinical. 2006;**129**(1–2):77-79

[5] Thornton WE, Bonato F. Space motion sickness and motion sickness: Symptoms and etiology. Aviation, Space, and Environmental Medicine. 2013;**84**(7):716-721

[6] Williams D, Kuipers A, Mukai C, Thirsk R. Acclimation during space flight: Effects on human physiology. CMAJ: Canadian Medical Association Journal (Journal de l'Association Medicale Canadienne). 2009;**180**(13):1317-1323

[7] Bagian JP, Hackett P. Cerebral blood flow: Comparison of ground-based and spaceflight data and correlation with space adaptation syndrome. Journal of Clinical Pharmacology. 1991;**31**(10):1036-1040

[8] Friedman DI, Jacobson DM. Diagnostic criteria for idiopathic intracranial hypertension. Neurology. 2002;**59**(10):1492-1495

[9] Mader TH, Gibson CR, Pass AF, Kramer LA, Lee AG, Fogarty J, et al. Optic disc edema, globe flattening, choroidal folds, and hyperopic shifts observed in astronauts after long-duration space flight. Ophthalmology. 2011;**118**(10):2058-2069

[10] Alexander DJ, Gibson CR, Hamilton DR, Lee SMC, Mader TH, Otto C, et al. Risk of spaceflight-induced intracranial hypertension and vision alterations. Lyndon B. Houston, Texas: National Aeronautics and Space Administration, Johnson Space Center; 2012

[11] Kramer LA, Sargsyan AE, Hasan KM, Polk JD, Hamilton DR. Orbital and intracranial effects of microgravity: Findings at 3-T MR imaging. Radiology. 2012;**263**(3):819-827

[12] Liu D, Kahn M. Measurement and relationship of subarachnoid pressure of the optic nerve to intracranial pressures in fresh cadavers. American Journal of Ophthalmology. 1993;**116**(5):548-556

[13] Otto C. Risk of Microgravity-Induced Visual Impairment and Elevated Intracranial Pressure (VIIP). NASA Johnson Space Center: Scientific and Technical Information Program; 2011, http://ntrs.nasa.gov/archive/nasa/casi.ntrs.nasa.gov/20110012449.pdf

[14] Lee AG, Mader TH, Gibson CR, Tarver W. Space flight-associated neuro-ocular syndrome. JAMA Ophthalmology. 2017;**135**(9):992-994

[15] Mader TH, Gibson CR, Hart SF, Lee AG. Asymmetric papilledema in idiopathic intracranial hypertension: Comment. Journal of Neuro-Ophthalmology. 2016;**36**(1):111-112

[16] Fogarty JA, Otto C, Kerstman E, Oubre C, Wu J. The Visual Impairment Intracranial Pressure Summit Report. http://ston.jsc.nasa.gov/collections/trs/_techrep/TP-2011-216160.pdf2011

[17] Taibbi G, Cromwell RL, Kapoor KG, Godley BF, Vizzeri G. The effect of microgravity on ocular structures and visual function: A review. Survey of Ophthalmology. 2013;**58**(2):155-163

[18] Mader TH, Gibson CR, Caputo M, Hunter N, Taylor G, Charles J, et al. Intraocular pressure and retinal vascular changes during transient exposure to microgravity. American Journal of Ophthalmology. 1993;**115**(3):347-350

[19] Mader TH, Taylor GR, Hunter N, Caputo M, Meehan RT. Intraocular pressure, retinal vascular, and visual acuity changes during 48 hours of 10 degrees head-down tilt. Aviation, Space, and Environmental Medicine. 1990;**61**(9):810-813

[20] Berdahl JP, Yu DY, Morgan WH. The translaminar pressure gradient in sustained zero gravity, idiopathic intracranial hypertension, and glaucoma. Medical Hypotheses. 2012;**79**(6):719-724

[21] Thornton WE, Moore TP, Pool SL. Fluid shifts in weightlessness. Aviation, Space, and Environmental Medicine. 1987;**58**(9 Pt 2):A86-A90

[22] Parazynski SE, Hargens AR, Tucker B, Aratow M, Styf J, Crenshaw A. Transcapillary fluid shifts in tissues of the head and neck during and after simulated microgravity. Journal of Applied Physiology. 1991;**71**(6):2469-2475

[23] Thornton WE, Moore TP, Pool SL, Vanderploeg J. Clinical characterization and etiology of space motion sickness. Aviation, Space, and Environmental Medicine. 1987;**58**(9 Pt 2):A1-A8

[24] Murthy G, Marchbanks RJ, Watenpaugh DE, Meyer JU, Eliashberg N, Hargens AR. Increased intracranial pressure in humans during simulated microgravity. The Physiologist. 1992;**35**(1 Suppl):S184-S185

[25] Tatebayashi K, Asai Y, Maeda T, Shiraishi Y, Miyoshi M, Kawai Y. Effects of head-down tilt on the intracranial pressure in conscious rabbits. Brain Research. 2003;**977**(1):55-61

[26] Aaslid R. Cerebral autoregulation and vasomotor reactivity. Frontiers of Neurology and Neuroscience. 2006;**21**:216-228

[27] Taylor CR, Hanna M, Behnke BJ, Stabley JN, McCullough DJ, Davis RT 3rd, et al. Spaceflight-induced alterations in cerebral artery vasoconstrictor, mechanical, and structural properties: Implications for elevated cerebral perfusion and intracranial pressure. FASEB Journal: Official Publication of the Federation of American Societies for Experimental Biology. 2013;**27**(6):2282-2292

[28] Kawai Y, Murthy G, Watenpaugh DE, Breit GA, Deroshia CW, Hargens AR. Cerebral blood flow velocity in humans exposed to 24 h of head-down tilt. Journal of Applied Physiology. 1993;**74**(6):3046-3051

[29] Iwasaki K, Levine BD, Zhang R, Zuckerman JH, Pawelczyk JA, Diedrich A, et al. Human cerebral autoregulation before, during and after spaceflight. The Journal of Physiology. 2007;**579**(Pt 3):799-810

[30] Geary GG, Krause DN, Purdy RE, Duckles SP. Simulated microgravity increases myogenic tone in rat cerebral arteries. Journal of Applied Physiology. 1998;**85**(5):1615-1621

[31] Wilson MH, Imray CH, Hargens AR. The headache of high altitude and microgravity—Similarities with clinical syndromes of cerebral venous hypertension. High Altitude Medicine & Biology. 2011;**12**(4):379-386

[32] Zhang LN, Zhang LF, Ma J. Simulated microgravity enhances vasoconstrictor responsiveness of rat basilar artery. Journal of Applied Physiology. 2001;**90**(6):2296-2305

[33] Sun B, Zhang LF, Gao F, Ma XW, Zhang ML, Liu J, et al. Daily short-period gravitation can prevent functional and structural changes in arteries of simulated microgravity rats. Journal of Applied Physiology. 2004;**97**(3):1022-1031

[34] Wilkerson MK, Lesniewski LA, Golding EM, Bryan RM Jr, Amin A, Wilson E, et al. Simulated microgravity enhances cerebral artery vasoconstriction and vascular resistance through endothelial nitric oxide mechanism. American Journal of Physiology. Heart and Circulatory Physiology. 2005;**288**(4):H1652-H1661

[35] Stevens SA, Lakin WD, Penar PL. Modeling steady-state intracranial pressures in supine, head-down tilt and microgravity conditions. Aviation, Space, and Environmental Medicine. 2005;**76**(4):329-338

[36] Lakin WD, Stevens SA, Penar PL. Modeling intracranial pressures in microgravity: The influence of the blood-brain barrier. Aviation, Space, and Environmental Medicine. 2007;**78**(10):932-936

[37] Herault S, Fomina G, Alferova I, Kotovskaya A, Poliakov V, Arbeille P. Cardiac, arterial and venous adaptation to weightlessness during 6-month MIR spaceflights with and without thigh cuffs (bracelets). European Journal of Applied Physiology. 2000;**81**(5):384-390

[38] Alperin N, Lee SH, Sivaramakrishnan A, Hushek SG. Quantifying the effect of posture on intracranial physiology in humans by MRI flow studies. Journal of Magnetic Resonance Imaging: JMRI. 2005;**22**(5):591-596

[39] Alperin N, Lee SH, Sivaramakrishnan A, Lichtor T. Relationship between total cerebral blood flow and ICP measured noninvasively with dynamic MRI technique in healthy subjects. Acta Neurochirurgica. Supplement. 2005;**95**:191-193

[40] Wiener TC. Space obstructive syndrome: Intracranial hypertension, intraocular pressure, and papilledema in space. Aviation, Space, and Environmental Medicine. 2012;**83**(1):64-66

[41] Gisolf J, van Lieshout JJ, van Heusden K, Pott F, Stok WJ, Karemaker JM. Human cerebral venous outflow pathway depends on posture and central venous pressure. The Journal of Physiology. 2004;**560**(Pt 1):317-327

[42] Buckey JC, Gaffney FA, Lane LD, Levine BD, Watenpaugh DE, Blomqvist CG. Central venous pressure in space. The New England Journal of Medicine. 1993;**328**(25):1853-1854

[43] Foldager N, Andersen TA, Jessen FB, Ellegaard P, Stadeager C, Videbaek R, et al. Central venous pressure in humans during microgravity. Journal of Applied Physiology. 1996; **81**(1):408-412

[44] Buckey JC Jr, Gaffney FA, Lane LD, Levine BD, Watenpaugh DE, Wright SJ, et al. Central venous pressure in space. Journal of Applied Physiology. 1996;**81**(1):19-25

[45] Ainslie PN, Duffin J. Integration of cerebrovascular CO_2 reactivity and chemoreflex control of breathing: Mechanisms of regulation, measurement, and interpretation. American Journal of Physiology. Regulatory, Integrative and Comparative Physiology. 2009;**296**(5): R1473-R1495

[46] James JT. The headache of carbon dioxide exposures. In: International Conference on Environmental Systems; Houston, TX, United States: NASA Johnson Space Center; 2007

[47] CH S, JL Z, CH L. Investigation of Airflow and Accumulation of Carbon Dioxide in the Service Module Crew Quarters. Society of Automotive Engineers. 2002;2002–01-2341

[48] Ursino M, Lodi CA. A simple mathematical model of the interaction between intracranial pressure and cerebral hemodynamics. Journal of Applied Physiology. 1997;**82**(4):1256-1269

[49] Zuj KA, Arbeille P, Shoemaker JK, Blaber AP, Greaves DK, Xu D, et al. Impaired cerebrovascular autoregulation and reduced CO_2 reactivity after long duration spaceflight. American Journal of Physiology. Heart and Circulatory Physiology. 2012;**302**(12):H2592-H2598

[50] Zwart SR, Gibson CR, Mader TH, Ericson K, Ploutz-Snyder R, Heer M, et al. Vision changes after spaceflight are related to alterations in folate- and vitamin B-12-dependent one-carbon metabolism. The Journal of Nutrition. 2012;**142**(3):427-431

[51] Sanzari J, Muehlmatt A, Savage A, Lin L, Kennedy A. Increased intracranial pressure in mini-pigs exposed to simulated solar particle event radiation. Acta Astronautica. 2014; **94**(2):807-812

[52] Leszczynski D, Joenvaara S, Reivinen J, Kuokka R. Non-thermal activation of the hsp27/ p38MAPK stress pathway by mobile phone radiation in human endothelial cells:

Molecular mechanism for cancer- and blood-brain barrier-related effects. Differentiation; Research in Biological Diversity. 2002;**70**(2–3):120-129

[53] Leszczynski D, Nylund R, Joenvaara S, Reivinen J. Applicability of discovery science approach to determine biological effects of mobile phone radiation. Proteomics. 2004; **4**(2):426-431

[54] Nittby H, Brun A, Eberhardt J, Malmgren L, Persson BR, Salford LG. Increased blood-brain barrier permeability in mammalian brain 7 days after exposure to the radiation from a GSM-900 mobile phone. Pathophysiology: The Official Journal of the International Society for Pathophysiology/ISP. 2009;**16**(2–3):103-112

[55] Lempert P, Cooper KH, Culver JF, Tredici TJ. The effect of exercise on intraocular pressure. American Journal of Ophthalmology. 1967;**63**(6):1673-1676

[56] Vieira GM, Oliveira HB, de Andrade DT, Bottaro M, Ritch R. Intraocular pressure variation during weight lifting. Archives of Ophthalmology. 2006;**124**(9):1251-1254

[57] Cavanagh PR, Licata AA, Rice AJ. Exercise and pharmacological countermeasures for bone loss during long-duration space flight. Gravitational and Space Biology Bulletin. 2005;**18**(2):39-58

[58] Junqueira LF Jr. Teaching cardiac autonomic function dynamics employing the Valsalva (Valsalva-Weber) maneuver. Advances in Physiology Education. 2008;**32**(1):100-106

[59] Brimioulle S, Moraine JJ, Norrenberg D, Kahn RJ. Effects of positioning and exercise on intracranial pressure in a neurosurgical intensive care unit. Physical Therapy. 1997;**77**(12): 1682-1689

[60] Lane HW, Bourland C, Barrett A, Heer M, Smith SM. The role of nutritional research in the success of human space flight. Advances in Nutrition. 2013;**4**(5):521-523

[61] Newborg B. Pseudotumor cerebri treated by rice reduction diet. Archives of Internal Medicine. 1974;**133**(5):802-807

[62] Intracranial Hypertension Research Foundation. 2015. Available from: http://www.ihrfoundation.org/

[63] Yri HM, Fagerlund B, Forchhammer HB, Jensen RH. Cognitive function in idiopathic intracranial hypertension: A prospective case-control study. BMJ Open. 2014;**4**(4):e004376

[64] McGeeney BE, Friedman DI. Pseudotumor cerebri pathophysiology. Headache. 2014; **54**(3):445-458

[65] Kharkar S, Hernandez R, Batra S, Metellus P, Hillis A, Williams MA, et al. Cognitive impairment in patients with pseudotumor cerebri syndrome. Behavioural Neurology. 2011;**24**(2):143-148

[66] Sorensen PS, Thomsen AM, Gjerris F. Persistent disturbances of cognitive functions in patients with pseudotumor cerebri. Acta Neurologica Scandinavica. 1986;**73**(3):264-268

[67] Wall M, Kupersmith MJ, Kieburtz KD, Corbett JJ, Feldon SE, Friedman DI, et al. The idiopathic intracranial hypertension treatment trial: Clinical profile at baseline. JAMA Neurology. 2014;**71**(6):693-701

[68] Hoffmann J, Huppertz HJ, Schmidt C, Kunte H, Harms L, Klingebiel R, et al. Morphometric and volumetric MRI changes in idiopathic intracranial hypertension. Cephalalgia: An International Journal of Headache. 2013;**33**(13):1075-1084

[69] Roberts DR, Antonucci MU. Effects of spaceflight on astronaut brain structure. The New England Journal of Medicine. 2018;**378**(6):582-583

[70] Koppelmans V, Bloomberg JJ, Mulavara AP, Seidler RD. Brain structural plasticity with spaceflight. NPJ Microgravity. 2016;**2**:2

[71] Alperin N, Bagci AM, Oliu CJ, Lee SH, Lam BL. Role of cerebrospinal fluid in spaceflight-induced ocular changes and visual impairment in astronaut. Radiology. 2017;**285**(3):1063

[72] McGuire S, Sherman P, Profenna L, Grogan P, Sladky J, Brown A, et al. White matter hyperintensities on MRI in high-altitude U-2 pilots. Neurology. 2013;**81**(8):729-735

[73] Balldin UI, Pilmanis AA, Webb JT. Central nervous system decompression sickness and venous gas emboli in hypobaric conditions. Aviation, Space, and Environmental Medicine. 2004;**75**(11):969-972

[74] Jersey SL, Baril RT, McCarty RD, Millhouse CM. Severe neurological decompression sickness in a U-2 pilot. Aviation, Space, and Environmental Medicine. 2010;**81**(1):64-68

[75] NASA. International Space Station Research Projects Life Sciences Data Archive at Johnson Space Center. 2014. Available from: http://lsda.jsc.nasa.gov/docs/research/research_detail.aspx?experiment_type_code=21&research_type=current

[76] Motschmann M, Muller C, Kuchenbecker J, Walter S, Schmitz K, Schutze M, et al. Ophthalmodynamometry: A reliable method for measuring intracranial pressure. Strabismus. 2001;**9**(1):13-16

[77] Samuel M, Burge DM, Marchbanks RJ. Tympanic membrane displacement testing in regular assessment of intracranial pressure in eight children with shunted hydrocephalus. Journal of Neurosurgery. 1998;**88**(6):983-995

[78] Rajajee V, Vanaman M, Fletcher JJ, Jacobs TL. Optic nerve ultrasound for the detection of raised intracranial pressure. Neurocritical Care. 2011;**15**(3):506-515

[79] Miller N. Do Astronauts Develop Papilledema. Review of NASA's Evidence Reports on Human Health Risks. 2013 Letter Report. Washington, DC: The National Academies Press; 2014

[80] Barr YR. Lumbar puncture during spaceflight: Operational considerations, constraints, concerns, and limitations. Aviation, Space, and Environmental Medicine. 2014;**85**(12):1209-1213

[81] Otto C, editor. NASA's visual impairment & intracranial pressure risk: Utilizing the ISS for risk reduction. In: 1st Annual ISS Research & Development Conference; Denver Marriot City Center, CO; 2012

Chapter 6

The Effect of Gravity on the Nervous System

Florian P.M. Kohn, Claudia Koch and
Ramona Ritzmann

Additional information is available at the end of the chapter

http://dx.doi.org/10.5772/intechopen.74715

Abstract

Gravity affects the nervous system of living organisms. This book chapter reviews historical and recent findings on how changes in gravity affect cellular and subcellular parameters of human and animal cells as well as the timing and shaping of complex sensorimotor responses. With an emphasize on weightlessness, partial, and hypergravity conditions, the gravity dependencies of living organisms have been manifested on different levels of organization, ranging from changes in biophysical properties of single cells to the intact nervous system. An effort has been made to integrate the various findings into a consistent model for a better understanding of how the components of the nervous system interact as a response to acute and long-term gravitational variation. Especially with planned long-term manned missions to Mars and beyond, knowledge about the impact of increased and decreased gravity on the nervous system is essential for the physical and cognitive preparation to assure the success of space missions and human survival in space.

Keywords: gravity, microgravity, hypergravity, adaptation, reflex, sensorimotor function, biophysical properties, electrophysiology

1. Introduction

The gravitational force on Earth has remained constant in direction and magnitude since the formation of the planet [1]. Therefore, living species including plants, animals, and humans have evolved to cope with and rely upon gravity equal to 1 g. Throughout the history of the Earth, all living organisms adapted their cellular and behavioral function to this particular physical environment characteristic for our home planet. Gravity—as a permanent and constant vector-calibrated stimulus—led to various gravity-perceiving systems in organisms that control growth or influence movement and behavior. But what happens if this constant stimulus is changed?

Future challenges in terms of long-term interplanetary manned space missions moved the adaptability of living organisms and their vital systems to heterogravitational habitats into scientific focus [2]. With emphasis on our astronomical neighbors Mars and Moon with a reduced gravitational force of approximately a third and a fifth of the Earth's gravitation, it became apparent that orbital or interplanetary space explorations require knowledge about gravity-perceiving systems, which determine movement, cognition, and survival [3]. In the past decades, space research manifested a significant gravity dependency for various biological processes and vital systems. A special focus lies on the animal and human nervous system (NS) as it is crucial for integration of sensory input, for example, from the vestibular system, movement control, and terrestrial locomotion on Earth. The NS governs muscle contraction enabling the body to counteract the gravitational force and controlling locomotor patterns and reflexes during the evolutionary shift from aqueous to terrestrial life. For interplanetary and orbital missions in future human space flight, knowledge about the gravity sensitivity of the NS is crucial to anticipate major challenges, train the astronauts, and prepare adequate countermeasures to conserve elementary sensorimotor skills during long-term partial-gravity exposure.

The NS is a network of neurons and fibers which transmits nerve impulses between parts of the body. It is composed of interconnected nerve and supporting glial cells. The mechanism of neuronal communication is based on electrochemical coupling, the modulation of intra- and extracellular ions to modify the electrical properties of a cell (intracellular signaling), and the controlled release of transmitters (intercellular communication). Resulting action potentials (APs) are the basic communication unit, and their conduction frequency serves as the coding for the stimulus' intensity.

One of the fundamental circuits within the central nervous system (CNS) to control muscle contraction is the monosynaptic reflex arch [4]. These reflexes are neuromuscular reactions in response to an external stimulus, which lead to fast muscle contractions. The magnitude of muscle contraction depends on the magnitude of sensory input. Allowing mobility of terrestrial life, sensory input from the vestibular and visual systems and proprioception is processed by the NS, and by means of muscle innervation, appropriate forces are generated to control simple posture or movement [5–10]. These sensorimotor competencies are crucial for life. Since the first manned spaceflight of Yuri Gagarin in 1961, the effect of microgravity on the human body has been intensively investigated. In the decades since his first spaceflight, many experiments have been performed which made gravity-induced changes on astronauts and cosmonauts apparent. With an emphasis on weightlessness and our astronomical neighbors Mars and Moon [2, 5], the authors found directly related health effects, among others a persistent modulation in the sensory [7, 11] and motor system [12] and the resulting structural loss of muscle [13] and bone mass [14]. In addition, there are modulations in the neuromuscular system underlying those health-related changes that open up many questions on how the variation of gravity influences the NS. These questions led to numerous experiments to investigate the effect of varying gravity conditions on the different levels of organization, from the molecular and cellular level, up to the whole NS and its interconnection with movement control and mobility. The functional properties of these levels were thoroughly investigated, however, with barely any interconnection.

This chapter systematically reviews results on how changes in gravity affect neurons of human and animal as well as temporal and spatial characteristics of complex sensorimotor responses. For that purpose, the subject of this chapter is divided in three subthemes: the gravity dependence of subcellular and cellular parameters associated with neuronal activation is followed by an outline of the sensitivity of the human NS to gravitational variation in the context of movement. To interconnect these transdisciplinary findings, a working model is introduced on how the effects observed on the molecular and biophysical level may impact the sensorimotor control of the NS. The chapter ends with a conclusive statement that refers to movement in terms of long-term interplanetary manned space missions.

2. Gravity and the nervous system

2.1. The gravity dependence of subcellular and cellular parameters

A variety of life science experiments executed in gravity conditions different from Earth gravitation, 1 g, have been executed in cellular model systems. With an emphasize on subcellular and cellular parameters and the associated biophysical attributes, most of the *in vitro* experiments have been conducted on short-term gravity-research platforms as drop towers and parabolic flights. Findings from the late twentieth century and recent findings manifest a significant gravity dependency of the basic cell function associated with changes in membrane and channel properties as well as the underlying biophysical characteristics. Results are outlined in the following subchapter.

2.1.1. Membrane parameters

From experiments with unicellular organisms [15] and various cell types as immune cells [16] and neuronal cells [17], it is well established that single cells react to changes in gravity even though they do not have dedicated gravity-sensing structures. One of the major components that all these cell types and organisms have in common is the cell membrane. These complex structures are mainly composed of proteins and lipids [18].

To communicate, cells of the nervous system are able to modify their membrane potential. This ability is based on the activity of integrated membrane proteins as ion channels and ion pumps. But it is well known that the physicochemical state of the lipid membrane can directly modify the function of membrane proteins [19, 20]. In non-space-related experiments, it was shown that the closed-state probability of nicotinic acetylcholine receptors (nAChRs) increased with a decreased membrane fluidity [21]. These nAChRs are a major player in the sensorimotor system as they are located in the motor end plates that form the interface between the neuronal system and the muscles.

Due to these findings, experiments have been performed to monitor the changes of membrane viscosity in micro- and hypergravity with several models (artificial asolectin vesicles and human neuronal SH-SY5Y cells). In all models, the membrane fluidity significantly increases in microgravity and decreases in hypergravity, but in a different distinctness [22]. The difference in distinctness might be explained with the absence of a cytoskeleton in artificial membranes or a different lipid composition.

Nevertheless, this finding, that the membrane fluidity is gravity-dependent, will have a huge impact on biological and medical gravity research, as this is a basic physical mechanism that affects every cell in an organism [23].

2.1.2. Ion channel parameters

Ion channels are crucial for neuronal communication. They form controllable pores through the cell membrane. Charged ions can diffuse through these pores, following electrical and chemical gradients, changing the electrical properties of the cell. Ion channel parameters as open- and closed-state probability have been investigated by using pore-forming peptides which can be used as ion channel analogs. Until now, no native ion channel proteins have been used for gravity research.

The open-state probability of porin channels from *Escherichia coli* is significantly decreased in microgravity, whereas in hypergravity, it is increased. No effect on conductance was found [24].

Similar findings have been made with alamethicin, a pore-forming peptide from *Trichoderma viride*. In microgravity, the activity of alamethicin is decreased, whereas in hypergravity, it is increased [25, 26].

The effect on ion channels is—similar to changes in membrane fluidity—fully reversible and fast. With the onset of a different gravity condition, the open-state probability is changed, returning to normal as soon as the experiment returns to normal 1 g gravity.

2.1.3. Electrophysiological properties of single cells

By having a stable-resting potential, a cell is able to communicate. By changing the activity of relevant ion channels, the membrane potential can be modulated. During parabolic flight, the resting potential of human neuronal cells is significantly depolarized in microgravity and it is hyperpolarized in hypergravity. During microgravity, the depolarization is about 3 mV [27]. This gravity dependence of resting potential is not limited to excitable cells as neuronal cells; it was also found during a drop-tower mission in SF21 cells, an ovary cell line from the insect *Spodoptera frugiperda* [17].

Again in parabolic flight, in microgravity, the transmembrane currents in oocytes from *Xenopus laevis* show a significant decrease at a holding potential of −100 mV, whereas in hypergravity, there is a tendency of increased currents [28].

2.1.4. Propagation of action potentials

Action potentials (APs) are the basic communication unit in the nervous system. The intensity of a stimulus is frequency-coded: while the amplitude of APs remains constant, their frequency differs dependent on the stimulus strength. In microgravity obtained by drop tower, the rate of action potentials triggered by spontaneous active leech neurons is significantly increased [29]. This means on the level of single cells, more action potentials are generated in weightlessness.

Simultaneously, the conduction velocity of APs on the axonal level is decreased in microgravity and increased in hypergravity. This was demonstrated in parabolic flight missions *in vitro* in isolated earthworm axons and isolated rat axons and *in vivo* in intact earthworms. [29]. Again, the changes are fast and fully reversible.

2.2. The gravity dependence of the human nervous system

In addition to the abovementioned molecular and cellular experiments, a variety of studies have been conducted to investigate the effect of gravity on the nervous system in humans [4, 10, 30–34]. In the context of movement control, it becomes apparent that the biophysical attributes underlying cell communication and the nervous capacity to inhibit and facilitate neural pathways are of fundamental importance to activate and control the skeletal muscle, allowing the living organisms to displace themselves. On the complex sensorimotor level, the gravitational force determines human movement control, and its impact is considered to be of major relevance for the astronaut's safety management in scenarios that require spontaneous or chronic adaptation to an astronomical environment different from the Earth. Not only are short-term platforms as parabolic flights and centrifuges used for this research, the experiments are also conducted during long-term space missions or exploration class missions (up to 1.5 years).

A frequently used technique is the peripheral nerve stimulation (PNS) as it is a noninvasive and reliable approach, providing information about nerve communication including temporal and spatial characteristics of direct motor (M-wave) and reflector responses (Hoffmann(H)-reflex) of the skeletal muscle [35, 36]. By external electrical stimulation, neurons, axons, or cell bodies are depolarized, and the bipolar potential difference of the muscle is measured and interpreted [4]. The nerve *tibialis posterior* and the muscle *soleus* have been established as a model for describing the adaptation processes of the neuromuscular system with emphasize on the temporal and spatial characteristics of the electromyographic signal.

2.2.1. Spatial attributes

The shaping of the potential difference includes peak-to-peak amplitudes normalized to the input stimulus and is associated with the magnitude of the muscle output [37]. Furthermore, the stimulation threshold corresponds to the threshold for axonal excitation with a minimal current evoking a muscle contraction [4].

2.2.2. Stimulation threshold of the H-reflex

The needed electrical stimulation to depolarize an axon to generate a constant muscle response can be interpreted as the responsiveness of a nerve to external stimuli. In reduced gravity conditions, similar to Moon (0.16 g) and Mars (0.36 g), generated in parabolic flights, higher stimulation currents for PNS were needed to depolarize the neurons. In hypergravity (1.8 g), the needed currents were smaller [4]. Although the respective partial-gravity level lasts only 24–33 s [10] and effects are reversible within seconds, it can be concluded that the stimulation threshold is acutely increased in reduced gravity and decreased in hypergravity.

2.2.3. *Amplitude of the H-reflex*

The H-reflex amplitude describes the neuronal output signal of the reflectory reaction of muscles and is proportional to the muscle contraction after peripheral electrical stimulation of sensory fibers in their innervating nerves. Gravity dependency has been reported in cross-sectional study designs with neuroplastic changes for amplitudes of H-reflexes and stretch reflexes [10, 30–34]. The peak-to-peak amplitudes increased during hypergravity, independently from the method of stimulation [10, 33].

In micro- and reduced gravity, the results are more inhomogeneous. Experiments in Mars and Moon gravity showed a gravity dependence in the decrease of peak-to-peak amplitudes of Hmax. Less gravity resulted in a higher decrease in Hmax amplitude [4]. Nevertheless, in microgravity, the H-reflex was either not changed [10, 34] or it was increased [30–33]. A long-term experiment on the International Space Station (ISS) revealed a decrease of H-reflexes in space [38]. This decrease was found for 5 months in space, but it was recovered shortly after the return to Earth.

The inhomogeneous findings might be explained by (1) active adaptation processes during long-term missions and (2) mainly due to differences in methodology [4].

The amplitudes of the different sections of the H-reflex depend on the stimulation threshold. As the threshold is gravity-dependent, this has to be taken into account when a constant stimulus intensity is used during the experiments [30–33]. H/M-wave recruitment curves are independent of stimulation threshold [10, 34]. As a consequence, gravity-induced changes in H-reflex amplitudes elicited with a constant and submaximal stimulus are rather attributed to threshold shifts than changes in gravity [30–33].

2.2.4. *Temporal attributes*

Temporal characteristics of motor and reflectory responses are characterized by latencies relying on the nerve's conduction velocity [39], duration, and inter-peak intervals (IPI) associated with the conduction speed along the muscle fibers at the neuromuscular junction where the nerve interconnects with the muscle [40].

2.2.5. *Neuromuscular latency*

Neuromuscular latency describes the time between a given stimulus and the measured muscle response. The latency of H-reflex and M-wave in the *Soleus* muscle was investigated in many experiments, short term [4, 32] and long term [40], but the results are again ambiguous, similar to the findings for the amplitudes of H-reflex. In eight subjects, Ritzmann et al. showed an increase in H-reflex latencies with gradually decreasing gravity (from hyper to 1 g to Mars to Lunar gravity) with a simultaneous tendency of an increase of M-wave latencies [4]. However, Ohira et al. showed that hyper- and microgravity had no immediate effect on the H-reflex and M-wave latencies; unfortunately, they did not give information about the sample size [32].

2.2.6. Inter-peak interval

By interpreting the IPI between the negative and the positive maxima of the biphasic amplitude, information about the conduction velocity from the motor end plate to the muscle fibers can be gained. The motor end plates (or neuromuscular junction) are the interface between the nervous system and the muscles. It could be showed that the IPIs of the peak *M. soleus* M-wave and H-reflex significantly increase with decreasing gravity from hyper- to 1 g to Mars to Moon gravity conditions [4]. This finding can be interpreted that the conduction velocity at the neuromuscular junction is decreasing in reduced gravity and is increasing in hypergravity. This effect occurs immediately and is fully reversible.

2.2.7. Duration

The duration of the H-reflex is established as the interval from the first rise of the electromyographic signal until return to baseline. Ritzmann et al. demonstrated a gradual decrease in H-reflex duration with increasing gravitation from lunar to Martian to earth gravitation to hypergravity [4]. Accordingly, the duration of the M-waves showed a strong tendency to decline with increasing gravitation. As the duration of the motor and reflectory responses cover information about the conduction velocity of signal transmission from the motor end plate to the muscle fibers, results indicate a major impact of gravity on the temporal characteristics of sensorimotor responses.

3. A model for the immediate adaptation of the nervous system to changes in gravity

The following model integrates the results from the various experiments that have been carried out in the past decades from cellular level up to the neuromuscular interface. To avoid long-term adaptation processes, only immediate effects have been taken into account. The model was designed in a bottom-up approach, starting at the very base level of gravity dependence. Therefore, it can be used as a framework for future—more complex data—as long-term adaptation processes and the gravity dependence of for example, the human brain.

3.1. Molecular level

Micro- and hypergravity change the biophysical properties of biological membranes in every cell in the body. This is not due to some biological effect or process, it is a change in thermodynamic properties of biological membranes [20]; therefore, this can be seen as the basic principle of how gravity affects cells as neuronal cells, for example.

On Earth, it is well known that the properties of membrane-integrated proteins as ion channels depend on the physical state of the membrane. Lateral pressure or membrane fluidity is an important component, for example, the open state of alamethicin pores clearly depends on the lateral pressure of the membrane [41], and the pore activity increases with an increased lateral pressure. An increased lateral pressure can be interpreted as decreased membrane fluidity. This was also shown for other ion channels, for example, the closed-state probability of nicotinic acetylcholine receptor channels increases (the open-state probability decreases) toward decreased membrane fluidity [21].

The pore activity of alamethicin and the open-state probability of ion channels is also gravity-dependent [24, 25]. In microgravity, the open-state probability decreases, whereas in hypergravity, it increases.

As membrane fluidity is affected by gravity and due to the fact that ion channels are affected by membrane fluidity, the first part of the model can be described as follows:

In microgravity, the membrane fluidity is increased. This changed membrane fluidity decreases the open-state probability of ion channels. This effect is inversed in hypergravity: membrane fluidity decreases and the open-state probability of ion channels increases (**Figure 1**).

3.2. Single cells

It was shown that cells slightly depolarize in microgravity—the membrane potential gets more positive—and they hyperpolarize in hypergravity. With a light depolarization of the resting potential, the threshold to trigger action potentials is reached more easily. This effect was demonstrated in spontaneous active leech neurons. The rate of APs increased in microgravity.

With these findings, the model of gravity dependence on the molecular level can be extended to explain the cellular gravity dependence of single (neuronal) cells (**Figure 2**).

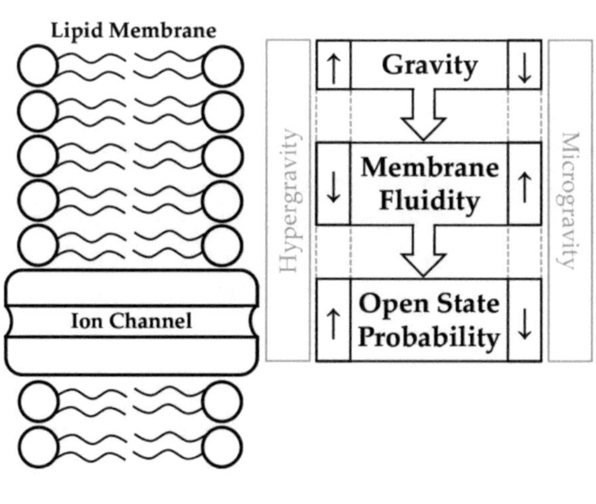

Figure 1. The biophysical gravity dependence of cell membranes and incorporated ion channel proteins. Modified from [42].

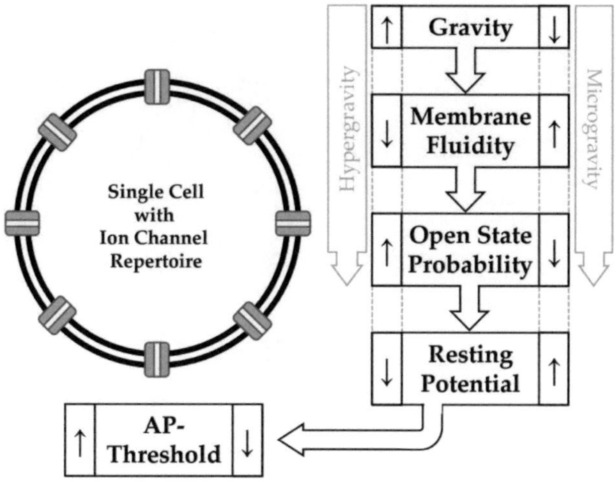

Figure 2. The gravity dependence of a single neuronal cell. Modified from [42].

3.3. Neuronal system: Sensorimotor system

The influence of different gravity conditions on neuronal tissue is clearly visible. In isolated single axons as well as in living animals and in human test subjects, the propagation velocity of APs is decreased in microgravity and it is increased in hypergravity.

Neuromuscular reflex arcs in humans are influenced by gravity. In microgravity, increased latencies can be measured. An increased latency can be explained with a decreased conduction speed—the APs are slower in microgravity.

In Mars and Moon gravity, a higher stimulus has to be given to get the same Hmax as in 1 g, and the peak-to peak amplitude of the H-reflex is decreased (with heterogeneous data at real microgravity). Unfortunately, as the methods of single-cell electrophysiology and peripheral nerve stimulation are different, their results cannot be compared directly. Nevertheless, a decreased propagation velocity of APs in the axons can also explain the decrease in Hmax in microgravity. Less APs per time arrive at the muscle, which leads to a reduced contraction. Two findings support this explanation: first, the decrease can be compensated with a higher stimulus. Due to the frequency coding of sensory input, a higher stimulus generates more APs per time. With more APs per time arriving at the muscle, the contraction force is increased. Second, the decrease in inter-peak intervals of the H-reflex indicates a decreased signal speed at the neuromuscular junction. In increased gravity, these effects are reversed (**Figure 3**).

In microgravity, the rate of action potentials is increased, while at the same time, the propagation speed of APs is decreased. This might look like an inconsistency, but it is not. It can be explained with a mathematical equation. Matsumoto and Tasaki developed a mathematical model to calculate the conduction speed of APs in unmyelinated axons [43]. This equation can also be used to estimate the conduction velocity of APs in myelinated axons

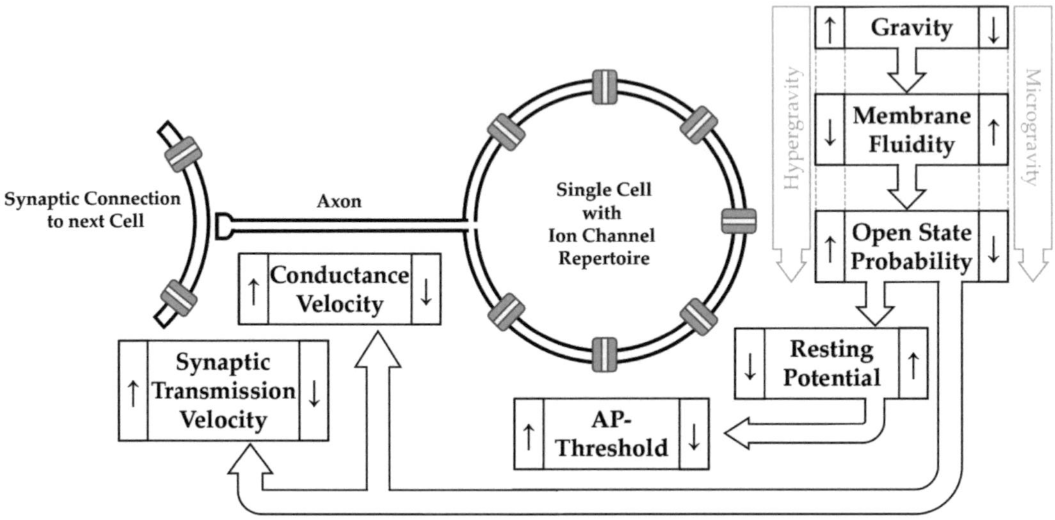

Figure 3. The gravity dependence of a multicellular network, connected via synapses as the sensorimotor system. Modified from [42].

$$v_{axon} \approx \sqrt{\frac{d}{8 \cdot \varrho C^2 \cdot R^*}} \tag{1}$$

where v_{axon} = conduction velocity, C = membrane capacity, d = diameter of the nerve, R^* = resistance of the membrane, and ϱ = axoplasmic resistance.

By integrating the data from gravity research and Matsumoto and Tasaki's model, at first view, the inconsistent findings from single-cell electrophysiology and the data from PNS can be brought together quite nicely to a working model on how the sensorimotor system adapts to changes in gravity.

The increased membrane viscosity in microgravity decreases the open-state probability of ion channels, leading to a slightly depolarized membrane potential. With a reduced open-state probability, the resistance of the membrane (R^*) is increased. If axoplasmic resistance (ϱ), membrane capacity (C), and the diameter of the axon (d) are treated as constant in changed gravity, the increased resistance of the membrane leads to a decreased conduction velocity of APs (v_{axon}) while simultaneously APs can be triggered more easily.

To sum up, the described effects are a gravity-dependent decrease in neuronal conduction velocity–or, more general, an increase in electrical and chemical time constants—under reduced gravity and vice versa in hypergravity.

4. Conclusion

In the decades since the first manned space mission, many *in vitro* and *in vivo* experiments have been conducted to investigate the effect of micro- and hypergravity on neuronal processes.

Adaptation processes occur on all levels of organization, from the subcellular level up to the neuromuscular system (and even up to the brain). Unfortunately, till date, the discrete results of these experiments were never brought together to see (1) whether they can be integrated to a working model of neuronal adaptation in varying gravity or (2) to reveal inconsistencies or (3) areas, which have not been investigated yet. This model aims at bringing insight to the short-term adaptation of the neuronal system to varying gravity conditions. Simultaneously—as some points still are based on reasoned assumptions [42]—it has to be seen as a framework, which should be fleshed out more in future experiments to include long-term adaptation processes and the adaptation of the human brain. A more interconnected and interdisciplinary analysis of all the data can serve as a "roadmap," aiming for giving more structure to ongoing and future research.

Findings are of major functional relevance in the application field of manned space flight as well as countermeasure development. As more and more space agencies and private space companies are planning long-term missions into space, for example, to Mars, the effect of gravity—and its absence—on the human organisms has to be understood overreaching all vital body systems to minimize the risks for space-faring humans [2]. Today, scientific outcomes of life science experiments executed in samples of astronauts and cosmonauts encompass a variety of long-term adaptation in regard to their sensory perception, motor execution, and planning as well as complex body motion. They are interrelated to neural adaptation to varying gravity and have been verified as follows (for review, see [44]): a recalibration of sensory perception, vestibular and proprioceptive dysfunction [7, 11], changes in muscle synergies and coordination, a decline of muscle force as well as deficits in posture control [6], locomotion, and functional mobility [8]. Reduced and delayed reflex responses and a decline in intramuscular and intermuscular function occur concomitantly with an increased muscle weakness, fatigue concomitant with a higher fall, and injury prevalence [40, 44]. With a persistency beyond the acute period of space flight, these adaptations are of clinical relevance as manifested by significant adverse effects which entail fragility and bone fractures [14, 44].

To reduce health and life risk throughout long-term exposure to low gravity during manned space explorations, scientists and space agencies developed intelligent exercise technologies and efficient interventions validated in cohorts of space crew members to prevent the human body from deconditioning [2]. Empirical outcomes subject to the NS and its adaptability to changes in gravity are included in the concepts of ancient and future countermeasures as manifested, for instance, for strength or jump exercises, vibration treatment, sensorimotor training, and artificial gravity [44].

Although great efforts have been made to optimize countermeasures, limitation on the cellular level such as changes in membrane fluidity as well as complex adaptations on the spinal level encompassing mechanisms of facilitating and inhibiting is of major relevance and cannot be diminished by countermeasures, only [4, 10, 23].

As astronauts traveling to Mars will live in the absence of gravity for approximately 2 years with transition between weightlessness and planetary gravitational forces at the beginning, middle, and end of the mission, further research and countermeasure development considering the gravity dependency of the NS will be obligate to assure a safe space travel and Earth return in the future [44].

Acknowledgements

The research was supported by the German Aerospace Center (DLR), the European Space Agency (ESA), and Novespace.

Author details

Florian P.M. Kohn[1*], Claudia Koch[1] and Ramona Ritzmann[2]

*Address all correspondence to: florian.p.m.kohn@uni-hohenheim.de

1 Department of Membrane Physiology (230b), University of Hohenheim, Stuttgart, Germany

2 Departments for Sports and Sport Science, University of Freiburg, Freiburg, Germany

References

[1] Montmerle T, Augereau JC, Chaussidon M, Gounelle M, Marty B, Morbidelli A. Solar system formation and early evolution: The first 100 million years. Earth, Moon and Planets. 2006;**98**(1):39-95

[2] Spudis PD. An argument for human exploration of the moon and Mars. American Scientist. 1992;**80**:269-277

[3] Lister A, Rothschild LJ, editors. Evolution on Planet Earth: The Impact of the Physical Environment. Amsterdam, Boston: Academic Press; 2003

[4] Ritzmann R, Krause A, Freyler K, Gollhofer A. Gravity and neuronal adaptation - Neurophysiology of reflexes from hypo to hyper gravity conditions. Microgravity Science and Technology. 2017;**29**(1-2):9-18. DOI: 10.1007/s12217-016-9519-4

[5] Margaria R, Cavagna GA. Human locomotion in subgravity. Aerospace Medicine. 1964;**22**: 1140-1146

[6] Homick JL, Reschke MF. Postural equilibrium following exposure to weightless space flight. Acta Oto-Laryngologica. 1977;**83**(5-6):455-464

[7] Paloski WH, Black FO, Reschke MF, Calkins DS, Shupert C. Vestibular ataxia following shuttle flights: Effects of microgravity on otolith-mediated sensorimotor control of posture. The American Journal of Otology. 1993;**14**:9-17

[8] Bloomberg JJ, Layne CS, McDonald PV, Peters BT, Huebner WP, Reschke MF. Effects of Space Flight on Locomotor Control. NASA Technical Report (20040201527)

[9] Layne CS, Mulavara AP, McDonald PV, Pruett CJ, Kozlovskaya IB, Bloomberg JJ. Effect of long-duration spaceflight on postural control during self-generated perturbations. Journal of Applied Physiology. 2001;**90**:997-1006

[10] Ritzmann R, Freyler K, Weltin E, Krause A, Gollhofer A. Load dependency of postural control—kinematic and neuromuscular changes in response to over and under load conditions. PLoS One. 2015;**10**(6):e0128400

[11] Reschke MF, Anderson DJ, Homick JL. Vestibulo-spinal response modification as determined with the H-reflex during the Spacelab-1 flight. Experimental Brain Research. 1986;**64**:367-379

[12] Blottner D, Salanova M. The NeuroMuscular System: From Earth to Space Life Science. Springer Publishing; 2015

[13] Di Prampero P, Narici M. Muscles in microgravity: From fibres to human motion. Journal of Biomechanics. 2003;**36**(3):403-412

[14] Loomer P. The impact of microgravity on bone metabolism in vitro and in vivo. Critical Reviews in Oral Biology and Medicine. 2001;**12**(3):252-261

[15] Häder D-P, Hemmersbach R, Lebert M. Gravity and the Behavior of Unicellular Organisms. Cambridge: Cambridge University Press; 2005

[16] Hauschild S, Tauber S, Lauber B, Thiel CS, Layer LE, Ullrich O. T cell regulation in microgravity – The current knowledge from in vitro experiments conducted in space, parabolic flights and ground-based facilities. Acta Astronautica. 2014;**104**:365-377. DOI: 10.1016/j.actaastro.2014.05.019

[17] Wiedemann M, Kohn FPM, Roesner H, Hanke W. Self-organization and Pattern-formation in Neuronal Systems Under Conditions of Variable Gravity. Springer Publishing; 2011

[18] Pollard TD, Earnshaw WC. Cell Biology. Philadelphia, USA: Elsevier; 2008

[19] Lee AG. How lipids affect the activities of integral membrane proteins. Biochimica et Biophysica Acta. 2004;**1666**:62-87

[20] Heimburg T. Thermal Biophysics of Membranes. Weinheim, Germany: Wiley-VCH Verlag GmbH & Co. KGaA; 2007. DOI: 10.1002/9783527611591

[21] Zanello LP, Aztiria E, Antollini S, Barrantes FJ. Nicotinic acetylcholine receptor channels are influenced by the physical state of their membrane environment. Biophysical Journal. 1996;**70**(5):2155-2164

[22] Sieber M, Hanke W, Kohn FPM. Modification of membrane fluidity by gravity. Open Journal of Biophysics. 2014;**4**(4):105-111

[23] Kohn F, Hauslage J, Hanke W. Membrane fluidity changes, a basic mechanism of interaction of gravity with cells? Microgravity Science and Technology. 2017;**151**:2529. DOI: 10.1007/s12217-017-9552-y

[24] Goldermann M, Hanke W. Ion Channel are sensitive to gravity changes. Microgravity Science and Technology. 2001;**13**(1):35-38

[25] Klinke N, Goldermann M, Hanke W. The properties of alamethicin incorporated into planar lipid bilayers under the influence of microgravity. Acta Astronautica. 2000;**47**:771-773

[26] Wiedemann M, Rahmann H, Hanke W. Gravitational impact on ion channels incorporated into planar lipid bilayers. In: TiTien and Ottova, editors. Planar lipid bilyers and their applications. Elsevier Sciences

[27] Kohn FPM. High throughput fluorescent screening of membrane potential under variable gravity conditions. Proceedings of "Life in Space for Life on Earth", 18-22 June 2012, Aberdeen, UK

[28] Schaffhauser DF, Andrini O, Ghezzi C, Forster IC, Franco-Obregón A, Egli M, Dittrich PS. Microfluidic platform for electrophysiological studies on Xenopus Laevis oocytes under varying gravity levels. Lab on a Chip. 2011;**11**:3471-3478

[29] Meissner K, Hanke W. Action potential properties are gravity dependent. Microgravity Science and Technology. 2005;**17**(2):38-43

[30] Nomura T, Kawano F, Ishihara A, Sato Y, Mitarai G, Iwase S, Kamiya A, Mano T, Ohira Y. Enhanced Hoffman-reflex in human soleus muscle during exposure to microgravity environment. Neuroscience Letters. 2001;**316**:55-57

[31] Sato T, Miyoshi T, Nakazawa K, Yano H, Takeoka H. Reflex response changes during hyper and microgravity. Journal of Gravitational Physiology. 2001;**8**:97-99

[32] Ohira Y, Nomura T, Kawano F, Soduh M, Ishihara A. Responses of Hoffman-reflex in human soleus to gravity and/or fluid shift. Journal of Gravitational Physiology. 2002;**9**:129-130

[33] Miyoshi T, Nozaki D, Sekiguchi H, Kimura T, Sato T, Komeda T, Nakazawa K, Yano H. Somatosensory graviception inhibits soleus H-reflex during erect posture in humans as revealed by parabolic flight experiment. Experimental Brain Research. 2003;**150**:109-113

[34] Kramer A, Gollhofer A, Ritzmann R. Acute exposure to microgravity does not influence the H-reflex with or without whole body vibration and does not cause vibration-specific changes in muscular activity. Journal of Electromyography and Kinesiology. 2013;**23**:872-878

[35] Crone C, Hultborn H, Mazieres L, Morin C, Nielsen J, Pierrot-Deseilligny E. Sensitivity of monosynaptic test reflexes to facilitation and inhibition as a function of the test reflex size: A study in man and the cat. Experimental Brain Research. 1990;**81**:35-45

[36] Zehr EP. Considerations for use of the Hoffmann reflex in exercise studies. European Journal of Applied Physiology. 2002;**86**(6):455-468

[37] Palmieri RM, Ingersoll CD, Hoffman MA. The Hoffmann reflex: Methodologic considerations and applications for use in sports medicine and athletic training research. Journal of Athletic Training. 2004;**39**(**3**):268-277

[38] Watt DG. Effects of Altered Gravity on Spinal Cord Excitability (Final Results). Galveston, TX: Bioastronautics Investigators' Workshop; 2003

[39] Soudmand R, Ward LC, Swift TR. Effect of height on nerve conduction velocity. Neurology. 1982;**32**:407-410

[40] Ruegg DG, Kakebeeke TH, Gabriel JP, Bennefeld M. Conduction velocity of nerve and muscle fiber action potentials after a space mission or a bed rest. Clinical Neurophysiology. 2003;**114**:86-93

[41] Hanke W, Schluhe W. R. Planar Lipid Bilayer Experiments: Techniques and Application. Oxford: Academic Press; 1993

[42] Kohn FPM, Ritzmann R. Gravity and neuronal adaptation, in vitro and in vivo-from neuronal cells up to neuromuscular responses: A first model. European Biophysics Journal: EBJ. DOI: 10.1007/s00249-017-1233-7

[43] Matsumoto G, Tasaki I. A study of conduction velocity in nonmyelinated nerve fibers. Biophysical Journal. 1977;**20**:1-13

[44] Hilbig R, Gollhofer A, Bock O, Manzey D. Sensory Motor and Behavioral Research in Space. Cham: Springer International Publishing; 2017

Chapter 7

Spaceflight: Immune Effects and Nutritional Countermeasure

Anil D Kulkarni, Marie-Francoise Doursout,
Asmita Kulkarni, Alamelu Sundaresan,
Takehito Miura, Koji Wakame and Hajime Fujii

Additional information is available at the end of the chapter

http://dx.doi.org/10.5772/intechopen.74709

Abstract

Microgravity is predicted to be a significant challenge to immune system during space travel. Consequences of weakened immune responses range from increased disease susceptibility to neoplastic growth. Degree of immune dysfunction is considered proportional to duration of stay in spaceflights. As a result of these risks, there is major concern over potential health risk for space travels that ultimately result in serious and considerable loss of mission objectives. Therefore, here is a need to explore the immune effects of spaceflight and its countermeasures. Several attempts have been made to develop effective measure to alleviate or prevent immune dysfunction due to microgravity. Among them, immunonutritional model has been shown to effectively modulate and upregulate immune system. This is further supported by our experiments demonstrating that supplementation of nutritional substrates like nucleotide and mushroom extracts active hexose-correlated compound (AHCC) effective in maintaining or restoring immunity in microgravity analog models.

Keywords: space travel, immune system, countermeasures, nutrition

1. Introduction

Microgravity and stress of space travel affect many organ systems and their functions in the body. Exposure to microgravity may produce changes in the performance of the immune system at the cellular level and in the major physiological systems of the body. Consequently, abnormal immune responses observed in microgravity may pose serious consequences, especially in future long-term space missions. Existing evidence suggests that spaceflight environment impairs immune system function in space travelers. Spaceflight environment has a cumulative effect on the body due to inherent stressors such as microgravity, cosmic radiation, and increase in corticosteroids [1]. A weakened immune system increases susceptibility to diseases and infectious pathogenesis. Immunosuppression puts hosts at risk for adverse effects such as infection from previously innocuous microorganisms that they are harboring or from microorganisms in their surroundings. Level of immune dysfunction is considered proportional to duration of stay in spaceflights. Longer the durations, such as missions to Moon and Mars and other deep space exploration flights, the effects are likely to be irreversible [2]. Maintenance of a healthy immune system is vital for resistance to infection and is essential in the homeostasis required for resistance to neoplastic disease, for prevention of autoimmune disease, tissue repair, and wound healing.

Based on the evidence, NASA in its roadmap has documented the immunological risks and consequences in space travel and exploration [3]. Among the highlighted risks are carcinogenesis caused by immune system changes, immunodeficiency, infections, altered wound healing, allergies and hypersensitivities, altered host-microbial interactions. As a result of these risks, NASA is concerned of major impact on health and mission objectives and irreversible potential loss of life which ultimately will result in serious and considerable loss of mission objectives. Therefore, it is essential to consider and highlight the immune effects of spaceflight and its preventive measures.

2. Spaceflight: stress and immune response

Space travelers are subjected to myriad of stressors of psychosocial, physical and environmental origin like microgravity, increased radiation, sleep deprivation, persistent circadian misalignment, and nutritional factors [4]. A common clinical observation is often the adverse relationship between stress and human disease. Stress such as injury or physical and physiological stress can result in metabolic stress and can cause severe impact on host health. Metabolic response to stress almost always results in adverse effects on the host defense mechanisms. Stress-related events cause breakdown in physical barrier, disrupt phagocytic cell function, and decreased antigen presentation altering cellular immune reactions. Stress is also suspected to play a role in morbidity and mortality in other immune-based diseases such as cancer, inflammatory bowel diseases, and even aging [5–13]. Although such dysfunctions have been thought of primarily as immunosuppressive, recent data have suggested immunoregulatory dysfunctions may play a more central role in stress-induced immune alterations [14]. Sleep alterations are suggested to modulate the stress-health relationship [15, 16]. Poor sleep, in turn, is associated with subsequent decrements in mental health including symptom reporting, incident cases of mood and anxiety disorders, and immune function [17–20]. Recent research suggests that stress is also associated with increased latent viral reactivation, upper respiratory tract infections, and increased wound healing time [21–26].

For instance, the decrease in immune cell function has been observed after flights of varying duration in the Soyuz, Skylab, Salyut, and Space Shuttle programs [27, 28]; these studies have also reported a reduction in lymphocyte proliferation. Reduced interferon-γ (IFN-γ) production in suspended mice has been correlated to increased susceptibility to viral infection similar to that observed in rats flown on the US Space Shuttle. Therefore, there is an increased risk of infections among crewmembers during spaceflight resulting from working and living in a crowded, closed environment with limited capabilities for air revitalization and disinfection. Consequently, alterations in the immune response during spaceflight, as well as stress, aerosols, and altered fluid distribution within the body could increase the incidence of infectious diseases during long-duration space missions. Importantly, bacterial infection can be a major cause of morbidity after any traumatic injury, but trauma during spaceflight may substantially increase the infectious risk. An increase in neutrophil counts has been reported in tail-suspended rats [29, 30]. It is well established that activation of large numbers of neutrophils is likely to result in excessive generation of free radicals and associated tissue damage. Neutrophils also produce nitric oxide (NO), another free radical that reacts with superoxide to form peroxynitrite, decomposing to the highly toxic hydroxyl radical.

Enclosed cabin of spacecraft and free-floating environment increase the potential for infections among the crewmembers. Like all the objects, microbes are also in free-floating state increasing the potential for inoculation by inhalation increasing the regional susceptibility of respiratory tract. Altered metabolism and virulence reported in *in vivo* models of simulatory microgravity [31–35]. Increased morbidity and slower rate of wound healing in *S. aureus* sepsis reported in experimental animals exposed to test environment. The evidence is significant as the possibility of sepsis increases in space cabin environment as well as after return to earth from space mission. Spaceflights has profound effect on ecologic control of the gastrointestinal tract as cosmonauts, upon returning to earth, were found to have their normal gut flora replaced by potential pathogenic microorganisms.

Oxidative stress is known to occur in disuse and in many pathological conditions, and is now widely considered a major trigger of the imbalance between protein synthesis and degradation leading to muscle atrophy [36]. Reactive oxygen species (ROS) and elevated proinflammatory cytokines, in particular, TNF-α, mediate muscle atrophy via the redox-sensitive transcription factor nuclear factor-κB (NF-κB). It has been suggested that the exposure of brain to simulated microgravity can induce expression of certain transcription factors, which are oxidative stress dependent [37]. We have reported that the regulation and production of free radicals, and the relationship between oxidative stress and production of inflammatory cytokines and their subsequent effects on the healing of traumatic skeletal injuries in animals as well as cells subjected to analog microgravity.

3. Space travel and immune response studies

US Apollo missions were the first to identify the immune dysfunction [38]. The study of spaceflight immunology is limited due to relative inaccessibility, difficulty of performing experiments in space and inadequate provisions in this area in the United States and Russian space programs [39]. Most of the immune studies performed in the early days of spaceflight era had astronauts and cosmonauts participated in such studies. These studies assessed the immune effects by in vitro analysis of blood samples that were obtained before and after spaceflights.

Most of the experimental studies are of a pre- and postflight nature involving both humans and experimental animals and are divided into categories of short duration (< 2 weeks) and long duration (> 2 weeks) missions. In short duration flights, the majority of the outcomes are from postflight period analysis showing decreased cellular response to mitogens, decreased T cell counts and somewhat variable leukocyte counts [39]. Long-duration studies (1–12 months) that are performed by Russians, on board the Mir space station, have documented a 50% reduction in lymphocytic response to phytohemagglutinin (PHA) on the day after the mission, as compared with the preflight response. Levels returned to normal by day seven postlanding. Other studies showed decreased graft-versus-host response to xenoantigens and mitogen-induced IL-2 production [40]. The limited in-flight studies of delayed-type hypersensitivity (DTH) using commercial kits for the assessment of cell-mediated immunity showed significant suppression in half the subjects of 3–5 months in space and upon landing [41]. There have been several studies reported from space shuttle missions indicating alterations in lymphocyte response and decreased production of cytokines including interferons-α, β and γ and interleukin-2 [42].

Several studies have indicated that spaceflight can adversely affect tissue repair in muscle and bone. Mechanical unloading and physical deconditioning, which are thought to be central components in the effects of microgravity on the human body, have also broader clinical applications on Earth, for example, as it relates to prolonged bed rest or inactive geriatric patients. As a result, ground-based animal models have been used to mimic the mechanical unloading and physical deconditioning associated with microgravity and bed rest in humans. Because the phenotype of skeletal muscle is importantly dependent on mechanical loading, muscle plasticity is highlighted by the severe loss of mass (atrophy) after a few days of reduced weight-bearing activity such as bed rest or spaceflight. Hind limb unloaded (HU) of rats is an established model for atrophy which produces many of the muscular and systemic changes seen in humans as a consequence of muscle disuse [43, 44]. Consequently, results indicated that microgravity adversely affects the capacity of wounds to heal and that this may be related to a diminished cellular response to growth factors known to be present at sites of wounding [45, 46]. However, one such area of biomedical research where little is known concerns the effects of mechanical unloading and physical deconditioning on bone fracture repair and wound healing.

These research studies suggest that stress-induced changes in psychological, behavioral, and/or physiological functioning can be harmful and may result in negative health consequences. The clinical significance of these stressors and immune system changes must be defined, evaluated, and identified in space travelers.

4. Ground-based simulated microgravity studies

Due to high cost of spaceflight experiments and infrequency of flights, ground-based models that mimic the effect of microgravity have been extensively used. Among the models developed were human analogs model such as bed rest, physical stress, academic stress, and confinement, which allow some aspects of the spaceflight stressors [39]. Mechanical unloading and physical deconditioning, which are thought to be central components in the effects of microgravity on the human body, have also broader clinical applications on Earth, for example, as it relates to prolonged bed rest or inactive geriatric patients. Exclusive use of human subjects for space research has severe limitations due to the ethical issues involved. Animal models provide more opportunities for research as it allows wider range of possible experiments. With variety of techniques available, rodents are preferred choice for space research studies. Several models were designed with specific effects to be studied in each individual model.

4.1. *In vivo* studies with rodent hind limb unloading model

Hind limb suspension of rodents was initially developed to study musculoskeletal system. In this system, the hind limbs of rodents are elevated to produce a 30° head-down tilt, which results in a cephalad-fluid shift and avoids weight-bearing by the hindquarters. When spaceflight effects were compared with ground-based weight unloading models, such as, bed rest studies and hind limb suspension model, there are many common features and effects [47–49]. These are shown in **Table 1**. Similar to many physiologic effects, the immune function and its dysfunction in both ground-based models is also very similar to spaceflight effects on the body. Many of the areas correlate with the spaceflight and its stress that have many consequential responses produced in body.

Antiorthostatic hind limb suspension of rodents, a ground-based model for simulation of microgravity, has summarized the physiologic and immunologic changes induced by antiorthostatic suspension and indicates a correlation with physiologic changes induced by spaceflight [50]. This position simulates the cephalad fluid and organ shift, a negative balance of water, nitrogen, and potassium; and increased metabolic turnover observed in astronauts during spaceflight. Studies using this model have shown interesting contradictory observations relative to organ-specific immunologic changes. Overall results of such antiorthostatic suspension models have shown a decrease in immunity.

	Spaceflight	Bed rest	Tail Suspension
Cephalic fluid shift	+	+	+
Redistribution of bones	+	+	+
Bone resorption	↓→	↓→	↓→
Calcium balance	↓	↓	↓
Fecal calcium	↑	↑	↑
Urine calcium	↑	↑	↑
Serum calcium	↑	↑→	↑→
PTH	↓	→	→
1, 25 (OH)2 D	↓	↓	↓
Serum osteocalcin	↓	→	↓
Bone strength	↓	ND	↓
Immune function	↓	↓→	↓

Table 1. Comparison of spaceflight to ground-based models of skeletal unloading.

4.2. *In vitro* studies in simulated microgravity using clinostat bioreactors

Numerous attempts to identify and separate the effects of microgravity and stress have met with difficult challenges, further raising the issue whether single cells are also affected by microgravity. Among the microgravity simulator models, an apparatus called a rotating wall vessel (RWV) developed by NASA is an ideal ground-based model system for examining the effects of microgravity on cells of the immune system without the presence of psychoneuroendocrine factors [51]. The RWV, based on clinostat technology, is a microgravity simulator Couette flow bioreactor. It consists of a zero-head space, aqueous filled culture vessel that suspends cells by rotating at low speed (10–60 rpm) around a horizontal axis. These conditions subject the cells to a randomized gravity field and low shear forces [52, 53]. Cells in the RWV are estimated to experience acceleration forces that simulate microgravity as low as 2 × 10–4 g. Using a Clinostat tissue culture apparatus, Cogoli [54] has shown that microgravity alters cell membrane permeability and thickness as well as cytoplasmic streaming. Several studies have reported the effect of microgravity on T lymphocyte activation. Clinostat culture studies showed that T cell responses to concanavalin-A (Con-A) were decreased by 50% [54]. Cooper and Pellis [55] have documented, using a clinostatic RWV bioreactor, that during polyclonal activation the signaling pathways leading to protein kinase C (PKC) activation are sensitive to simulated microgravity. Although several investigators used cell cultures subjected to analog microgravity to study potential impacts that space travel imposes on humans, the *ex vivo* has a serious lack of *in vivo* measurements of immune-physiological responses.

In the bioreactor microgravity cultured cells, there was a reproducibility of significantly suppressed PHA response when compared to static cultured cells as described by Cooper and Pellis [55]. In preliminary experiments, supplementation of the culture medium with nucleoside-nucleotide mixture or uridine (preferred nucleotides for solubility and bioavailability) significantly enhanced the PHA response in bioreactor microgravity. To our knowledge, this is the first observation documenting the reversal of decreased PHA response in simulated microgravity using the NASA bioreactor. Continuation of these studies using the biotechnology of *in vitro* modeled microgravity will provide additional data to support the hypothesis and prove the countermeasure effects of nucleotides.

During the countermeasure experiments, we studied the effects of housing environments on the production and expression of biologically and immunologically important molecules, namely cytokines, nitric oxide (NO), and inducible nitric oxide synthase (iNOS).

5. Countermeasure for prevention of immunosuppressive effects

Several attempts have been made to develop effective measure to alleviate or prevent immune dysfunction. There is definitely a need for countermeasures that will maintain normal immune system during spaceflight, especially when missions are prolonged. Almost all were found to be inadequate and presented adverse responses. For example, the use of immunomodulator agents and neurohormonal regulation was suggested to ameliorate the immune dysfunction in space [56]. However, the suggested methods of neurohormonal regulation by using agents that act upon the nervous system may have deleterious effects on systems besides the immune system. The use of immunomodifying preparations such as LPS, MDP, and proteoglycans had no beneficial effects. In clinical practice, these compounds exhibited toxicity. Use of growth factors and interleukins was beneficial to small extent but had no influence on the increased corticosteroid (due to microgravity environment) levels [57, 58]. Most of the microgravity studies have documented and analyzed the immunosuppressive effects of true or simulated microgravity; however, there are scanty reports of efforts to modulate the immune system, host defense system, and its function. Especially, scarce are the studies that approach the issue of the maintenance and restoration of immunosurveillance.

Nutrition has played a critical role throughout the history of exploration, and space exploration is no exception. Environmental impacts like radiation, and spacecraft and spacesuit atmospheres can alter nutritional status and nutritional requirements of spaceflight. The physiological changes that occur during spaceflight influence spaceflight nutritional requirements. Therefore, understanding the nutritional requirements of space travelers and the role of nutrition in human adaptation to microgravity are as critical to crew safety and mission success. Many potential targets for nutritional countermeasures proposed to counteract or mitigate some of the negative effects of spaceflight on the human body. Recently, immunenutritional model has been shown to effectively modulate and upregulate immune system where a nutritional substrate has benefits beyond basic nutrition. Based on our extensive experience in R&D of nutritional immunomodulation, we evaluated two nutrients, which we have been studying for several years.

6. Dietary nucleotides

Dietary nucleotides are reported to restore innate and adaptive T-cell mediated immunity both at the peripheral mature immune compartment and stem cell level (**Figure 1**) [59, 60]. Nutritional upregulation of the global host defense mechanisms would have the great advantage of being technically feasible and applicable in people and it would be economical without the untoward effects. Laboratory findings and progress in multidisciplinary emerging field of nutritional immunology justify emphatically the proposed novel approach of nutritional modulation of host defense system in space travel. With the experimental evidence and information of nucleotide nutrition on immunity, it is plausible to provide both the prophylactic and therapeutic approach to the modulation of host defense mechanisms during spaceflights (**Figure 2**) [61].

6.1. Nucleotide supplementation in microgravity experiments

6.1.1. Hind limb unloading (HLU) in vivo model

We have documented that nucleotide supplementation significantly reversed the immunosuppression observed *in vivo* HLU model and *in vitro* BIO model [62]. The results were dramatic in the HLU group where the control chow group had significantly lower popliteal lymph node (PLN) response as compared to other housing groups. This effect of immunosuppression was reversed by dietary supplementation of nucleotides with Uracil effect reported the highest and significant as compared to the chow group in HU. Thus, the antiorthostatic HLU model of modeled microgravity can be used successfully to document nutritional immunomodulatory countermeasure. We also assessed the stress effect by measuring the serum levels

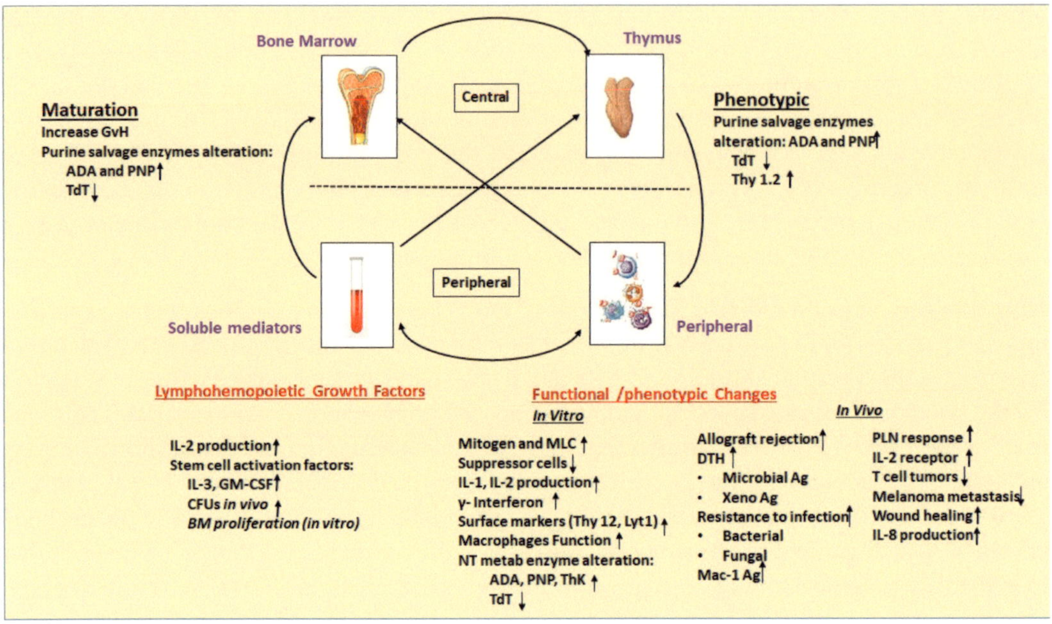

Figure 1. Pleiogenic immunologic responses by supplemental nucleotides.

of corticosterone (CORT) in experimental groups. The experimental evidence showed that in non-HLU animals, there was minimal effect of supplemented nucleotides (at the given dose) and did not encounter the immune depressive effects seen in HLU animals. Thus, nucleotide supplementation was beneficial for immune restoration in modeled microgravity environmental conditions. These results confirm our observations that RNA and Uracil are effective in maintaining or restoring immunity when the animals are under stressful situations (such as protein starvation, total starvation or dietary nucleotide deficiency, and HLU) or other trauma (such as sepsis, or inflammatory hypersensitivity stimulations). Our data also show the HU group had increased oxygen radicals (ROS) to 130% in the brain as compared to control mouse brain. This ROS increase was inversely proportional to glutathione levels (75%) in the brains. Therefore, our data confirm that oxidative stress is induced in animals subjected to hind limb unloaded.

6.1.2. Bone density after hind limb suspension or spinal cord injury in a rat model of osteoporosis

Space travelers are subjected to significant bone loss due to increased resorption and altered remodeling of bone tissue. In spite of calcium supplementation, increased excretion of calcium, reduced absorption of calcium from intestine, and diminished vitamin D synthesis due to space suit ultimately result in bone loss. Bone loss is proportional to the length of time in space. The changes in bone during spaceflight are similar to those seen in osteoporosis. Dietary nucleotides have long been known to positively affect the immune system and more recently have been shown to have beneficial effects in rapidly proliferating tissues.

http://dx.doi.org/10.5772/intechopen.74709

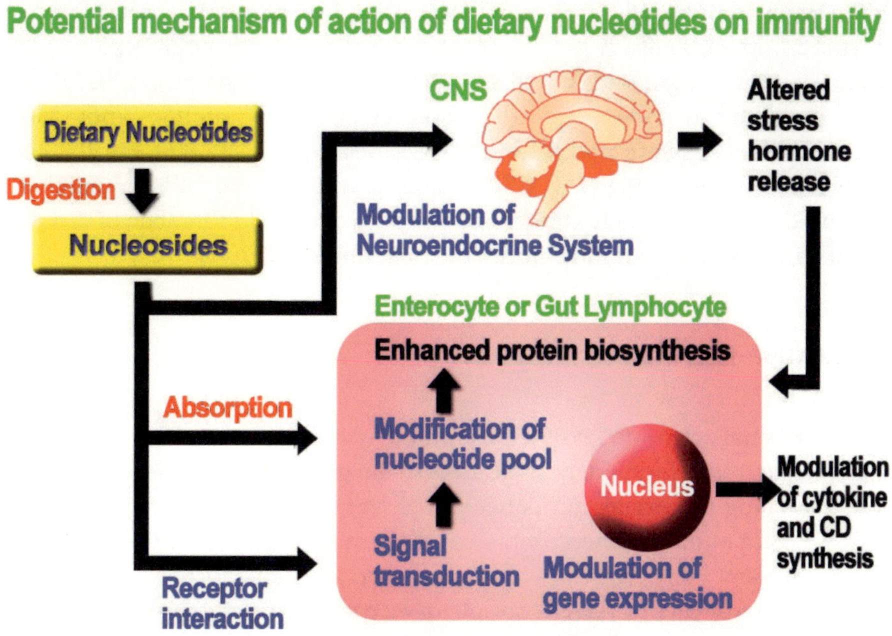

Figure 2. Potential mechanism of action of dietary nucleotides on immunity.

We have studied the effects of dietary nucleotides on bone loss after a disuse model of tail suspension in rats and have found that, in addition to exerting positive effects on the immune system, a nucleotide-enriched diet reduced the amount of bone loss seen in these animals. In a different model of bone demineralization, we have recently found that providing a diet of normal rat chow enriched with enhanced nucleotides reduces bone density loss in the femur when initiated immediately following spinal contusion injury [63].

6.1.3. Bioreactor in vitro modeled microgravity for T cell suppression and lymphocyte locomotion

In microgravity, immune suppression is a documented phenomenon in astronauts. It is also documented in *in vitro* and *in vivo* studies in modeled microgravity and the antiorthostatic rodent models of microgravity. In our earlier study, we reported that in the BIO microgravity cultured cells, there was a reproducibility of significantly suppressed phytohemaglutinin, a T cell mitogen response when compared to static cultured cells [1, 55]. We also reported that supplementation of the culture medium with nucleoside-nucleotide mixture or uridine (preferred nucleotides for solubility and bioavailability) significantly enhanced the PHA response in bioreactor microgravity.

Lymphocyte locomotion along the interstitium is integral to the immune response. Microgravity is a stressor that inhibits this phenomenon. The microgravity cell culture analog system also has the same effect on locomotion inhibition of lymphocytes [64]. Since locomotion is critical for an optimal immune response, countermeasure strategies for its restoration in lymphocytes were sought. When lymphocytes were treated with 0.5 ng/ml phorbol myristate acetate (PMA) after exposure to microgravity culture, recovery of locomotion through type I collagen was 87%. However, in the human setting, PMA is a tumor promoter and cannot be administered. Studies with hind limb suspended mouse splenocytes displayed immune suppression, which was mitigated by the use of nucleotides and nucleosides (NS/NT). In lymphocytes cultured in modeled microgravity using the NASA BIO model of microgravity, it is shown that the NS/NT mixture used was able to orchestrate locomotion recovery by more than 87% documented with PMA in lymphocytes from three normal human donors.

7. Active hexose-correlated compound (AHCC)

AHCC is a nutritional substrate known for immune enhancing properties in humans and in laboratory studies. It is being widely used around the globe as a nutritional supplement for health and well-being in normal and patients with various afflictions to improve quality of life. AHCC is a compound obtained from enzyme-fermented extract of the mycelia of basidiomycetes mushrooms. AHCC consists of oligosaccharides, amino acids, lipids, and minerals [65]. The main components of AHCC are oligosaccharides (~74% of AHCC), and approximately 20% of AHCC are partially acetylated α-1, 4-glucans with a mean molecular weight of 5 kDa. These oligosaccharides including α-1, 4-glucans are believed to be the active components of AHCC [66, 67].

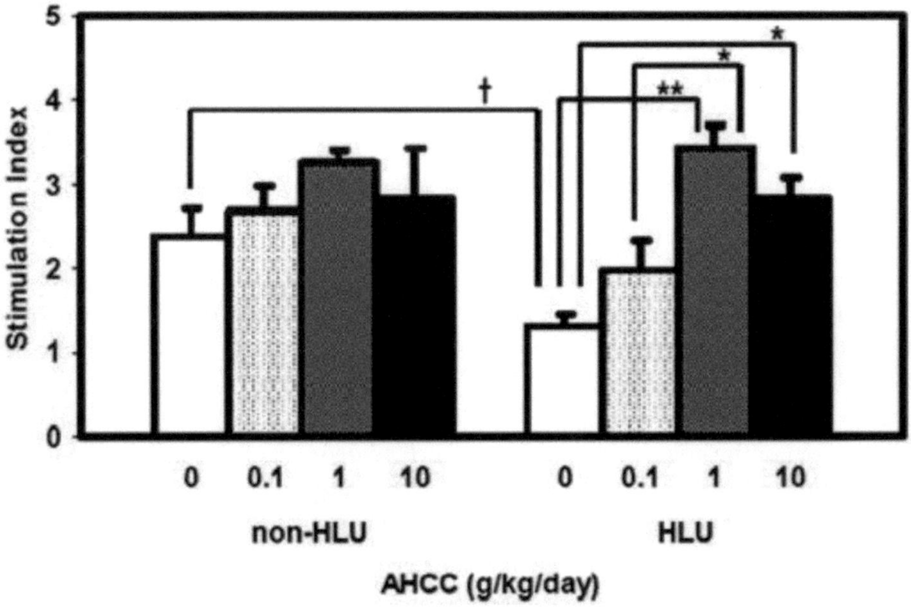

Figure 3. Stimulation index for *in vivo* PLN proliferative response vs. AHCC dose in control and HLU mice. means ± SEM; *p <0.05, **p =0.001, †P <0.05.

http://dx.doi.org/10.5772/intechopen.74709

We examined the effect of AHCC on microgravity-induced immune changes by using a hind limb unloading (HLU) of mice as a microgravity analog [68]. A beneficial effect of AHCC on T cells has been reported in various models [69–71]. We induced immune changes by using a hind limb unloading (HLU) of mice as a microgravity analog and accessed the effect of AHCC supplementation on various immune functions. To access the immune function, Popliteal lymph node (PLN) response was analyzed as it involves all phases of immune response, e.g., antigen processing and presentation, followed by proliferative phase of immune response. PLN response was significantly decreased in mice in the HLU group compared to that in mice in the control group, and AHCC supplementation significantly reversed this response (**Figure 3**). AHCC reversed HLU-induced T cell dysfunction in PLNs. Since T cells play an important role in acquired immunity, a countermeasure for T cell dysfunction is imperative.

Spaceflight environment is one of the serious immune-compromised conditions due to closed space, and recycling of air and water may increase the risk of microbial load and reactivation of opportunistic pathogens, latent bacteria, and viruses. Neutrophils-, macrophages-, or monocytes-mediated innate immunity is the first step to exclude pathogens. To assess over-

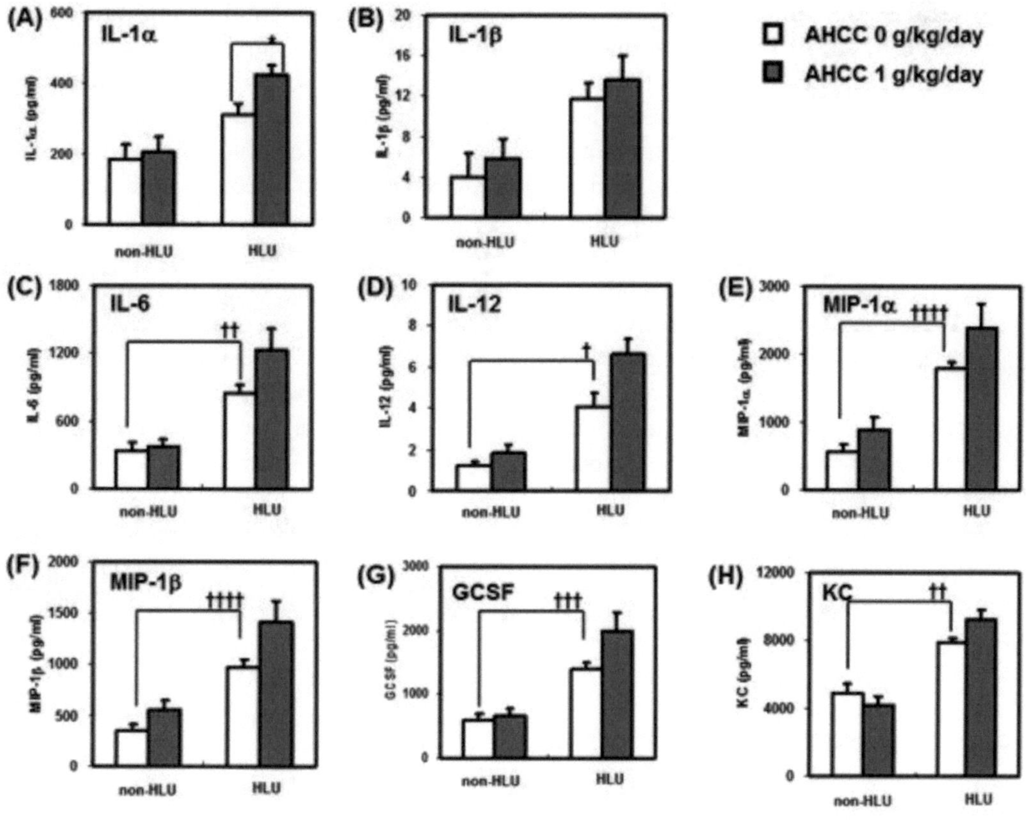

Figure 4. Production of cytokines and chemokines vs. AHCC dose in control and HLU mice.

all functions of these types of cells, we measured the levels of LPS-stimulated cytokine and chemokine production from splenocytes. Both inflammatory cytokines and chemokines were increased in mice in the HLU group and AHCC supplementation in HLU mice tended to further enhance the inflammatory cytokine and chemokine production (**Figure 4**). Increased interleukin like IL-6 is implicated in increased stimulated immune responses, e.g., during infection and after trauma, especially burns or other tissue damage leading to inflammation [72], while increased IL-12 (T cell stimulating factor) is known to be a stimulator of the TNF-α pathway and increases adaptive immunity. Enhancement of inflammatory cytokine and chemokine production by AHCC supplementation suggested to be effective for preventing infection [69, 73].

8. Summary and significance

Spaceflight observations thus far clearly document the adverse effects on the immune system, concomitant persistence of space environment stressors, and potential increase in virulence of infectious agents. It is imperative to design and develop effective countermeasures to secure health aspects of humans in space. The literature from experimental models and clinical human applications clearly documents that supplemental dietary nucleotides have beneficial effects on the immune system under stress conditions and environments. It is also known that supplemental dietary nucleotides had beneficial enhanced resistance to *Staphylococcus aureus* (SA), methicillin-resistant SA (MRSA), and *Candida albicans* infections in mice [74–77]. Similarly, use of a nucleotide containing formula in humans showed that there is a significant decrease in infectious complications in various patients. Immune enhancing nutritional supplements like AHCC have also found to be effective in restoring and maintaining immune system function in spaceflight analog animal model. These results emphasize the role of nucleotide nutrition and nutritional substrates like AHCC as a promising and plausible preventive measure to the immunologic consequences pre-/during /postspaceflight. A multipronged research will be an effective and safe countermeasure for spaceflight effects and to obviate stress and observed immune dysfunction. These studies should be of significant interest to NASA and other space agencies around the world by identifying profiles (immune, endocrine, and psychological) of individuals at risk for these immune dysfunctions with subsequent clinical manifestations and how nutritional countermeasures may impact such profiles. Such approaches should provide pragmatic clinical interventions for alleviation of stress and preservation of crew health particularly during long-term flights such as ISS, long-term interplanetary excursions, and deep space explorations.

Acknowledgements

NASA; Department of Surgery, UTMMS; Amino Up Chemical Company, Sapporo, Japan.

http://dx.doi.org/10.5772/intechopen.74709

Author details

Anil D Kulkarni*, Marie-Francoise Doursout[1], Asmita Kulkarni, Alamelu Sundaresan[2], Takehito Miura[3], Koji Wakame[4] and Hajime Fujii[3]

*Address all correspondence to: anil.d.kulkami@ uth.tmc.edu

1 Department of Surgery, The University of Texas Health McGovern Medical School, Houston, Texas, USA

2 Department of Anesthesiology, The University of Texas Health McGovern Medical School, Houston, Texas, USA

3 Texas Southern University, Houston, Texas, USA

4 Amino Up Chemical Co, Sapporo, Hokkaido, Japan

5 Hokkaido Pharmaceutical University, Hokkaido, Japan

References

[1] Kulkarni AD, Kogiso M, Wakame K. AHCC Newsletter. 2006;**3-4**:3-4

[2] Lang T, Van Loon JJWA, Bloomfield S, Vico L, Chopard A, Rittweger J, Kyparos A, Blottner D, Vuori I, Gerzer R, Cavanagh PR. Towards human exploration of space: The THESEUS review series on muscle and bone research priorities. NPJ Microgravity. 2017;**3**:8

[3] Crucian B, Kunz H, Sams CF. Risk of Crew Adverse Health Event Due to Altered Immune Response. Human Research Program Human Health Countermeasures Element. Evidence Report—Human Research Roadmap—NASA. 2005

[4] Frippiat JP, Crucian BE, de Quervain DJ, Grimm D, Montano N, Praun S, Roozendaal B, Schelling G, Thiel M, Ullrich O, Choukèr A. Towards human exploration of space: The THESEUS review series on immunology research priorities. NPJ Microgravity. 2016;**2**:16040

[5] Bergsma J. Illness, the mind and the body: Cancer and immunology: An introduction. Theoretical Medicine. 1994;**15**:337-347

[6] Chorot P, Sandín B. Life events and stress reactivity as predictors of cancer, coronary heart disease and anxiety disorders. International Journal of Psychosomatics. 1994;**41**(1-4):34-40

[7] Kusnecov AW, Rabin BS. Stressor-induced alterations of immune function: Mechanisms and issues. International Archives of Allergy and Immunology. 1994;**105**(2):107-121

[8] Levenstein S, Prantera C, Varvo V, Scribano ML, Berto E, Andreoli A, Luzi C. Psychological stress and disease activity in ulcerative colitis: A multidimensional cross-sectional study. The American Journal of Gastroenterology. 1994;**89**(8):1219-1225

[9] Cohen S, Tyrrell DA, Smith AP. Negative life events, perceived stress, negative affect, and susceptibility to the common cold. Journal of Personality and Social Psychology. 1993; **64**(1):131-140

[10] McEwen BS, Stellar E. Stress and the individual. Mechanisms leading to disease. Archives of Internal Medicine. 1993;**153**(18):2093-2101

[11] Cohen S, Tyrrell DA, Smith AP. Psychological stress and susceptibility to the common cold. The New England Journal of Medicine. 1991;**325**(9):606-612

[12] Garrett VD, Brantley PJ, Jones GN, McKnight GT. The relation between daily stress and Crohn's disease. Journal of Behavioral Medicine. 1991;**14**(1):87-96

[13] Bonneau RH, Kiecolt-Glaser JK, Glaser R. Stress-induced modulation of the immune response. Annals of the New York Academy of Sciences. 1990;**594**:253-269

[14] Marshall GD Jr. The adverse effects of psychological stress on immunoregulatory balance: Applications to human inflammatory diseases. Immunology and Allergy Clinics of North America. 2011;**31**(1):133-140

[15] Hall M, Baum A, Buysse DJ, Prigerson HG, Kupfer DJ, Reynolds CF 3rd. Sleep as a mediator of the stress-immune relationship. Psychosomatic Medicine. 1998;**60**(1):48-51

[16] Hall M, Buysse DJ, Dew MA, Prigerson HG, Kupfer DJ, Reynolds CF 3rd. Intrusive thoughts and avoidance behaviors are associated with sleep disturbances in bereavement-related depression. Depression and Anxiety. 1997;**6**(3):106-112

[17] Gamaldo CE, Shaikh AK, McArthur JC. The sleep-immunity relationship. Neurologic Clinics. 2012;**30**:1313-1343

[18] Bollinger T, Bollinger A, Oster H, Solbach W. Sleep, immunity, and circadian clocks: A mechanistic model. Gerontology. 2010;**56**:574-580

[19] Ramsawh HJ, Stein MB, Belik SL, Jacobi F, Sareen. Relationship of anxiety disorders, sleep quality, and functional impairment in a community sample. Journal of Psychiatric Research. 2009;**43**(10):926-933

[20] Strine TW, Chapman DP. Associations of frequent sleep insufficiency with health-related quality of life and health behaviors. Sleep Medicine. 2005;**6**(1):23-27

[21] Marucha PT, Kiecolt-Glaser JK, Favagehi M. Mucosal wound healing is impaired by examination stress. Psychosomatic Medicine. 1998;**60**(3):362-365

[22] Cohen S. Psychological stress and susceptibility to upper respiratory infections. American Journal of Respiratory and Critical Care Medicine. 1995;**152**(4 Pt 2):S53-S58

[23] Kiecolt-Glaser JK, Marucha PT, Malarkey WB, Mercado AM, Glaser R. Slowing of wound healing by psychological stress. Lancet. 1995;**346**(8984):1194-1196

[24] Krueger E, Krueger GR. How does the subjective experience of stress relate to the breakdown of the human immune system. In Vivo. 1991;**5**(3):207-215

[25] Kiecolt-Glaser JK, Glaser R, Shuttleworth EC, Dyer CS, Ogrocki P, Speicher CE. Chronic stress and immunity in family caregivers of Alzheimer's disease victims. Psychosomatic Medicine. 1987;**49**(5):523-535

[26] Glaser R, Kiecolt-Glaser JK, Speicher CE, Holliday JE. Stress, loneliness, and changes in herpes virus latency. Journal of Behavioral Medicine. 1985;**8**(3):249-260

[27] Levine DS, Greenleaf JE. Immunosuppression during spaceflight deconditioning. Aviation, Space, and Environmental Medicine. 1998;**69**(2):172-177

[28] Barger LK, Greenleaf JE, Baldini F, Huff D. Effects of space missions on the human immune system: A meta-analysis. Sports Medicine, Training and Rehabilitation. 1995; **5**:293-310

[29] Robinson RR, Doursout MF, Chelly JE, Powell MR, Little TM, Butler BD. Cardiovascular deconditioning and venous air embolism in simulated microgravity in the rat. Aviation, Space, and Environmental Medicine. 1996;**67**(9):835-840

[30] Tipton CM, Overton JM, Joyner MJ, Hargens AR. Local fluid shifts in humans and rats: Comparison of simulation models with actual weightlessness. Physiologist. 1987;**30**(1 Suppl):S117-S120

[31] Nickerson CA, Ott CM, Mister SJ, Morrow BJ, Burns-Keliher L, Pierson DL. Microgravity as a novel environmental signal affecting *Salmonella enterica* serovar *Typhimurium virulence*. Infection and Immunity. 2000;**68**(6):3147-3152

[32] Miller ES, Bates RA, Koebel DA, Sonnenfeld G. Antiorthostatic suspension stimulates profiles of macrophage activation in mice. Neuroimmunomodulation. 1999;**6**(3):160-167

[33] Fang A, Pierson DL, Koenig DW, Mishra SK, Demain AL. Effect of simulated microgravity and shear stress on microcin B17 production by *Escherichia coli* and on its excretion into the medium. Applied and Environmental Microbiology. 1997;**63**(10):4090-4092

[34] Fang A, Pierson DL, Mishra SK, Koenig DW, Demain AL. Gramicidin S production by *Bacillus brevis* in simulated microgravity. Current Microbiology. 1997;**34**(4):199-204

[35] Miller ES, Sonnenfeld G. Influence of antiorthostatic suspension on resistance to murine *Listeria monocytogenes* infection. Journal of Leukocyte Biology. 1994;**55**(3):371-378

[36] Jackson MJ. Redox regulation of adaptive responses in skeletal muscle to contractile activity. Free Radical Biology & Medicine. 2009;**47**(9):1267-1275

[37] Wise KC, Manna KS, Yamauchi K, Ramesh V, Wilson BL, Thomas RL, Sarkar S, Kulkarni AD, Pellis NR, Ramesh GT. Activation of nuclear transcription factor kappa-β in mouse brain induced by a simulated microgravity environment. In Vitro Cellular & Developmental Biology—Animal. 2005;**41**:118-123

[38] Kimzey SL, Fisher CL, Johnson PC, Ritzmann SE, Mengel CE. Hematology and immunology studies. In: Johnston RS, Dietlein LF, Berry CA, editors. Biomedical Results of Apollo. Washington DC: National Aeronautics and Space Administration; 1975 SP-368. 1975. pp. 197-226

[39] Taylor GR. Overview of spaceflight immunology studies. Journal of Leukocyte Biology. 1993;**54**(3):179-188

[40] Konstantinova IV, Rykova MP, Lesnyak AT, Antropova EA. Immune changes during long-duration missions. Journal of Leukocyte Biology. 1993;**54**:189-201

[41] Gmünder FK, Konstantinova I, Cogoli A, Lesnyak A, Bogomolov W, Grachov AW. Cellular immunity in cosmonauts during long duration spaceflight on board the orbital MIR station. Aviation, Space, and Environmental Medicine. 1994;**65**(5):419-423

[42] Lesnyak AT, Sonnenfeld G, Rykova MP, Meshkov DO, Mastro A, Konstantinova I. Immune changes in test animals during spaceflight. Journal of Leukocyte Biology. 1993; **54**(3):214-226

[43] Narici MV, de Boer MD. Disuse of the musculo-skeletal system in space and on earth. European Journal of Applied Physiology. 2011;**111**:403-420

[44] Däpp C, Schmutz S, Hoppeler H, Flück M. Transcriptional reprogramming and ultrastructure during atrophy and recovery of mouse soleus muscle. Physiological Genomics. 2004;**20**(1):97-107

[45] Delp MD. Unraveling the complex web of impaired wound healing with mechanical unloading and physical deconditioning. Journal of Applied Physiology (1985). 2008; **104**(5):1262-1263

[46] Radek KA, Baer LA, Eckhardt J, LA DP, Wade CE. Mechanical unloading impairs keratinocyte migration and angiogenesis during cutaneous wound healing. Journal of Applied Physiology (1985). 2008;**104**(5):1295-1303

[47] Carpenter RD, Lang TF, Bloomfield SA, Bloomberg JJ, Judex S, Keyak JH, et al. Effects of long-duration spaceflight, microgravity, and radiation on the neuromuscular, sensorimotor, and skeletal systems. Journal of Cosmology and Astroparticle Physics. 2010; **12**:3778-3780

[48] Morey-Holton E, Globus RK, Kaplansky A, Durnova G. The hindlimb unloading rat model: Literature overview, technique update and comparison with space flight data. Advances in Space Biology and Medicine. 2005;**10**:7-40

[49] Morey-Holton ER, Globus RK. Hindlimb unloading rodent model: Technical aspects. Journal of Applied Physiology. 2002;**92**:1367-1377

[50] Chapes SK, Woods KM, Armstrong JW. Ground-based experiments complement microgravity flight opportunities in the investigation of the effects of space flight on the immune response: Is protein kinase C gravity sensitive? Transactions of the Kansas Academy of Science. 1993;**96**(1-2):74-79

[51] Schwarz RP, Goodwin TJ, Wolf DA. Cell culture for three-dimensional modeling in rotating-wall vessel: An application of simulated microgravity. Journal of Tissue Culture Methods. 1992;**14**:51-58

[52] Jessup JM, Goodwin TJ, Spaulding G. Prospects for use of microgravity-based bioreactors to study three-dimensional host-tumor interactions in human neoplasia. Journal of Cellular Biochemistry. 1993;**51**:290-300

[53] Tsao YD, Goodwin TJ, Wolf DA, Spaulding GF. Responses of gravity level variations on the NASA/JSC bioreactor system. Physiologist. 1992;**35**(s):s49-s50

[54] Cogoli A. The effect of hypogravity and hypergravity on cells of the immune system. Journal of Leukocyte Biology. 1993;**54**:259-268

[55] Cooper D, Pellis NR. Suppressed PHA activation of T lymphocytes in simulated microgravity is restored by direct activation of protein kinase C. Journal of Leukocyte Biology. 1998;**63**(5):550-562

[56] Fuchs SS, Medvedev AE. Countermeasures for ameliorating in-flight immune function. Journal of Leukocyte Biology. 1993;**54**(3):245-252

[57] Armstrong JW, Balch S, Chapes SK. Interleukin-2 therapy reverses some immuno-suppressive effects of skeletal unloading. Journal of Applied Physiology. 1994;**77**(2):584-589

[58] Armstrong JW, Kirby-Dobbles K, Chapes SK. The effects of rM-CSF and rIL-6 therapy on immunosuppressed antiorthostatically suspended mice. Journal of Applied Physiology. 1995;**768**(3):968-975

[59] Kulkarni AD, Yamauchi K, Hales NW, Ramesh V, Ramesh GT, Sundaresan A, Andrassy RJ, Pellis NR. Nutrition beyond nutrition: Plausibility of immunotrophic nutrition for space travel. Clinical Nutrition. 2002;**21**(3):231-238

[60] Kulkarni AD, Fanslow WC, Rudolph FB, Van Buren CT. Immunohemopoietic effects of dietary nucleotide restriction in mice. Transplantation. 1992;**53**(2):467-472

[61] Kulkarni AD, Rudolph FB, Van Buren CT. The role of dietary sources of nucleotides in immune function: A review. The Journal of Nutrition. 1994;**124**(8 Suppl):1442S-1446S

[62] Yamauchi K, Hales NW, Robinson SM, Niehoff ML, Ramesh V, Pellis NR, Kulkarni AD. Dietary nucleotides prevent decrease in cellular immunity in ground-based microgravity analog. Journal of Applied Physiology. 2002;**93**(1):161-166

[63] Kulkarni A, Richardson M, Ono K, Chandwani V, Johnson E, Kogiso M, Wakame K, Ambrose C, Martinez-Puig D. Nutritional nucleotides improve bone integrity in skeletal disuse. In: American Society for Parenteral and Enteral Nutrition 2008 Annual Conference; Chicago; USA. 2008

[64] Sundaresan A, Yamauchi K, Kulkarni AD, Pellis NR. Microgravity and modeled microgravity effects on lymphocyte signal transduction: Comparisons between human and mouse lymphocyte signaling. In: Proceedings of the 23rd ISTA Conference, Matsue, Japan. May 2002; p. 2

[65] Hales NW, Yamauchi K, Martinez AA, Sundaresan A, Pellis NR, Kulkarni AD. A countermeasure to ameliorate immune dysfunction in in vitro simulated microgravity

environment: Role of cellular nucleotide nutrition. In Vitro Cellular & Developmental Biology—Animal. 2002;**38**(4):213-217

[66] Kidd PM. The use of mushroom glucans and proteoglycans in cancer treatment. Alternative Medicine Review. 2000;**5**(1):4-27

[67] Ritz BW, Nogusa S, Ackerman EA, Gardner EM. Supplementation with active hexose correlated compound increases the innate immune response of young mice to primary influenza infection. The Journal of Nutrition. 2006;**136**(11):2868-2873

[68] Kogiso M, Wakame K, Sakai T, Yamamoto S, Sundaresan A, Kulkarni AD. Active hexose correlated compound and T cell response in hind limb—Unloaded BALB/c mice. International Journal of Surgical Research. 2015;**2**(5):32-38

[69] Aviles H, Belay T, Vance M, Sun B, Sonnenfeld G. Active hexose correlated compound enhances the immune function of mice in the hindlimb-unloading model of spaceflight conditions. Journal of Applied Physiology. 2004;**97**:1437-1444

[70] Gao Y, Zhang D, Sun B, Fujii H, Kosuna K, Yin Z. Active hexose correlated compound enhances tumor surveillance through regulating both innate and adaptive immune responses. Cancer Immunology, Immunotherapy. 2006;**55**:1258-1266

[71] Burikhanov RB, Wakame K, Igarashi Y, Wang S, Matsuzaki S. Suppressive effect of active hexose correlated compound (AHCC) on thymic apoptosis induced by dexamethasone in the rat. Endocrine Regulations. 2000;**34**:181-188

[72] van der Poll T, Keogh CV, Guirao X, Buurman WA, Kopf M, Lowry SF. Interleukin-6 gene-deficient mice show impaired defense against pneumococcal pneumonia. The Journal of Infectious Diseases. 1997;**176**(2):439-444

[73] Aviles H, Belay T, Fountain K, Vance M, Sun B, Sonnenfeld G. Active hexose correlated compound enhances resistance to *Klebsiella pneumoniae* infection in mice in the hindlimb-unloading model of spaceflight conditions. Journal of Applied Physiology. 2003;**95**:491-496

[74] Adjei AA, Takamine F, Yokoyama H, Shiokawa K, Matsumoto Y, Asato L, Shinjo S, Imamura T, Yamamoto S. The effects of oral RNA and intraperitoneal nucleoside-nucleotide administration on methicillin-resistant *Staphylococcus aureus* infection in mice. JPEN Journal of Parenteral and Enteral Nutrition. 1993;**17**(2):148-152

[75] Adjei AA, Yamamoto S. A dietary nucleoside-nucleotide mixture inhibits endotoxin-induced bacterial translocation in mice fed protein-free diet. The Journal of Nutrition. 1995;**125**(1):42-48

[76] Fanslow WC, Kulkarni AD, Van Buren CT, Rudolph FB. Effect of nucleotide restriction and supplementation on resistance to experimental murine candidiasis. JPEN Journal of Parenteral and Enteral Nutrition. 1988;**12**(1):49-52

[77] Kulkarni AD, Fanslow WC, Rudolph FB, Van Buren CT. Effect of dietary nucleotides on response to bacterial infections. JPEN Journal of Parenteral and Enteral Nutrition. 1986;**10**(2):169-171

Chapter 8

Countermeasure Development for Lumbopelvic Deconditioning in Space

Andrew Winnard, Dorothee Debuse and
Nick Caplan

Additional information is available at the end of the chapter

http://dx.doi.org/10.5772/intechopen.72881

Abstract

Physical inactivity and lumbopelvic deconditioning have been linked to increased incidence of non-specific low back pain (LBP) and spinal injury in those who are exposed to microgravity (e.g. astronauts and individuals on long-duration bed rest) and in the general population. Astronauts have an increased risk of experiencing moderate to severe LBP during microgravity exposure and herniated intervertebral discs within 1 year following spaceflight. Atrophy and reduced motor control of the lumbar multifidus (LM) and transversus abdominis (TrA) muscles resulting from periods of deconditioning are linked to non-specific LBP and spinal injury risk in both post-flight astronauts and general populations. However, voluntary recruitment of these two key muscles is difficult and presents a rehabilitation challenge. This chapter reviews the concept of spinal stability as it relates to microgravity, discusses how existing exercise countermeasures used in space do not successfully maintain lumbopelvic muscle size, and introduces the functional readaptive exercise device (FRED) that shows potential to activate the LM and TrA muscles automatically and in a tonic fashion, which has relevance to rehabilitation of both astronaut and terrestrial populations.

Keywords: spinal, lumbopelvic, deconditioning, rehabilitation, astronaut

1. Introduction

There is a 53–68% risk of experiencing moderate to severe low back pain (LBP) during microgravity exposure [1] and fourfold increased risk of herniated intervertebral discs within 1 year following the spaceflight [2], which demonstrate a need to understand the underlying mechanisms of LBP and spinal changes that result from exposure to microgravity. Developing an effective rehabilitation programme to address and rehabilitate spaceflight-related spinal changes is also required. Atrophy and reduced motor control of the lumbar multifidus (LM) and transversus abdominis (TrA) muscles, resulting from periods of deconditioning, are linked to non-specific LBP and spinal injury risk in both post-flight astronauts and general populations [3–6]. However, voluntary recruitment of these two key muscles is difficult and presents a rehabilitation challenge [7].

This chapter discusses the key factors contributing to lumbopelvic deconditioning in space-flight and shows how exercise countermeasures against deconditioning can be developed to aid both astronauts and people on Earth with LBP.

2. Link between segmental spinal stability and upright sagittal spinal motor control on Earth

Non-specific LBP is experienced in the lower region of the spine and is not attributable to a known cause or specific pathology such as infection, systemic disease, fracture or cauda equina syndrome [8]. Non-specific LBP is often multifactorial in its origin; this makes it complex to treat. Panjabi [9] was one of the first to recognise that abnormal spinal mechanics may be a common factor in people with back pain. He identified several potential triggers and causes of abnormal mechanics, including inflammation, biochemical and nutritional changes, immunological factors, structural changes in discs and endplates.

Atrophy [4, 10–13] and altered motor control [4] of the LM and TrA muscles are linked with the common symptom of altered mechanics. Both muscles have a substantial body of evidence linking their dysfunction and atrophy with LBP on Earth [4, 6, 14–18]. Similar patterns of muscle atrophy and LBP have also been observed in those who are exposed to microgravity [5, 6, 19, 20].

2.1. Deep and superficial lumbopelvic muscles in spinal stability

The paraspinal muscles can be divided into deep and superficial muscles based on a structural model of the spine presented by Bergmark [21] who provided the following definitions. **Deep muscles** have their origin and/or insertion at the vertebrae and have an action that includes controlling the curvature and/or structural stiffness of spine. Deep muscles include the LM and TrA muscles. The LM muscle controls and stabilises lumbar lordosis [22] during force transfer through the spine [23, 24] and provides segmental stiffness [25, 26]. The TrA muscle provides a transverse force, therefore increasing stiffness and extrinsic stability of the spine [4] by increasing intra-abdominal pressure [14, 27]. **Superficial muscles** control the large spinal movements and transfer loads between the thorax and pelvis, they do not directly increase stiffness or stability of the spine at a segmental level [21] but can increase global trunk stiffness [28]. Superficial muscles include erector spinae, the internal and external oblique muscles, rectus abdominis, quadratus lumborum and psoas.

Bergmark [21] also defined stability in engineering terms as the ability of a structure to maintain its equilibrium under loading. This definition was extended to define clinical spinal stability as the ability of the spine, under physiological loads, to limit structural displacement so as to prevent damage to spinal structures including the discs, ligaments and neural structures. The spine gains passive stability from bones, ligaments, tendons and fascia, while it was suggested that active stability is provided by deep muscles [21]. Studies using in vitro cadaveric specimens of human spinal segments found that the specimens became mechanically unstable at loads much less than those experienced by in vivo spines [25]. This finding highlighted the importance of the stabilising force provided by the LM and TrA muscles in allowing the spine to function under everyday loading.

2.2. Spinal stabilising system and motor control

To achieve spinal stability, the deep muscles must be controlled by precise coordination of activation and timing. The complete spinal stabilising system was, therefore, conceptualised by Panjabi [29] as having a neural control element, a passive spinal column (and ligaments) and an active system of deep muscles. The control system assesses and directs the deep muscles to provide varying levels of extrinsic stability while the passive elements of the spinal column provide intrinsic stability. To successfully provide control, actions are based on feedback from both the active and passive components. Mechanoreceptors in the passive structures indicate levels of force and stress, while feedback on muscle activation patterns and stretch are provided by the active system. In addition to the muscle feedback system, there is now strong evidence that LM and TrA are ideally activated in a feedforward mechanism, that is, they act in anticipation of changing loads. Importantly, the dysfunction of this feedforward control system has been linked with LBP [28].

2.3. Segmental stability and the neutral zone

During dynamic loads into spinal flexion and extension, there is displacement of each vertebra, which allows flexibility. At low loads, the spine was observed to be highly flexible and then stiffening as loads increased. A neutral zone was defined as the range of segmental displacement within which there is a minimal resistance to the displacement [29]. This is represented graphically in **Figure 1** with the neutral zone being represented by a ball in a bowl. The motion of the ball represents the displacement motion of the vertebral segment, while the steepness of the sides represents varying stability with steeper sides demonstrating increased resistance to displacement. As segmental spinal stability increases, the neutral zone becomes smaller, demonstrated by placing the ball in a wine glass. As segmental spinal stability decreases, the neutral zone gets larger, demonstrated by placing the ball in a flat bowl (**Figure 1**).

It was hypothesised that decreased stability may be caused either by damage to the passive stability system and/or abnormal activity or control of the active system that leads to a larger neutral zone [29]. An increase in the neutral zone is likely to be associated with increased stress on spinal structures so, it results in pain. Therefore, interventions were suggested for unstable painful spines which aimed at reducing the neutral zone through retraining control of the active stability system or through the use of spinal fusion [29].

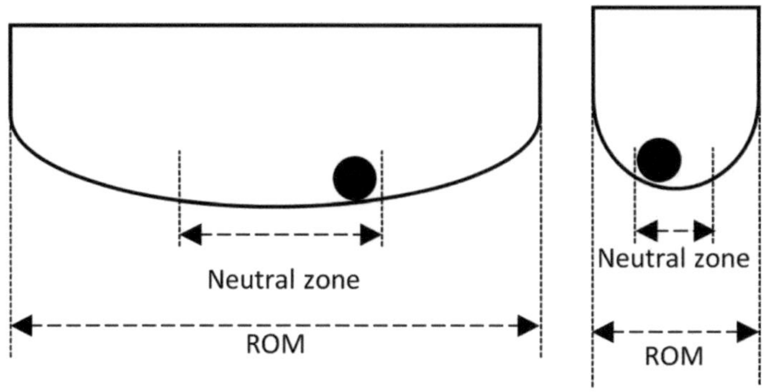

Figure 1. A visual representation of the neutral zone as a flat dish with less stability (left) and a wine glass with more stability (right) (Adapted from Panjabi [29]).

2.4. Theory linking low back injury with altered motor control and low back pain

The theory of how low back injuries can cause altered motor control and lead to low back pain was summarised by Panjabi [9] as follows:

1. Initial trauma occurs in spinal structures such as ligaments. This can be either a long-term build-up of microtrauma or an acute injury.
2. During dynamic loading of the injured spine, mechanoreceptor signals sent to the neural control system, produced by the injured tissue are now corrupted due to injury.
3. The motor control area of the brain finds a mismatch between expected signals and those actually being received. This causes control unit output to the active stability system in response to dynamic loading to also become corrupted.
4. Corrupted output from the control unit leads to the changes in the activation of the deep muscles in response to the dynamic load. These changes lead to abnormal activation and timing of the active stabilising deep muscles—LM and TrA. This then causes altered spinal mechanics.
5. Abnormal activation patterns of the deep muscles causes their returning feedback to also become corrupted, causing further mismatch in signals being received by the control unit.
6. Increased corruption of control unit output occurs in response to continued dynamic loading. This has great potential to lead to segmental instability, increased segmental neutral zone and higher stresses on spinal structures.
7. Inflammation of stressed spinal tissues around unstable segments is then likely to occur and nociceptive pain signals produced.
8. If left unchecked, chronic non-specific LBP may develop.

Evidence supporting these hypotheses exists from several experimental studies. Danneels et al. [11] conducted a comparison study of chronic LBP and matched non-symptomatic participants, which found reduced cross-sectional area of LM in the lower lumbar spine. In the study, 32 clinical participants were compared with 23 matched no-LBP volunteers, and the LM cross-sectional area was measured using CT scans. A study in pigs by Hodges et al. [12] found that induced L4 spinal disc lesions resulted in LM cross-sectional area decreases at the same level of the injury within 3 days, compared to no change in no-LBP controls. Injury to the L3 nerve root resulted in LM cross-sectional area reduction at the affected level and down to L4, L5 and S1 levels in 15 induced injury pigs compared to six controls. The controls were, however, still subjected to a sham surgical procedure that involved all the same steps as the injured pigs apart from the inducing of the injury. A comparison study by Hides et al. [3] of 26 first episode acute unilateral low back pain patients with 51 healthy controls, found that LM asymmetry in the people with back pain, isolated to the symptomatic level compared to symmetrical LM muscles in the no-LBP controls. A comparison study by Hodges and Richardson [4] of 15 LBP patients with 15 no-LBP matched controls used electromyography to assess the activation and timing of TrA in response to upper limb movements. It was observed that TrA activation was consistently delayed in the people with back pain. A comparison study by Ferreira and Hodges [13], in which 10 low back pain patients compared with 10 healthy matched controls, found consistently reduced changes in TrA thickness in their group with back pain during lower limb exercises, which was measured using ultrasound imaging.

While Panjabi's theories were seminal to improve our understanding of lumbopelvic pain, they missed one important factor, and that is the fact that, unlike the feedback control of many superficial muscles, there is a feedforward control of the deep spinal muscles LM [24, 30] and TrA [31, 32]. In other words, the LM and TrA muscles work in anticipation of loads and movements, not in response to them, thus, providing spinal stability. Importantly, there has also been evidence that the anticipatory activation of the deep spinal muscles is impaired in people with LBP compared to non-symptomatic controls [4, 31, 33] and that this is reversible with certain exercise approaches [34]. Based on a considerable body of literature, key authors in the field [16, 35] also suggest that secondary compensatory postural mechanisms are likely to contribute further to LBP, which would go some way towards explaining chronic LBP in the face of minimal tissue abnormality. It should be noted that the studies used to have small sample sizes may not be strongly statistically powered.

2.5. Lumbopelvic adaptations to microgravity

Astronauts returning from long-duration space missions (~6-months duration) [36] and participants following long-duration bed rest studies, which are commonly used to simulate microgravity exposure [37], have a range of muscular and postural problems. Human spaceflight results in exposure to an altered gravity state, mostly eliminating weight bearing and axial loads, resulting in physiological changes and potentially increased injury risk [36, 38, 39]. Buckey [36] grouped these changes into broad themes allowing them to be listed briefly as follows: bone loss, psychosocial, radiation biological, muscle loss, balance and postural control, cardiovascular and nutritional. These changes include decreased balance and proprioception, decreased muscle mass, force and power with increased loss of technique (specifically affecting lower limb antigravity muscles and lumbopelvic segmental control muscles) [36], decreased ability to control posture—specifically, the ability to achieve a balanced pelvic tilt and normal spinal curves in the sagittal plane, increased risk of spinal injury from poor spinal positioning during everyday activities—especially involving trunk flexion, increased chance of poor global movement patterns, and risk of injury from musculoskeletal weakness and atrophy [5, 19, 40, 41]. Those in microgravity also experience lengthening of the spine due to swelling and hyper-hydration of the intervertebral discs which, in turn, become deconditioned resulting in increased risk of disc injury [19].

Gernand [38] reported the implications of these physiological changes on subsequent safe functioning when returning to a gravity-loaded environment, highlighting the need for both countermeasure interventions during spaceflight and rapid and effective rehabilitation following spaceflight. For spaceflight of around 6 months, Gernand [38] noted significant bone and muscle loss, as well as altered postural control, leaving the body susceptible to fractures, muscle injury and the potential to develop osteoporosis. Muscle atrophy and altered motor control have been specifically observed in the lumbopelvic region [42]. A European Space Agency (ESA) report by Snijders et al. [43] reported LBP in 12 out of 20 astronauts during spaceflight. The report highlighted the importance of maintaining spinal movements, as end range flexion and extension exercises were anecdotally noted as being employed to ease pain during spaceflight. A relationship was also highlighted between LBP and atrophy of deep spinal muscles, particularly LM, during bed rest studies [44]. Wing et al. [1] reported that 53–68% of astronauts experienced moderate to severe back pain when in space. On landing after a shuttle mission, a US astronaut reported severe LBP which was later linked with a herniated nucleus pulposus at the L4-5 intervertebral (IV) disc and required surgical intervention [2]. Johnston et al. [2] also reported that astronauts had a more than fourfold increased risk of herniated disc pulposus within the first year following spaceflight, compared with controls. Sayson and Hargens [42] suggested that this back pain and disc injury could be caused by a range of factors linked to spinal lengthening and reduced loading. A review by Belavy et al. [19] supported this, suggesting the increased lumbar IV disc herniation risk in the astronaut population was most likely caused by long-term disc tissue deconditioning resulting from swelling of the discs due to unloading during spaceflight. However, the review only considered IV discs in isolation and did not refer to any potential predisposing factors such as spinal motor control. It should be noted that data from actual astronaut studies are usually from small samples, and therefore, statistical power is often low. Earth-based simulation studies, such as bed rest studies, are therefore, useful to increase the overall sample size on which to base conclusions.

Lumbopelvic adaptations to microgravity include adoption of a flexed posture (**Figure 2**) [36], spinal lengthening, increased intervertebral disc height and disc deconditioning, altered spinal curvatures [42] and atrophy of the lumbopelvic musculature. Selective atrophy of spinal extensors without corresponding atrophy of the psoas muscle was also seen in terrestrial individuals with LBP compared to no-LBP controls by Danneels et al. [11]. Atrophy and motor control changes in the LM muscle have been linked with LBP [3, 10] and development of poor intersegmental control of the lumbar spine [15, 21, 27, 45], which can potentially cause increased stress on spinal structures, resulting in pain [25, 29, 46].

Humans exposed to sustained microgravity develop a risk of significant spinal injury as a result of microgravity-induced poor intersegmental control of the lumbar spine combined with loaded activities, such as extra-vehicular activity, physically demanding medical procedures, landing

http://dx.doi.org/10.5772/intechopen.72881

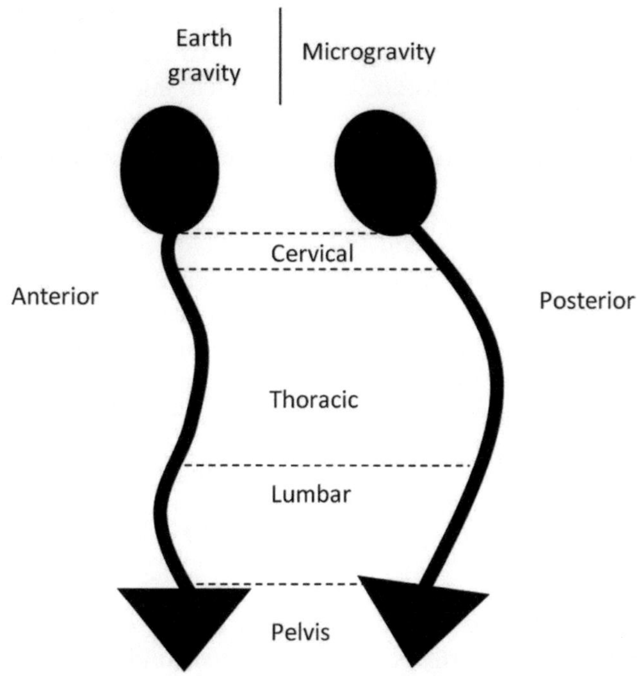

Figure 2. Postural adaptation to microgravity, showing loss of normal spinal curvature and increased flexion of the spinal column.

and return to a g-loaded environment, which have the potential to be at least as demanding as those undertaken in normal Earth gravity [38]. It is necessary, therefore, to know what physiological changes occur, which could lead to increased injury risk, and which interventions, both preventative and rehabilitative, can be used to minimise and effectively rehabilitate physiological compromise. The current evidence also suggests that interventions to address lumbopelvic physiological adaptations are likely to be a required element of any rehabilitation programme following exposure to microgravity. While Evetts et al. [47] indicated that European post-flight rehabilitation includes specific training for lumbopelvic posture and spinal muscles involved in intersegmental control of the lumbar spine, they highlighted a need to compare the effectiveness of interventions to advance the treatments given to astronauts. Such improvements are also likely to aid terrestrial healthcare with more effective interventions for people with LBP and post-bed rest rehabilitation [47].

3. Management of low back pain on Earth using the motor control approach to improve spinal stability and control

Management of segmental instability using specific motor control exercises aimed at normalising the recruitment patterns of the deep muscles was first summarised by O'Sullivan [48]. The first stage of training is learning to isolate and correctly voluntarily contract the deep muscle system. The voluntary contractions are intended to be low level and at 30–40%

maximal voluntary contraction. Contractions are taught in postures such as supine, prone and four-point-kneeling while patients are asked to perform abdominal drawing in using TrA while maintaining a neutral lumbar lordosis. In addition to this, patients are taught: differentiation of lumbar, pelvic and hip movements and diaphragmatic breathing and maintenance of neutral lordosis in different postural sets such as sitting and standing.

Live biofeedback with use of palpation, ultrasound imaging or possibly electromyography can be included to help isolate TrA and LM activation [49]. Treatment is then progressed to the second stage where the deep muscle recruitment learned in stage one is incorporated into functional movement, and compensatory muscle strategies are discouraged. Patients are taught movements such as sit to stand, walking, bending and twisting while maintaining activation of deep muscles. The third and final stage of training is for patients to carry the newly learned and stable functional movements into their activities of daily life.

The 'bare bones' have evolved since 2000 into an evidence-based and integrated approach which is summarised by Hodges et al. [28] and too complex to discuss here. However, it is based on a considerable amount of good evidence, one of which will be summarised in the following paragraph.

Hides et al. [49] assessed LM size in athletes with LBP and determined the effectiveness of a motor control intervention. Ten participants with back pain underwent a 6-week intervention programme of learning to correctly activate TrA and LM. Live biofeedback using ultrasound imaging was used during muscle activation teaching. Abdominal drawing in exercises were used to teach recruitment of TrA while maintaining a normal, relaxed, breathing pattern, followed by participants attempting to swell the LM muscle while holding a breath out and keeping the spine still with a neutral lumbar lordosis. Initially, activation was taught in lying and then progressed to upright sitting and standing, all the while maintaining a neutral lumbar lordosis. Further progression to functional movements was then performed. By the end of the programme, pain scores had dropped from an average of 4.3–2.3 ($p < 0.05$). Before treatment, asymmetry had been observed in LM cross-sectional area, which also significantly decreased, while overall muscle size increased. This is just one study that provides evidence that motor control exercises including recruitment of deep muscles can improve clinical outcomes.

4. In-flight countermeasures for lumbopelvic deconditioning in space

On the International Space Station (ISS), astronauts take part in up to 2.5 h of exercise each day including running, cycling and strength training. These exercise countermeasure programmes are known to be relatively successful at preventing bone loss and loss of muscle mass in some regions of the body. However, as mentioned earlier, they are not specifically targeted at preventing lumbopelvic muscle loss.

A number of bed rest studies have investigated potential inflight countermeasures, reporting on their effects on lumbopelvic musculoskeletal parameters. However, no studies have tested the operational countermeasures currently in use on the ISS [50]. Tested countermeasures included lower body negative pressure treadmill running, resistance exercises with external vibration, resistance exercise alone, rowing like exercises using a flywheel device and self-performed exercises designed to mobilise the spine. In a recent systematic review of these bed rest studies, Winnard et al. [50] identified that no single potential countermeasure can successfully prevent all lumbopelvic musculoskeletal adaptations to simulated microgravity. For example, resistive vibration exercise was the only countermeasure, which is able to protect against lumbopelvic muscle adaptations, but it did not prevent spinal morphology changes such as loss of lumbar lordosis. No other countermeasure tested was able to do more than partially prevent (at best) the lumbopelvic muscle adaptations. This demonstrates a need for further research into new interventions to better protect the spine during microgravity exposure. It might be possible to translate ground-based interventions into new countermeasures or develop new ones based on the current lumbopelvic deconditioning rehabilitation theory. Any new interventions will need testing in ground-based microgravity simulations before incurring costs associated with actual spaceflight testing. Lower body negative pressure treadmill was the most effective currently researched countermeasure against lumbopelvic deconditioning. As this countermeasure is not yet used in operational spaceflight and no countermeasure is fully effective, there remains a need for rehabilitation.

5. Rehabilitation following actual and simulated spaceflight

Due to the lack of effective in-flight countermeasures targeted at preventing lumbopelvic musculoskeletal adaptations, astronauts require significant rehabilitation to reduce injury risk on their return to Earth. During spaceflight, a general pattern of selective extensor muscle atrophy has been seen throughout the body [51]. Decrease in spinal extensor volume has been reported as being greater than hip flexor (psoas muscle) decline in astronauts [52]. Anecdotal accounts also appear to show selective atrophy of trunk extensor muscles concomitant with improved flexor muscle performance immediately post mission [53]. This muscle imbalance results in temporary loss of lumbopelvic posture, flexion of thoracic spine and hyperextension of cervical spine, the centre of gravity is moved anteriorly and increases the risk of musculoskeletal injury [47]. Hides et al. [40] suggested that deep spinal muscle changes such as atrophy of LM and TrA muscles, along with selective hypertrophy of spinal flexors over extensors [5], may impact on the ability of the spine to distribute loads appropriately shortly after spaceflight simulation via bed rest. The European approach to post-space mission rehabilitation addresses the muscle imbalance and uses motor control training in a way very similar to that described for people with LBP on Earth and based mostly on the existing terrestrial evidence transferred to a post-spaceflight setting. Initially, postural control, muscle control and muscle balance are restored, followed by the use of strategies to normalise muscle recruitment. Astronauts are then supported to redevelop postural alignment in line with the centre of gravity and to develop adequate motor control before they start to exercise with elements of loading and strength training. The latter is only started after the astronauts have regained correct postural alignment and control [54, 55].

No studies have investigated lumbopelvic rehabilitation approaches following actual spaceflight, and only one study has investigated rehabilitation approaches following simulated spaceflight (bed rest) [40]. The study assessed specific motor control (SMC) exercises compared to a control group performing trunk and general strength exercise (TFS) programme in a supine position. Results favoured SMC for restoring spinal length and posterior disc height, suggesting it may reduce the risk of IV disc injury during rehabilitation. However, TFS was favoured for training LM muscle and restoring lordosis angle and overall disc volume. Overall, it was suggested that SMC is favourable over TFS because SMC is expected to place less force on the discs and is associated with the lower rate of change in disc volume and anterior disc height [40]. Lower forces on the discs during rehabilitation—at a time when the discs may be deconditioned and vulnerable to injury—is expected to help restore posture and motor control with reduced risk of damage to the discs in the process. Therefore, in line with current ESA rehabilitation practice, a training programme starting with SMC when disc injury risk is high, then progressing to general trunk strengthening once lumbar postural control is restored would seem to be indicated. Other rehabilitation methods that train the LM muscle and maintain lordosis angle, without high axial loading, would also be worth investigating. Additionally, as noted previously, due to the low sample sizes from the terrestrial evidence on which these methods are based, further studies to improve statistical power would be useful to ensure a robust evidence-based approach. While motor control training has been shown to be useful in LBP rehabilitation, and is already used in the rehabilitation of European astronauts [47, 54, 55], many people have difficulty in recruiting LM, in particular, voluntarily [7]. This presents a challenge to physiotherapists involved in evidence-based practice for LBP. Many of the exercises used early in motor control training also lack functional relevance to activities of daily living, and there is a drive to make rehabilitation more functional [45, 56, 57]. As such, new interventions must be developed to address these challenges.

6. Developing a new countermeasure for lumbopelvic deconditioning in space

6.1. Early development of the functional readaptive exercise device (FRED)

In an attempt to address the challenges discussed earlier relating to motor control training, Debuse et al. [53] investigated the effects of a new exercise device, the functional readaptive exercise device (FRED) (**Figure 3**), that aims to recruit the LM and TrA muscles.

FRED exercise constitutes a combination of weight-bearing, an unstable base of support (at the feet), an upright 'standing' posture with a relatively stable lumbopelvic area, and functional lower limb movement, combined with real-time visual feedback of performance. As the FRED offers no resistance to lower limb movement, it requires good balance and coordination in order to achieve a smooth, controlled cyclical motion. Exercise on the FRED has been shown to recruit LM and TrA automatically (i.e. with no conscious effort by participants) and

Figure 3. Current prototype of the functional readaptive exercise device.

to recruit them differentially [53]. More recently, FRED exercise has been shown to promote tonic activity of LM, assessed through measurement of superficial muscle activity using surface electromyography [58], as well as the deep lumbopelvic muscles using intramuscular electromyography [59], which is considered the most rigorous way of investigating muscle activity [60]. FRED exercise was shown to result in more selective activation of the LM and TrA muscles than over-ground walking [59], and it was found to reduce lumbopelvic movement when compared to over-ground walking, especially axial rotation of the spine [61]. To date, these studies are all based on normal terrestrial gravity, and the next step will be a clinical trial of the FRED following bed rest as a simulation of space flight. A musculoskeletal modelling study that examined the potential role of the FRED in the recruitment of lumbopelvic muscles in both +1 and 0 Gz environments, Lindenroth et al. [62] predicted that FRED exercise is able to facilitate lumbopelvic muscle recruitment in microgravity similar to how it recruits the same muscles in Earth gravity.

Based on the early research findings relating to the early and current prototypes of FRED, it can be hypothesised that the device uses several mechanisms in combination, to produce rehabilitation effects on several of the problems found in spinal instability simultaneously within one intervention [58], as presented in **Table 1**.

These potential mechanisms show how the FRED has already demonstrated the ability to automatically activate both LM and TrA in an asymptomatic population without need for conscious muscle recruitment. This might have potential to solve the LM and TrA conscious recruitment difficulties found in traditional spinal motor control rehabilitation [7]. The exercise is dynamic, functional, weight-bearing, in an upright posture and relevant to common daily activities such as walking. These are all elements of motor control exercises covered in **Section 4**. It appears, therefore, that the device might be a useful intervention to train the LM and TrA muscles and segmental spinal stability.

Problem	FRED mechanism
Poor lumbopelvic motor control of deep spinal muscles	Exercising using a pattern of moving the feet in a quasi-elliptical path in antiphase with minimal resistance from the device or support from the upper limbs Exercising while maintaining a stable pelvis and upright trunk while having to maintain an even speed within one revolution. The abovementioned points create a need for greater control of the lower limbs and pelvis during an unstable dynamic movement. Greater control is particularly needed in resisting a fast descent of the foot in the forward-most position of the cycle. The movement is functional and similar to over-ground walking. Therefore, muscle activation training is learned in a functional movement, hoped to produce carry over into other functional daily activities. Clinical observations seem to indicate that relatively greater rear foot loading in standing results in greater recruitment of LM, whereas relatively greater front foot loading in standing has a deactivating effect on LM. It is hypothesised that correct exercise on FRED results in reduced front foot loading. FRED provides visual feedback that encourages users to exercise at a constant, controlled speed and frequency ratio, which is hypothetically the most energy efficient movement [63]. Additional feedback encourages users to maintain even movements throughout the exercise, training control of the lumbopelvic area and lower limbs during dynamic functional movements. It is thought that efficient and smooth controlled movement on FRED may improve LM and TrA neuromotor control. The exercise has already been shown to activate LM and TRA without the need to consciously trigger the activation in non-symptomatic populations [53]. In addition to this, LM was shown to have constant tonic activity throughout exercise cycle on the device in an electromyography study. The muscle was active for more time than during over-ground walking [58].
Reduced ability to control spinal posture and balance	Previous kinematic research has shown FRED exercise promotes an increased degree of anterior pelvic tilt during upright posture [61]. Increased anterior pelvic tilt, within a range where the thoracolumbar junction remains the inflexion point between lumbar lordosis and thoracic kyphosis, has been shown to create a well-balanced sagittal spinal posture [64]. Electromyography data have also shown that this type of posture produces the highest LM and TRA recruitment [22], though this study investigated sitting postures. Additionally, users of the device are required to exercise in an upright posture. It is hoped that these elements together mean FRED exercise promotes a balanced upright sagittal posture, with recruitment of LM and TrA. Having improved control of balanced posture is also hoped to improve overall balance.
Atrophy of spinal extensors	EMG data from FRED exercise show that it promotes increased activation of spinal extensors over flexors [58]. This may be relevant to the rehabilitation of astronauts who show increased flexion postures when in space [36]
Weakness of lower limb anti-gravity muscles	Previous kinematic research shows FRED exercise involves constant hip and knee flexion in a dynamic and gravity-loaded exercise, therefore, constantly loading lower limb extensor muscles [61]. This loading is expected to improve strength in the lower limb extensors, which is a common aim of traditional interventions for reducing falls risk in older people [65]

Table 1. FRED mechanisms.

6.2. Developing FRED for rehabilitation from lumbopelvic deconditioning in space

Following the early research on the early prototype of the FRED, a range of mechanistic studies were completed on the prototype shown in **Figure 3** in order to develop it for use as a rehabilitation intervention in groups with lumbopelvic muscle deconditioning (e.g. low back pain, astronauts). Winnard et al. [63] compared the thickness of LM and TrA at a range of movement amplitudes in an asymptomatic sample. A large body of evidence has linked dysfunction and/or atrophy of TrA and LM to lumbopelvic deconditioning and LBP [3, 4, 9, 11, 12]. Importantly, these muscles are difficult to recruit voluntarily, this presents a challenge in terms of their rehabilitation [7]. Therefore, the automatic recruitment of TrA and LM during FRED exercise would appear to offer an advantage over current practice. Increasing crank amplitude was observed to increase movement variability, the range of TrA and LM thickness peaks, as well as mean TrA muscle thickness [63]. These outcomes are all measures of motor control of either global movement or muscle recruitment.

In a large sample of both symptomatic (LBP) and asymptomatic participants, FRED exercise promoted increased lumbar extension and anterior pelvic tilt compared to over-ground walking [64]. Attaining a lordosis throughout the lumbar spine below the thoracolumbar junction is a common goal of current interventions [48]; it is the sagittal spinal position where LM tends to be most effectively recruited [22, 24, 65]. Although this finding alone does not indicate that the correct lordosis is promoted by FRED exercise, when combined with the finding that FRED exercise appears to recruit key lumbopelvic muscles automatically, there is increased likelihood that the spinal position promoted during FRED exercise is more conducive to LM recruitment than walking.

In the same sample, the FRED caused increased anteroposterior and mediolateral centre of mass variation compared to walking [64]. This suggests an increased challenge to balance and, therefore, motor control during FRED exercise. This may form part of the motor control mechanism of FRED exercise and adds to the overall evidence that FRED exercise is in line with current motor control interventions and adds weight to the justification for a clinical trial.

7. Conclusion

Astronauts undergo significant lumbopelvic musculoskeletal deconditioning following their exposure to microgravity. Many experience low back pain, and there is a fourfold increase in the incidence of intervertebral disc injury on their return to Earth as compared to their non-astronaut peers. It is known that the spine lengthens, normal posture is lost, intervertebral discs change their morphology, LM and TrA muscles atrophy and a flexor-extensor lumbopelvic muscle imbalance occur during spaceflight. Current in-flight countermeasures aim to generally prevent physiological adaptations to microgravity. However, they are not specifically targeted enough to do so for the lumbopelvic spaceflight adaptations. Astronauts, therefore, require rehabilitation upon return to Earth's gravity to reduce injury risk. The European rehabilitation interventions follow current evidence-based practice for treating people with LBP in whom dysfunction of TrA and LM is a key contributing factor to their symptoms such as motor control training.

FRED exercise shares many of the characteristics of motor control exercise. The findings of a range of mechanistic studies show that it results in automatic and tonic activation of LM and TrA, promotes normal lumbopelvic positioning against gravity, works trunk and lower limb extensors more than flexors, challenges balance and motor control, and is a functional progressive exercise. On this basis, as well as the results of a very recent clinical study on people with chronic LBP, we are confident that FRED exercise would complement and enhance current astronaut rehabilitation practice. Future research should now investigate the effectiveness of the FRED in larger terrestrial populations with low back pain as well as following simulated and actual spaceflight.

Author details

Andrew Winnard[1], Dorothee Debuse[2] and Nick Caplan[1]*

*Address all correspondence to: nick.caplan@northumbria.ac.uk

1 Northumbria University, Newcastle upon Tyne, UK

2 Lunex University, Differdange, Luxembourg

References

[1] Wing P, Tsang I, Susak L, Gagnon R, Potts J. Back pain and spinal changes in microgravity. Orthopedic Clinics of North America. 1991;**22**(2):255-262

[2] Johnston S, Campbell M, Scheuring R, Feiveson A. Risk of hernated nucleus pulposus among US astronauts. Aviation, Space and Environments Medicine. 2010;**81**(6):566-574

[3] Hides JA, Stokes MJ, Jull GA, Cooper DH. Evidence of lumbar multifidus muscle wasting ipsilateral to symptoms in patients with acute/subacute low back pain. Spine. 1994; **19**(2):165-172

[4] Hodges P, Richardson C. Inefficient muscular stabilization of the lumbar spine associated with low back pain: A motor control evaluation of transversus abdominis. Spine. 1996;**21**(22):2640-2650

[5] Hides JA, Belavy DL, Stanton W, Wilson SJ, Rittweger J, Felsenberg D, et al. Magnetic resonance imaging assessment of trunk muscles during prolonged bed rest. Spine. 2007; **32**(15):1687-1692

[6] Hides J, Lambrecht G, Stanton W, Damann V. Changes in multifidus and abdominal muscle size in response to microgravity: Possible implications for low back pain research. European Spine Journal. 2016;**25**(Suppl 1):175-182

[7] Van K, Hides J, Richardson C. The use of real-time ultrasound imaging for biofeedback of lumbar multifidus muscle contraction in healthy subjects. The Journal of Orthopaedic and Sports Physical Therpay. 2006;**36**(12):920-925

[8] Balague F, Mannion AF, Pellise F, Cedraschi C. Non-specific low back pain. Lancet. 2012; **379**:482-491

[9] Panjabi MM. A hypothesis of chronic back pain: Ligament subfailure injuries lead to muscle control dysfunction. European Spine Journal. 2006;**15**:668-676

[10] Hides J, Gilmore C, Stanton W, Bohlscheid E. Multifidus size and symmetry among chronic LBP and healthy asymptomatic subjects. Manual Therapy. 2008;**13**:43-49

[11] Danneels LA, Vanderstraeten GG, Cambier DC, Witrouw EE, De Cuyper HJ. CT imaging of trunk muscles in chronic low back pain patients and health control subjects. European Spine Journal. 2000;**9**:266-272

[12] Hodges P, Holm AK, Hannson T, Sten H. Rapid atophy of the lumbar multifidus follows experimental disc or nerve root injury. Spine. 2006;**25**:2926-2988

[13] Ferreira PH, Ferreira ML, Hodges PW. Changes in recuitment of the abdominal muscles in people with low back pain. Spine. 2004;**29**(22):2560-2566

[14] Hides J, Stanton W, Mendis M, Sexton M. The relationship of transversus abdominis and lumbar multifidus clinical muscle tests in patients with chronic low back pain. Manual Therapy. 2011;**16**(6):573-577

[15] Hodges P, Moseley G. Pain and motor control of the lumbopelvic region: Effect and possible mechanisms. Journal of Electromyography and Kinesiology. 2003;**13**:361-370

[16] Macdonald D, Moseley L, Hodges P. Why do some patients keep hurting their back? Evidence of ongoing back muscle dysfunction during remission from recurrent back pain. Pain. 2009;**142**(3):183-188

[17] Saunders S, Coppieters M, Hodges P, editors. Reduced tonic activity of the deep trunk muscle during locomotion in people with low back pain. In: World Congress of Low Back Pain and Pelvic Pain; 2004

[18] Wallwork T, Stanton W, Freke M, Hides J. The effect of chronic low back pain on size and contraction of the lumbar multifidus muscle. Manual Therapy. 2009;**14**(5):496-500

[19] Belavy DL, Adams M, Brisby H, Cagnie B, Danneels L, Fairbank J, et al. Disc herniations in astronauts: What causes them, and what does it tell us about herniation on earth? European Spine Journal. 2015;**24**:251

[20] Evetts S. Anecdotal Accounts of Selective Atrophy of Trunk Extensors over Flexors in an Astronaut Population. Personal Communication 2015

[21] Bergmark A. Stability of the lumbar spine – A study in mechanical engineering. Copenhagen: Department of Solid Mechanics, Lund Institute of Technology; 1989

[22] Claus AP, Hides JA, Moseley LG, Hodges PW. Different ways to balance the spine. Spine. 2009;**34**:208-214

[23] Macintosh J, Valencia F, Bogduk N, Munro R. The morphology of the human lumbar multifdus. Clinical Biomechanics. 1986;**1**(4):196-204

[24] Moseley L, Hodges P, Gandevia S. Deep and superficial fibers of the lumbar multifidus muscle are differentially active during voluntary arm movements. Spine. 2002;**27**(2): 29-136

[25] Panjabi MM. The stabilizing system of the spine. Part 1. Function, dysfunction, adaptation, and enhancment. Journal of Spinal Disorders & Techniques. 1992;**5**(4):383-289

[26] Kiefer A, Shirazi-Adl A, Parnianpur M. Synergy of the human spine in neutral postures. European Spine Journal. 1998;**7**(6):471-479

[27] Hodges P. Lumbopelvic stability: A functional model of the biomechanics and motor control. In: Richardson C, Hodges P, Hides J, editors. Therapuetic Exercise for Lumbopelvic Stabilization. Edinburgh: Churchull Livingston; 2004. pp. 13-28

[28] Hodges P, Cholewicki J, Van Dieen J. Spinal Control: The Rehabilitation of Back Pain State of the Art and Science. Edinburgh: Churchill Livingstone Elsevier; 2013

[29] Panjabi MM. Clinical spinal instabiilty and low back pain. Journal of Electromyography and Kinesiology. 2003;**13**:371-379

[30] Hodges PW, Moseley GL, Gabrielsson A, Gandevia SC. Experimental muscle pain changes feedforward postural responses of the trunk muscles. Experimental Brain Research. 2003;**151**(2):262-271

[31] Hodges PW, Richardson CA. Altered trunk muscle recruitment in people with low back pain with upper limb movement at different speeds. Archives of Physical Medicine and Rehabilitation. 1999;**80**(9):1005-1012

[32] Hodges P, Cresswell A, Thorstensson A. Reparatory trunk motion precedes upper limb movement. Experimental Brain Research. 1999;**124**:69-79

[33] Ferreira PH, Ferreira ML, Maher CG, Refshauge K, Herbert RD, Hodges PW. Changes in recruitment0020of transversus abdominis correlate with disability in people with chronic low back pain. British Journal of Sports Medicine. 2010;**44**(16):116-172

[34] Hides JA, Stanton WR, McMahon S, Sims K, Richardson CA. Effect of stabilization training on multifidus muscle cross-sectional area among young elite cricketers with low back pain. The Journal of Orthopaedic and Sports Physical Therapy. 2008;**38**:101-108

[35] Van Dieen J, Kingsma I. Spine function and low back pain: interactions of active and passive structures. In: Hodges P, Cholewicki J, Van Dieen J, editors. Spinal Control: The Rehabilitation of Back Pain. Edinburgh: Churchill Livingstong; 2014. pp. 41-59

[36] Buckey JC. Space Physiology. New York: Oxford University Press; 2006

[37] Pavy-Le Traon A, Heer M, Narici MV, Rittweger J, Vernikos J. From space to Earth: Advances in human physiology from 20 years of bed rest studies (1986-2006). European Journal of Applied Physiology. 2007;**191**:143-194

[38] Gernand JM, editor. Risk Assessment and Control Through Countermeasure System Implementation for Long-Term Crew Exposure to Microgravity; 2004. Anaheim: IMECE

[39] di Prampero PE, Narici MV. Muscles in microgravity: From fibres to human motion. Journal of Biomechanics. 2003;**36**:403-412

[40] Hides JA, Lambrecht G, Richardson CA, Stanton WR, Armbrecht G, Pruett C, et al. The effects of rehabilitation on the muscles of the trunk following prolonged bed rest. European Spine Journal. 2011;**20**:808-818

[41] Belavy DL, Armbrecht G, Richardson CA, Felsenberg D, Hides JA. Muscle atrophy and changes in spinal morphology, is the lumbar spine vulnerable after prolonged bed-rest? Spine. 2011;**36**(2):132-145

[42] Sayson J, Hargens A. Pathophysiology of low back pain during exposure to microgravity. Aviation, Space and Environmental Medicine. 2008;**79**:365-373

[43] Snijders CJ, Richardson CA, Hides JA, Pool-Doudzwaard AL, Belavy DL. Low back pain during flight – 1 year post-flight science report. European Space Agency Report. 2011. ESA Proposal number 2003-DSM-MUS-100

[44] Pool-Goudzwaard AL, Belavý DL, Hides JA, Richardson CA, Snijders CJ. Low back pain in microgravity and bed rest studies. Aerospace Medicine and Human Performance. 2015;**86**(6):541-547

[45] Hodges P, Cholewicki J. Functional control of the spine. In: Vleeming A, Mooney V, Stoeckart R, editors. Movement, Stability and Lumbopelvic Pain: Integration of Research and Therapy. 2nd ed. Edinburgh: Churchill Livingstone; 2007. pp. 489-512

[46] Panjabi MM. The stabilizing system of the spine. Part 2. Neutral zone and instability hypothesis. Journal of Spinal Disorders. 1992;**5**(4):390-397

[47] Evetts S, Caplan N, Debuse D, Lambrecht G, Damann V, Petersen N, et al. Post space mission lumbo-pelvic neuromuscular reconditioning: A European perspective. Aviation, Space and Environmental Medicine. 2014;**85**(7):764-765

[48] O'Sullivan PB. Lumbar segmental "instability": Clinical presentation and specific stabilizing exercise management. Manual Therapy. 2000;**5**(1):2-12

[49] Hides J, Stanton W, McMahon S, Sims K, Richardson C. Effect of stabilisation training on multifidus muscle cross-sectional area among young elite cricketers with low back pain. Journal of Orthopaedic & Sports Physical Therapy. 2008;**38**(3):101-108

[50] Winnard A, Nasser M, Debuse D, Stokes M, Evetts S, Wilkinson M, Caplan N, et al. Systematic review of countermeasures to minimise physiological changes and risk of injury to the lumbopelvic area following long-term microgravity. Musculoskeletal Science and Practice. 2017;**27**:S5-S14

[51] Edgerton VR, McCall GE, Hodgeson JA, Gotto J, Goulet C, Fleischmann K, et al. Sensorimotor adaptations to microgravity in humans. The Journal of Experimental Biology. 2001;**204**:3217-3224

[52] LeBlanc A, Rowe R, Schneider V, Evans H, Hedrick T. Regional muscle loss after short duration spaceflight. Aviation, Space, and Environmental Medicine. 1995;**66**(12):1151

[53] Debuse D, Birch O, Gibson ASC, Caplan N. Low impact weight-bearing exercise in an upright posture increases the activation of two key local muscles of the lumbo-pelvic region. Physiotherapy Theory and Practice. 2013;**29**(1):51-60

[54] Lambrecht G, Petersen N, Weerts G, Pruett C, Evetts S, Stokes M, et al. The role of physiotherapy in the European Space Agency strategy for preparation and reconditioning of astronauts before and after long duration space flight. Musculoskeletal Science and Practice. 2017;**27**:S15-S22

[55] Petersen N, Lambrecht G, Scott J, Hirsch N, Stokes M, Mester J. Postflight reconditioning for European astronauts – A case report of recovery after six months in space. Musculoskeletal Science and Practice. 2017;**27**:S23-S31

[56] Richardson C, Hides J. Open chain segmentral control and progression into function. In: Richardson C, Hodges P, Hides J, editors. Therapeutic Exercise for Lumbopelvic Stabilization. 2nd ed. Edinburgh: Churchill Livingstone; 2004. pp. 233-246

[57] Richardson C, Hides J. Closed chain segmental control. In: Richardson C, Hodges P, Hides J, editors. Therapeutic Exercise for Lumbopelvic Stabilization. 2nd ed. Edinburgh: Churchill Livingstone; 2004. pp. 221-232

[58] Caplan N, Gibbon KC, Hibbs A, Debuse D. Phasic-to-tonic shift in tunk muscle activity during low-impact weight bearing exercise. Acta Astronautica. 2014;**104**:388-395

[59] Weber T, Debuse D, Salomoni S, Elqueta Cancino E, De Martino E, Caplan N, et al. Trunk muscle activation during movement with a new exercise device for lumbo-pelvic reconditioning. Physiological Reports. 2017;**5**(6):e13188

[60] Richardson CA, Jull GA. Muscle control – Pain control. What exercises would you prescribe. Manual Therapy. 1995;**1**:2-10

[61] Gibbon KC, Debuse D, Caplan N. Low impact weight-bearing exercise in an upright posture achieves greater lumbopelvic stability than overground walking. Journal of Bodywork & Movement Therapies. 2013;**17**:462-468

[62] Lindenroth L, Caplan N, Debuse D, Salomoni S, Evetts S, Weber TA. Novel approach to activate deep spinal muscles in space – Results of a biomechanical model. Acta Astronautica. 2015;**116**:202-210

[63] Winnard A, Debuse D, Wilkinson M, Samson L, Weber T, Movement NC. Amplitude on the functional re-adaptive exercise device: Deep spinal muscle activity and movement control. European Journal of Applied Physiology. 2017;**117**:1597-1606

[64] Winnard A, Debuse D, Wilkinson M, Tahmosybayat R, Caplan N. The immediate effects of exercise using the functional re-adaptive exercise device on lumbopelvic kinematics in people with and without low back pain. Musculoskeletal Science and Practice. 2017;**27**:S47-S53

[65] O'Sullivan PB, Dankaerts W, Burnett AF, Farrell GT, Jefford E, Naylor CS, et al. Effect of different upright sitting postures on spinal pelvic curvature and trunk muscle activation in a pain-free population. Spine. 2006;**31**(19):707-712

Chapter 9

Tumor Cells in Microgravity

Jun Chen

Additional information is available at the end of the chapter

http://dx.doi.org/10.5772/intechopen.77214

Abstract

The excessive proliferation and metastasis of tumor cells are due to frequent genetic alterations and subsequent stimulation of abnormal signal transduction pathways. Inventing and improving novel therapeutic strategies are critically needed. However, it remains unknown which of these pathways is essential to tumor initiation and progression. A weightless environment on Earth is a rare phenomenon, achieved using various simulations, but brings about changes of internal cellular structure and interactions among cells not normally seen under normal terrestrial gravitational conditions. For this reason, spaceflight experiments are of great value for cell biology research in general and for cancer research in particular. Many experiments indicate that microgravity, more so actual spaceflight as opposed to simulations, induces changes in the expression and secretion of genes as well as proteins involved in cancer cell proliferation, metastasis, and survival, shifting the cells toward a less aggressive phenotype. Therefore, studies on the biological features and gene expression of tumors cells under microgravity conditions may underline new clues to the tumor initiation, process, diagnosis, and therapy.

Keywords: space, microgravity, morphology, apoptosis, migration, tumor cells

1. Introduction

In the past 40 years, the development of the space industry has made people aware of the effects of microgravity on biological life, including cerebrospinal fluid flow change, body fluid electrolyte loss, muscle atrophy, bone demineralization, and immune system function decline [1]. Similarly, microgravity has been shown to alter some properties of cells, including cell morphology, function, and the cellular response to the environment. Observations from cells in the space environment provide inspiration for our research, in particular, the use of microgravity simulation studies for growth of cancer cells.

The interest in cellular response to microgravity started early in the 1970s, aboard the US Skylab. Research on Skylab included studying the response of humans as well as their cells to microgravity exposure. Similar research continued on future missions and space stations. However, like all astronaut, biological and physiological system research in space, including studying cells and their growth in space, requires precious resources and astronauts' time, which are scarce and in short supply. As a result, ground analogs were developed to test different physiological systems in simulated microgravity, so space agencies developed ways to replicate the microgravity environment in order to culture cells on Earth.

In 1995, Dr. Jessup supplied and successfully cultured cancer cells aboard STS-70 and a few years later aboard STS-80 using the bioreactor to confirm his results. Jessup and his group showed that their flown samples of colon carcinoma cells grew bigger and aggregated better than ground controls, 30 times the volume of the ground controls, and were more representative of cancer seen when growing in the body. These initial studies showed that microgravity was an environment that is favorable to cell growth and differentiation in addition to being more representative of in vivo growth [2].

Earth-based research groups have used the rotating wall vessel (RWV) bioreactor to support the development of several models of cancer cell lines. These include models of breast cancer, cervical cancer, colon cancer, hepatocellular carcinoma, neuroblastoma, melanoma, ovarian cancer, and prostate cancer. The bioreactor was not only a technological innovation, but was also a unique system benefiting biomedical research by allowing cells to be cultured in a 3-D environment on Earth [3]. Other systems have been developed in an attempt to simulate microgravity, such as the random positioning machine (RPM) and magnetic levitation.

Through ground-based simulations and spaceflight research, we know that simulated microgravity has been emerging as a new tool to develop potential therapy for tumor treatment [4]. With the developments from this research, more interest is being paid to the effects of microgravity on tumor cells. Increasing number of investigations has indicated that microgravity has evident effects on the morphology, proliferation, apoptosis, invasion, and migration, along with inhibiting cancerous cell growth and invasion. However, the details of the exact mechanism still elude us and are still being studied.

2. Changes of morphology

It has been found that after microgravity intervention, mesenchymal stem cell morphology changed significantly from spindle to round. "Spirit composing body," shape change will inevitably lead to changes in their function. Under the guidance of the traditional Chinese medicine "Circumference philosophical," the cells with round change are primitive state. The round bone marrow mesenchymal stem cells have greater differentiation potential in microgravity [5] (**Figure 1**). This study uses ground rotary cultivating device (2-D-clinostat) simulation of cells in space (**Figure 2**), using 1G gravity (normal gravity, NG) as a control with simulated microgravity (SMG) intervention of bone marrow mesenchymal stem cells to observe the differentiation into neural cells and endothelial cells. At the same time, cytoskeleton as well as

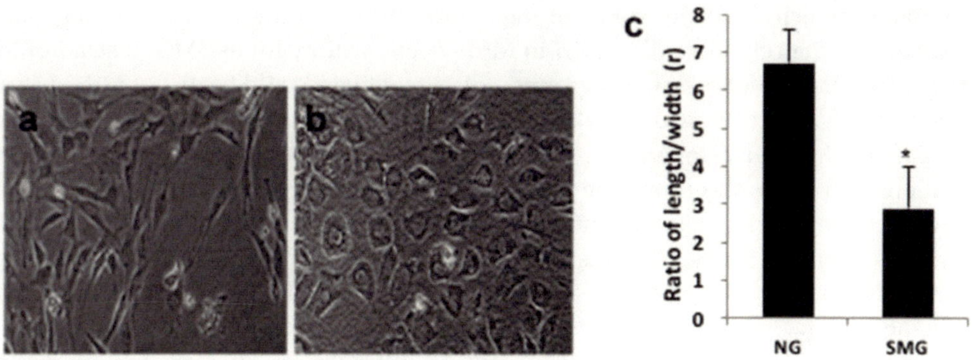

Figure 1. Phase-contrast microscopic analysis of the effect of SMG (b) on the morphology of BMSCs compared to NG (a). (c) The changes of the ratio of the width/length of BMSCs in the different groups. * denotes $P < 0.05$, versus the NG group.

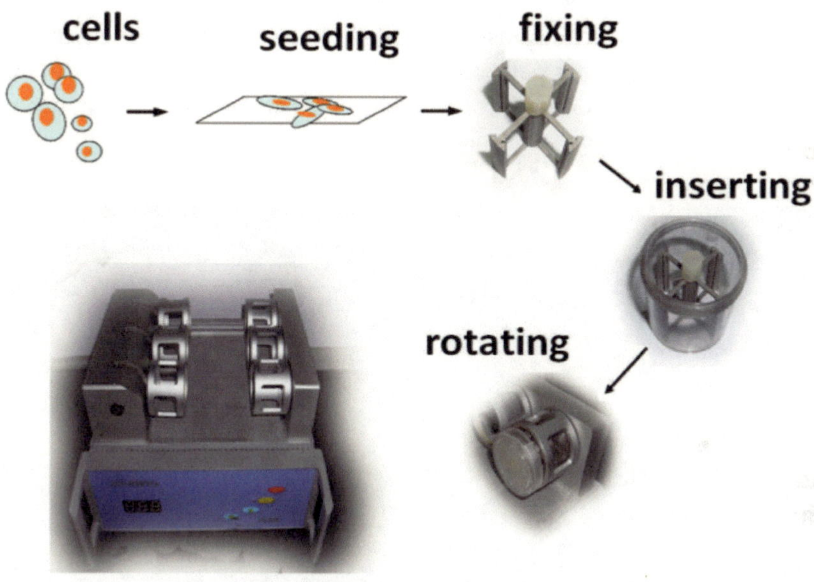

Figure 2. A clinostat or rotating vessel was used to form the weightless environment to prevent the cell from feeling gravity; the machines are based on the theory that sensing no gravity would have similar effects as being microgravity. The clinostat model system (clinorotation) used in this study tends to provide an average of zero vector in the apparent gravity on the cell culture.

the key molecule of RhoA activity is observed, thereby directing the possible differentiation mechanism of stem cells.

Other than stem cells, real and simulated microgravity induces early alterations of the cytoskeleton in different kinds of human cells, such as thyroid cancer cells [6, 7], endothelial cells [8], and glial cells [9].

Our previous works have concluded that modeled microgravity causes changes in the cytoskeleton and focal adhesions in malignant human MCF-7 cells [10].

Microtubules, which are components of the cytoskeleton that are involved in maintaining the structure of the cell, were disrupted in MCF-7 cells within 4 h of SMG. Instead of long, strongly labeled microtubules in NG groups, only a few fibers could be distinguished against the strong background (**Figure 3**). Self-organization of microtubules into stationary macroscopic patterns is gravity-dependent, and the patterns correspond to different microtubule orientations [11]. Also, the formation of radial microtubule arrays depends mostly on the activity of centrosome. Lewis et al. reported that microtubule organizing centers (MTOCs) were poorly defined in SMG [12]. It is obvious that the anchoring between microtubule and the centrosome is very complicated; so, many studies targeted the function of cytoplasmic dynein and its cofactor dynactin. Dynein transfers some centrosomal proteins to the centrosome and reorganizes radial microtubule arrays for cell division. Thus, we hypothesize that the disorganization of microtubule fibers and their reestablishment might be associated with MTOCs modification.

The cytoskeleton is made up of actin microfilaments, tubulin microtubules, and vimentin intermediate filaments [13]. The F-actin cytoskeleton has been involved in changes in cell morphosis and function as well as signaling path under weightless conditions. It was revealed that the amount of F-actin in A431 epidermoid carcinoma cells increased under real microgravity for

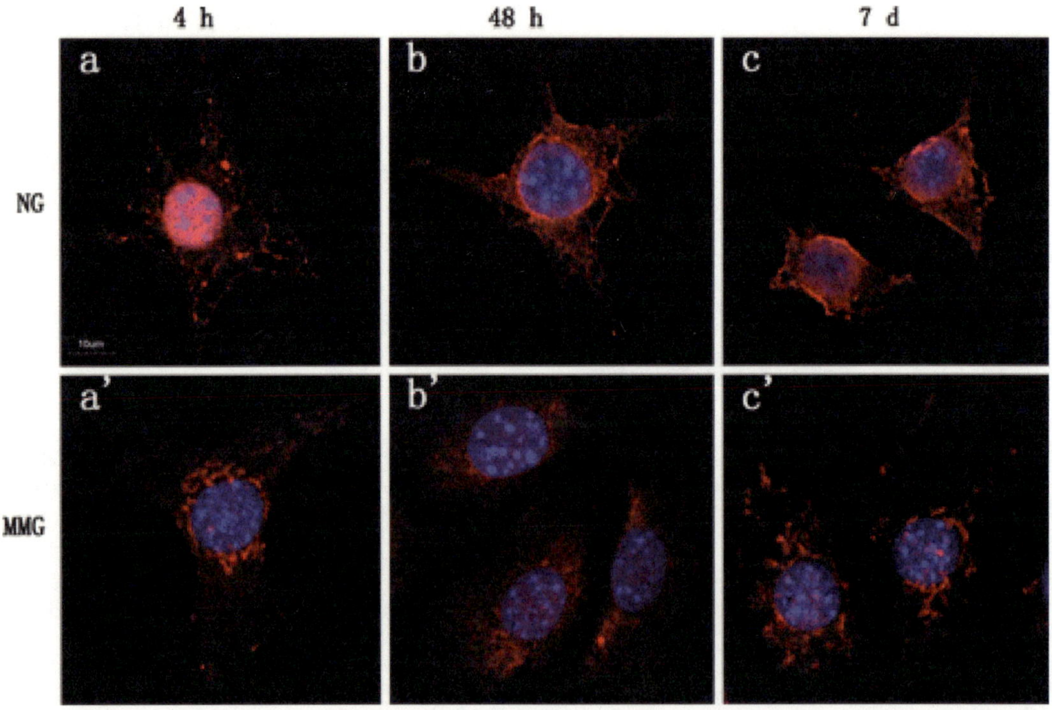

Figure 3. Microtubules formation of MCF-7 cells is changed in MMG. Scale bar is 10 μm. Visualization of β-tubulin (TRITC; red) and nuclear chromatin (Hochest; blue). It is shown that microtubules keep radiation from the perinuclear area toward the cell periphery in control cells as time varying. (a–c) (4 h, 48 h, and 7 days, respectively). While in the MMG groups, altered fibers and short microtubules filaments could be observed. (a'–c') (4 h, 48 h, and 7 days, respectively).

7 min [14]. This leads us to conclude that the actin microfilament structure is sensitive to gravity and that rebuilding of cytoskeleton may affect signal transduction [14]. Thyroid cancer cells were flown in a parabolic flight mission and they detected early alterations in the actin microfilaments. After 22-s microgravity stimulation, F-actin changed significantly, and the human beta actin (ACTB) expression was strikingly upregulated after the 1st and 31st parabolas [15].

Actin monomers polymerize only onto the existing barbed ends of the actin array, so that the actin filaments can be elongated to the cell periphery. In contrast to NG samples, microfilaments did not form preferential orientation with less labeled lamellipodia. Cell microfilament bundles are assembled primarily by bundling of preexisting actin filaments [16]. It has been reported that tension development in the preexisting actin cytoskeleton is critical for the formation of stress fibers [17]. Tension would act on mechanical connections of the actin filaments to reorganize a kind of meshwork. However, spaceflight or stimulated weightlessness would destroy the intracellular prestress and tension balance, and eventually cells show irregular formation of cytoskeletal actin filaments [18].

3. Changes of cell growth, cycle, and apoptosis

Glioma is the most common and aggressive form of cancer of the central nervous system with a median survival time of 15 months and a 5-year survival about 5% after initial diagnosis [19]. Despite the standard of therapy available, including maximal safe surgical resection, radiotherapy, and temozolomide (a form of chemotherapy) [20], nearly all patients relapse. The excessive proliferation and metastasis of glioma is due to frequent genetic alterations and subsequent stimulation of abnormal signal transduction pathways [21, 22]. Inventing and improving novel therapeutic strategies are critically needed. However, it remains unknown which of these pathways is essential to glioma initiation and progression.

To assess the effect of SMG on glioma growth, U251 cells were cultured under either ground condition or SMG (SM-31 random locator applied) for different time periods (0, 12, 16, 20, 24, 36, 48, 72, and 96 h). CCK8 measurements, a chemical kit employed to measure cell proliferation, showed that SMG inhibited U251 cells activity in a time-dependent manner. The more time handled, the less activity the U251 cells had. For 48~96 h, SMG markedly induced cell death of U251, cell activity decreased maximum about 45%. Therefore, we chose 72-h timepoint to use in the following experiments. This was further confirmed by our data derived from fluorescence-activated cell sorter (FACS) analysis via annexin V-FITC and PI double staining to investigate apoptosis of U251 cells exposed to SMG for 72 h, which showed 2.4% of control cells stained positive for both annexin V and PI, representing a minor subpopulation undergoing a spontaneous apoptosis. After being cultured in SMG for 72 h, apoptotic subpopulation increased to 29.3%, indicating that SMG promoted cell death via apoptosis. Western blot (WB) test confirmed that apoptotic promoting proteins cleaved-caspase 3 and cleaved-caspase 9 were markedly upregulated at 48 h by SMG compared with NG condition. Meanwhile, U251 cells' metabolic activity was detected by β-Gal Staining Set, blue-stained cells significantly increased by SMG, about 30% increase compared with NG condition. FACS analysis further demonstrated that SMG treatment led to 14.11 ± 1.73% of U251 cells arrested at G2/M phase (**Figure 4**).

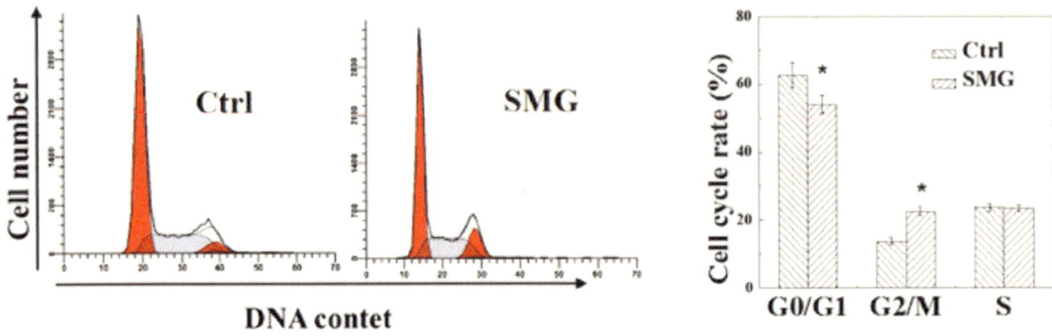

Figure 4. Cell cycle was analyzed by flow cytometry after PI staining. The bar graph represents the number of the cells in different phases.

Caspases are a family of cysteine proteases. Most caspases play a central role in activating cell apoptosis by cleaving selected target substrates in a cysteine-dependent aspartate-directed manner. To further dissect molecular mechanisms of SMG-induced glioma apoptosis, WB analysis was performed to confirm the level of apoptotic protein, Bcl2, Bnip3, pro-caspase 3, cleaved-caspase 3, pro-caspase 9, and cleaved-caspase 9. Zhao et al. found that SMG promotes BL6-10 melanoma cell apoptosis through downregulating Bcl2 and Bnip3 and upregulating caspases 3/7/8 [23]. We showed that two antiapoptotic proteins Bcl2 and Bnip3 were down-regulated, while cleaved-caspase 3 and cleaved-caspase 9 were upregulated in cells under microgravity conditions, respectively, leading to enhance glioma cell death.

Never in mitosis gene A (NIMA)-related kinase 2 (Nek2) is one of multiple cell cycle-regulated protein kinases that localizes to the centrosome and is required for mitotic progression and correct bipolar spindle formation [24]. Upregulated Nek2, which is confirmed in a number of neoplastic diseases, including prostate cancer, lung cancers, colorectal cancer [25], and myeloma, also exhibits adverse correlation with overall survival of multiple malignancies [26]. All of this suggests that Nek2 may regulate the metastasis of glioma, although few reports have shown this. In our unpublished study, Nek2 may play a crucial function in SMG-treated U251cells, which arrested at the mitotic phase (chromosome separation).

Nek2 is implicated in centrosome separation and is reported to displace linker proteins from centrosomes through phosphorylation at the beginning of mitosis, and Nek2 phosphorylates the centrosomal linker proteins C-Nap1 and rootletin resulting in linker dissociation. hSav1-Mst2-Nek2 centrosome disjunction pathway becomes essential for bipolar spindle formation, Nek2 kinase to regulate centrosome disjunction. We found SMG suppressed expression of Nek2 and distances between two centrioles by γ-tubulin staining. The experimental results were consistent as Di Agostino et al. [27].

4. Changes of migration

The formation of a protrusion initiates the cell migration cycle process, but the protrusions need to be stabilized to the substratum so as to move forward. The process of continuous coordinated formation and disassembly of adhesions is crucial for migration. These adhesion sites, named as FAs, serve as traction points to propel the cell forward. Therefore, the important mechanism for creating cell movement includes the maturation of FAs. FAs become mature during the binding and clustering of integrins and function physically as a link of the cytoskeleton to the extracellular matrix (ECM) [28]. This correlation between cytoskeleton and FAs in structure underlines the possibility that FAs formation might be changed during microgravity. In order to explore the possible mechanism, we examined FAs formation by vinculin immunofluorescence. Vinculin is one of the most prevalent elements in FAs. It seems that vinculin facilitates the assembly of FAs by interacting and recruiting its various partners [29]. We demonstrated that microgravity disorganized FAs of MCF-7 cells via quantifying FAs parameters. After 7 days stimulation of SMG, the amount of FAs was still low by detecting topographical adhesion parameters and showed no change as time varied. We therefore could conclude that FAs created in microgravity were less mature than those established in normal gravity. Fewer and smaller FAs can lead to the weaker cell spreading and migrating. Thus, we speculated that abnormal FAs structure in MCF-7 cells under SMG may contribute to the change of cell migration [10].

Since FAs also contain a lot of growth factor receptors, kinases, and signaling proteins, FAs have been referred as localized sites converging growth factor and adhesion signaling. The integrin family is the major transmembrane ECM receptors in these sites. FAK, PYK2, and ILK are well known as the key effectors in FAs signaling and a potential integrator of inhibition of MCF-7 cell migration by modeled microgravity. Because the changes of morphological and topographical cytoskeletal structures of FAs were observed in MCF-7 cells, we speculated that activities of the kinases would be altered under clinorotation conditions. Here, our hypothesis was confirmed by observing a decrease in FAs kinases phosphorylation level (FAK, PYK2, and ILK) in contrast to NG controls, while there was no significant change in total FAK, ILK, and PYK2 protein expression in both the NG controls and SMG groups. The effects of microgravity on the suppression of FAs kinases activity in our experiments were in accordance with other reports, though they were observed in different cells by different devices, which could suggest that SMG might suppress the FAs kinases activity in various kinds of cells [30].

As we know, integrins play a critical role in cell adhesion and migration. We explored that SMG could decrease the expression of integrins, a downregulation of integrins (integrinβ1 and integrinβ4) at both protein and mRNA levels after SMG compared to controls. Downregulated integrins in SMG are not a new finding, and it has been reported that SMG suppressed integrins (αv, α5,β1) expression in MG-63 cell. Via binding to the ECM, phosphorylation of signaling proteins at FAs was triggered, such as FAK, ILK, and PYK, whereby protein kinase C (PKC) and GTPases pathways were activated. The small G proteins, Rho family regulate the rebuilding of actin fibers and FAs formation, by which determines cell movement. In fact, the previous studies have reported that rho activation in the endothelial cells and MSCs could be affected by simulated microgravity. Also, it has been found repeatedly that microgravity modulates PKC signaling in neurons. Activation of PKC could directly induce cells' motility and migration. Furthermore, it is generally accepted that the releasing from the RhoGDI-1 molecule is required for activation of RhoGTPases. PKCα is confirmed to phosphorylate RhoGDI-1 and then catalyze the release of bound GTPases [31]. No doubt, the pathways regulating cell migration are very complicated and seem to vary in different cells and different species. In general, it seems that decreased integrin expression and downregulated FAs kinase activity are an essential step in suppression of cell migration by SMG.

5. Changes of genes

To investigate the involved mechanisms in more detail, gene level researches are expected to yield novel targets for cancer therapy, which may then be exploited in the form of new chemotherapeutics.

Dr. Xiao Ma cultured thyroid cancer cells in space (Shenzhou 8 space mission) on a random positioning machine (RPM) for 10 days to evaluate differences between real and SMG. About 2881 genes were regulated during 10 days of cell exposure to microgravity [32]. These genes were subdivided into different clusters (**Figure 5**) that allowed us to distinguish the difference between SMG (RPM) and spaceflight effects. However, two clusters of genes expressed similarly under either real or SMG. This research demonstrated that the effects of RPM and spaceflight both exert 3-D growth, but may not change gene expression in the same direction.

For the sake of the very limited amount of cells returned from space, the combination of gene expression and secretion analysis was detected on the cytokine, protease, and kinases factors that may play a pivotal role in the development of metastases. The list was made up of IL6, IL8, IL15 (interleukin family), OPN (osteopontin), VEGFA, VEGFD (vascular endothelial growth factor), and FGF17 (fibroblast growth factor).

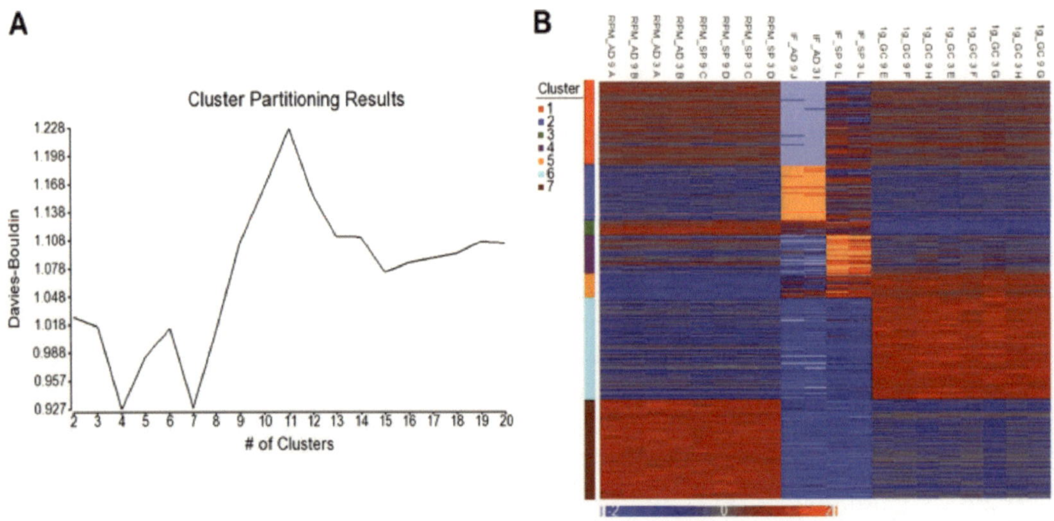

Figure 5. (A) Gene array analyses on FTC-133 cells cultured for 10 days under both conditions. ANOVA resulted in 2881 twofold-regulated probes. The Davies-Bouldin cluster estimation resulted in a local minimum of seven clusters. (B) RPM-specific effects (clusters 5 and 7), spaceflight-specific AD cell and MCS effects (clusters 1, 2, and 4), and general gravity effects (clusters 3 and 6) were shown in gene map. An outstanding expression level of spaceflight samples was found in contrast to the moderate levels in RPM samples and controls.

In this research, VEGFA and VEGFD, which are members of the vascular endothelial growth factor (VEGF) family, were also observed. It is well known that VEGFA is the most important growth factor driving angiogenesis and is also implicated in many processes like tumor vascularization, wound healing, and burn injury. VEGFA expression has been found to increase in papillary thyroid carcinoma in contrast to that in healthy patients [33]. While on the other hand, VEGFD serum levels were found to decrease in patients with thyroid cancer [34]. In this study, an obvious decrease in expression of VEGFA was confirmed in the RPM as well as the spaceflight samples, while no difference was observed between adherent (AD) cells or multicellular spheroids (MCSs). Similarly, VEGFA secretion was also reduced on the RPM, but no influence in real microgravity. Furthermore, VEGFD gene expression upregulated in cells cultivated on the RPM as well as in space. All these results seem to get a conclusion that microgravity exerts a shift of the thyroid tumor cells to a more benign, less metastatic phenotype.

Interestingly, the similar tendency was observed with the interleukins in microgravity. IL6 and IL8, encoded by the IL6 and IL8 genes, are involved in tumor cell growth, metastasis, and angiogenesis. IL15, on the contrary, is able to activate several antitumor mechanisms, such as activating CD8 T cells to kill tumor cells [35]. Furthermore, it has been found that IL15 was effective in several tumor therapy experiments [36]. Their results explored a significant downregulation of IL8 gene expression in both simulated and real microgravity cells, which is also in accordance with the IL8 secretion pattern. The weightless effect of reducing IL6 gene expression was only observed in adherent cells, the same to that described for endothelial cells [37].

Of note, both IL6 and IL8 gene expression were strikingly upregulated during a parabolic flight. The study also showed earlier that IL6 and IL8 were most possibly implicated in the gravity-sensitive signaling required for spheroid formation. It seems that this signal pathway is attenuated after exposure to microgravity and that the tumor cells are shifted toward a less aggressive biological behavior. This hypothesis was also confirmed by the observations for IL15. IL15 gene expression increased only slightly on the RPM and during parabolic flight, but they explored a strong increase in the MCSs during spaceflight. Thus, we speculate that microgravity triggers some antitumor pathways involving IL15 and, as MCSs resemble tumors in their 3-D structure, the cell-cell interactions should be affected.

6. Conclusion

Taken together, it is no doubt that spaceflight experiments are of great value for cell biology research, especially for cancer research. The previous studies indicate that microgravity, both actual and simulated microgravity, induces changes in cancer cell proliferation, metastasis, and survival, bringing the cells toward a less aggressive phenotype. This effect is greater in actual spaceflight; however, ground-based simulation has been shown to be an essential tool in our understanding of gravity and it effects on a cellular level in cancer cells. The above body shows that not only does the architecture of tumor cells change in microgravity, but the cell function and gene expression also are different. The changes of some gene expression reorganize the cytoskeleton which influences the cell growth, migration, and apoptosis. And some other genes impress tumor cell by regulating the immune antitumor pathway.

However, most of the studies are in vitro, and the evidence from human or animals is much rare. Is the microgravity strategy safe for human? How to create microgravity environment in body? Those are the thorny tissue for clinical translation. With the recent discoveries of nanomagnetic fluids, it suggests an innovative method of treating tumors using magnetic fluid-modeled microgravity. Magnetic fluids are delivered by outside magnetic field to tumor tissue either intravenously or through direct injection, and this is followed by application of a uniform external magnetic field that causes microgravity. The concept of magnetic fluid-modeled microgravity to treat tumor is novel, and the technology involved is simple, economical, and might be suitable for clinical applications in future.

All in all, as new information about the biology of cancer emerges, treatments will be developed and modified to increase effectiveness, precision, survivability, and quality of life. It is still a long shot to take advantage of microgravity as a suitable way to treat tumor.

Acknowledgements

The present chapter was funded by the National Natural Science Foundation of China (grant no. 81402055), and Youth Science and Technology Innovation Personnel Training Project of Shaanxi Province (grant no.2017KJXX-84).

Author details

Jun Chen

Address all correspondence to: chj2002819@163.com

Department of Encephalopathy, Traditional Chinese Medicine Hospital of Shaanxi Province, Xi'an, Shaanxi, China

References

[1] Sonnenfeld G. Space flight, microgravity, stress and immune responses. Advances in Space Research. 1999;**23**:1945-1953

[2] Jessup JM, Goodwin TJ, Garcia R, Pellis N. STS-70: First flight of EDU-1. In Vitro Cellular & Developmental Biology. 1996;**32**:13A

[3] Hjelm BE, Berta AN, Nickerson CA, Arntzen CJ, Herbst-Kralovetz MM. Development and characterization of a three-dimensional organotypic human vaginal epithelial cell model. Biology of Reproduction. 2009;**82**:617-627

[4] Jun C, Zhiqiang Y, Rongrong L, Nanding W, Jing L, Zongren W. Magnetic fluid-modeled microgravity: A novel way to treat tumor. Medical Hypotheses. 2011;**77**:953-955. DOI: 10.1016/j.mehy.2011.07.053

[5] Wang ND, Wang H, Chen J, Zhang XF, Xie J, Ma J, Wang W, Wang ZR. The simulated microgravity enhances multipotential differentiation capacity of bone marrow mesenchymal stem cells. Cytotechnology. 2014;**66**:119-131. DOI: 10.1007/s10616-013-9544-8

[6] Grimm D, Kossmehl P, Shakibaei M, Schulze-Tanzil G, Pickenhahn H. Effects of simulated microgravity on thyroid carcinoma cells. Journal of Gravitational Physiology. 2002;**9**:253-256. DOI: 10.1096/fj.01-0673fje

[7] Grimm D, Bauer J, Kossmehl P, Shakibaei M, Scho ̈nberger J. Simulated microgravity alters differentiation and increases apoptosis in human follicular thyroid carcinoma cells. The FASEB Journal. 2002;**16**:604-606. DOI: 10.1096/fj.01-0673fje

[8] Grimm D, Bauer J, Ulbrich C, Westphal K, Wehland M. Different responsiveness of endothelial cells to vascular endothelial growth factor and basic fibroblast growth factor added to culture media under gravity and simulated microgravity. Tissue Engineering. Part A. 2010;**16**:1559-1573. DOI: 10.1089/ten

[9] Uva BM, Masini MA, Sturla M, Prato P, Passalacqua M. Clinorotation-induced weightlessness influences the cytoskeleton of glial cells in culture. Brain Research. 2002;**934**:132-139. DOI: 10.1016/s0006-8993(02)02415-0

[10] Li J, Zhang S, Chen J, Du TY, Wang YC, Wang ZR. Modeled microgravity causes changes in the cytoskeleton and focal adhesions, and decreases in migration in malignant human MCF-7 cells. Protoplasma. 2009;**238**:23-33. DOI: 10.1007/s00709-009-0068-1

[11] Papaseit C, Pochon N, Tabony J. Microtubule self-organization is gravity-dependent. Proceedings of the National Academy of Sciences. 2000;**97**:8364-8368. DOI: 10.1073/pnas.140029597

[12] Lewis ML, Reynolds JL, Cubano LA, Hatton JP, Lawless BD, Piepmeier EH. Spaceflight alters microtubules and increases apoptosis in human lymphocytes (Jurkat). The FASEB Journal. 1998;**12**:1007-1018

[13] Benjamin M, Archer CW, Ralphs JR. Cytoskeleton of cartilage cells. Microscopy Research and Technique. 1994;**28**:372-377

[14] Boonstra J. Growth factor-induced signal transduction in adherent mammalian cells is sensitive to gravity. The FASEB Journal. 1999;**13**:35-42

[15] Ulbrich C, Pietsch J, Grosse J, Wehland M, Schulz H. Differential gene regulation under altered gravity conditions in follicular thyroid cancer cells: Relationship between the extracellular matrix and the cytoskeleton. Cellular Physiology and Biochemistry. 2011;**28**:185-198. DOI: 10.1159/000331730

[16] Hirata H, Tatsumi H, Sokabe M. Dynamics of actin filaments during tension-dependent formation of actin bundles. Biochimica et Biophysica Acta. 2007;**1770**:1115-1127. DOI: 10.1016/j.bbagen.2007.03.010

[17] Kolega J. Effects of mechanical tension on protrusive activity and microfilament and intermediate filament organization in an epidermal epithelium moving in culture. The Journal of Cell Biology. 1986;**102**:1400-1411

[18] Ingber DE. Tensegrity-based mechanosensing from macro to micro. Progress in Biophysics and Molecular Biology. 2008;**97**:163-179. DOI: 10.1016/j.pbiomolbio.2008.02.005

[19] Ostrom QT, Gittleman H, Fulop J. CBTRUS statistical report: Primary brain and central nervous system tumors diagnosed in the United States in 2008-2012. Neuro-Oncology. 2015;**17**:1-62. DOI: 10.1093/neuonc/nov189

[20] Stupp R, Mason WP, van den Bent MJ, et al. Radiotherapy plus concomitant and adjuvant temozolomide for glioblastoma. The New England Journal of Medicine. 2005;**352**:987-996

[21] Dong Z, Lei Q, Yang R, Zhu S, Ke XX, Yang L, Cui H, Yi L. Inhibition of neurotensin receptor 1 induces intrinsic apoptosis via let-7a-3p/Bcl-w axis in glioblastoma. British Journal of Cancer. 2017;**116**:1572-1584. DOI: 10.1038/bjc.2017.126

[22] Sturm D, Bender S, Jones DT, Lichter P, Grill J, Becher O, Hawkins C, Majewski J, Jones C, Costello JF, Iavarone A, Aldape K, Brennan CW, Jabado N, Pfister SM. Paediatric and adult glioblastoma: Multiform (epi)genomic culprits emerge. Nature Reviews. Cancer. 2014;**14**:92-107. DOI: 10.1038/nrc3655

[23] Zhao T, Tang X, Umeshappa CS, Ma H, Gao H, Deng Y, Freywald A, Xiang J. Simulated microgravity promotes cell apoptosis through suppressing uev1a/ticam/traf/nf-κb-regulated anti-apoptosis and p53/PCNA- and ATM/ATR-Chk1/2-controlled DNA-damage response pathways. Journal of Cellular Biochemistry. 2016;**117**:2138-2148. DOI: 10.1002/jcb.25520

[24] Bayliss R, Fry A, Haq T, Yeoh S. On the molecular mechanisms of mitotic kinase activation. Open Biology. 2012;**2**:120136. DOI: 10.1098/rsob.120136

[25] Xia J, Franqui Machin R, Gu Z, Zhan F. Role of Nek2A in human cancer and its therapeutic potentials. BioMed Research International. 2015;**2015**:862461. DOI: 10.1155/2015/862461

[26] Fang Y, Zhang X. Targeting Nek2 as a promising therapeutic approach for cancer treatment. Cell Cycle. 2016;**15**:895-907. DOI: 10.1080/15384101.2016.1152430

[27] Di Agostino S, Fedele M, Chieffi P, Fusco A, Rossi P, Geremia R, Sette C. Phosphorylation of high-mobility group protein A2 by Nek2 kinase during the first meiotic division in mouse spermatocytes. Molecular Biology of the Cell. 2004;**15**:1224-1232. DOI: 10.1091/mbc.E03-09-0638

[28] Zamir E, Geiger B. Molecular complexity and dynamics of cell-matrix adhesions. Journal of Cell Science. 2001;**114**:3583-3590

[29] Cox EA, Huttenlocher A. Regulation of integrin-mediated adhesion during cell migration. Microscopy Research and Technique. 1998;**43**:412-419

[30] Meyers VE, Zayzafoon M, Gonda SR, Gathings WE, McDonald JM. Modeled microgravity disrupts collagen I/integrin signaling during osteoblastic differentiation of human mesenchymal stem cells. Journal of Cellular Biochemistry. 2004;**93**:697-707. DOI: 10.1002/jcb.20229

[31] Mehta D, Rahman A, Malik AB. Protein kinase C-alpha signals rho-guanine nucleotide dissociation inhibitor phosphorylation and rho activation and regulates the endothelial cell barrier function. The Journal of Biological Chemistry. 2001;**276**:22614-22620. DOI: 10.1074/jbc.M101927200

[32] Ma X, Pietsch J, Wehland M, Schulz H, Saar K, Hübner N, Bauer J, Braun M, Schwarzwälder A, Segerer J, Birlem M, Horn A, Hemmersbach R, Waßer K, Grosse J, Infanger M, Grimm D. Differential gene expression profile and altered cytokine secretion of thyroid cancer cells in space. The FASEB Journal. 2014;**28**:813-835. DOI: 10.1096/fj.13-243287

[33] Hsueh C, Lin JD, Wu IC, Chao TC, Yu JS, Liou MJ, Yeh CJ. Vascular endothelial growth factors and angiopoietins in presentations and prognosis of papillary thyroid carcinoma. Journal of Surgical Oncology. 2011;**103**:395-359. DOI: 10.1002/jso.21844

[34] Nersita R, Matrone A, Klain M, Scavuzzo F, Vitolo G, Abbondanza C, Carlino MV, Giacco V, Amato G, Carella C. Decreased serum vascular endothelial growth factor-D levels in metastatic patients with differentiated thyroid carcinoma. Clinical Endocrinology. 2012;**76**:142-146. DOI: 10.1111/j.1365-2265

[35] Barry M, Bleackley RC. Cytotoxic T lymphocytes: All roads lead to death. Nature Re views. Immunology. 2002;**2**:401-409. DOI: 10.1038/nri819

[36] Zhang M, Yao Z, Dubois S, Ju W, Müller JR, Waldmann TA. Interleukin-15 combined with an anti-CD40 antibody provides enhanced therapeutic efficacy for murine models of colon cancer. Proceedings of the National Academy of Sciences of the United States of America. 2009;**106**:7513-7518. DOI: 10.1073/pnas.0902637106

[37] Grenon SM, Jeanne M, Aguado-Zuniga J, Conte MS, Hughes-Fulford M. Effects of gravitational mechanical unloading in endothelial cells: Association between caveolins, inflammation and adhesion molecules. Scientific Reports. 2013;**3**:1494. DOI: 10.1038/srep01494

Chapter 10

Plants in Space

Bratislav Stankovic

Additional information is available at the end of the chapter

http://dx.doi.org/10.5772/intechopen.74230

Abstract

Plants will play a critical role in the survival of human beings on long-duration space missions, probably beginning pretty soon with a mission to Mars. Plants can adapt to extreme environments on Earth, and model plants have been shown to grow and develop through a full life cycle in microgravity. In space, long-term human space exploration missions require a life support system in which higher plants play a vital role. Growing crops in space is as much about developing the humans' technological capacity to provide plants with adequate growth conditions in the unique microgravity environment, as is about the symbiotic relationship between plants and space travelers. After several decades of research, we have learned a lot about the impediments to growing plants in microgravity, in outer space, and on other planets. As human space exploration advances, we should feel confident about our ability to grow plants on board spacecraft during long-term space missions, on the Moon, and on other planets. Plants will require specialized environments for growth and development in microgravity, but – at least on a small scale – we already know how to produce such growth chambers and greenhouses.

Keywords: space biology, gravitational biology, microgravity, plants, spaceflight, international Space Station

1. Introduction

The phrase "plants in space" refers to plants that are grown in the physical universe known as outer space, a region beyond the Kármán line in the Earth's atmosphere, at an altitude of approximately 200–450 km above sea level, which is the typical orbit range of the Space Shuttle missions and of the International Space Station, where most of human spaceflight and research has taken place [1]. Outer space represents a challenging environment for human exploration for a number of reasons, including the lethal hazards of extreme temperatures, high vacuum, electromagnetic radiation, particle radiation, and magnetism. A deep understanding of the biological consequences of exposure to the space environment is required to design efficient countermeasures to minimize their negative impact on living organisms, humans and plants alike. In addition, the economic cost of sending anything into space is very high. In outer space, plants are typically grown in a microgravity (often referred to as weightlessness) controlled environment, in specific space plant growth chambers.

Plant space biology has been closely associated with human space exploration in that plants are considered as key parts of biologically based life support. Learning to grow plants in space is an essential goal for long duration space missions since crop growth in space will be beneficial in a variety of ways, aiding with air regeneration, food production, and water recycling [2–6]. The logistical challenges of the long-term human space exploration missions require a self-sustainable life support system. Traveling in a spacecraft to other worlds will put constraints on the quantity and weight of commodities that could be brought along. In that context, higher plants are of paramount importance for providing *in situ* resource utilization through a continuous supply of fresh food, atmosphere revitalization, and clean water for humans.

The many challenges of spaceflight research have logistical and resource constraints, including significant limitations on available space, power, crew time, cold stowage, and data downlinks. Additional issues are related to hardware development, safety concerns, and the engineering versus science culture in space agencies. There is not much space for growing plants in space. Concerning research, the difficulties of publishing the results from spaceflight research stem from the lack of adequate controls, limited sample size, the frequent impossibility for verification of the obtained data, and the indirect effects of the spaceflight environment.

The concept of growing crops in space is as much about developing the humans' technological capacity to provide plants with adequate growth conditions in the unique microgravity environment, as it is about the symbiotic relationship between plants and space travelers. Plants in space provide numerous benefits to the humans that accompany them. They improve the quality of indoor air by helping control humidity levels, and by removing and converting the carbon dioxide from air into essential oxygen that humans can breathe. Central to the concept of regenerative life support systems for space exploration is the use of photosynthetic organisms and light to generate oxygen and food. It is also axiomatic that plants can be consumed as food, providing a nutritive value to organisms throughout the food chain. Growing plants in space may also provide psychological and neurocognitive benefits to the human spaceflight crews, in the form of therapeutic people-plant interactions [7].

2. Fundamentals

Humans began the physical exploration of space during the 20th century with the advent of high-altitude balloon flights, followed by rocket launches. Plants have been used in space experiments since the early days of the space program. The early suborbital launches saw the first organisms in space, as unspecified "specially developed strains of seeds", sent to 134 km altitude above Earth on July 9, 1946, on a U.S. launched V-2 rocket; these samples were not recovered. The first seeds launched into space and successfully recovered were "ordinary corn seeds", launched on July 30, 1946; see [8] for a descriptive chronology of the early biological experiments in rockets. These early and very brief biological experiments were primarily concerned with the effects of radiation exposure on living tissue, including seeds. Some years later, the first plant materials taken into a microgravity environment for a longer ride in orbit were seeds of wheat, pea, maize, and onion, flown on board of Sputnik 4 in 1960 [2]. The first full life cycle of a plant (*Arabidopsis thaliana*) in space was completed on Salyut-7, resulting in clearly observable developmental alterations and in some viable seed, but mostly in seed having nonviable embryos [9].

Opportunities for space experiments greatly increased by the initiation of scientific operations in orbital laboratories, the latest being the International Space Station (ISS). Building upon accumulated knowledge, researchers took advantage of well-developed plant growth chambers for microgravity, which in general provided a very good environment for growing plants on the ISS [10]. These proved that it is possible to have plants pass the full cycle of ontogenesis in space (on the ISS), and to produce plants and viable seed similar to the ground controls. The first example of seed-to-seed-to-seed (i.e., two consecutive life cycles) of a plant (*Arabidopsis thaliana*) in space was completed in 2000–2001 [11]. With advanced plant growth chambers that in general provided a well-regulated environment for growing plants in microgravity on the ISS, most of the problems seen in previous plant spaceflight experiments were successfully eliminated. It turns out that gravity is not necessary for seed-to-seed growth of plants, though it plays a direct role in plant form, and may influence seed reserves [11].

The effort and resources allocated to plant cultivation in space have revealed many answers, while at the same time raising new research questions. Periodic literature updates on the status of plant space biology have reviewed the documented influence of gravity on both plant growth and development, and specifically on a myriad of cellular and molecular responses, including cell cycle, embryogenesis and seed development, photosynthesis and gas exchange, gravitropic sensing and response, phototropism, cell wall development, and gene expression changes [12–15]. More recent and also more sophisticated plant experiments during the Space Shuttle and the ISS era produced key science insights on the molecular and cellular mechanisms underlying biological adaptation to spaceflight, and especially to plant growth, development, tropisms, and stress responses in microgravity [16–19].

3. Particulars

The first experiments with higher plants grown in space were intended to assess whether plants could grow outside Earth and to determine what differences there were between spaceflight-grown and Earth-grown plants. As plant-growth hardware started to adapt to spaceflight, opportunities were created for more sophisticated plant experiments. Direct microgravity effects started being differentiated from confinement effects, and Earth orbit became a laboratory where plants could be grown without the influence of Earth gravity.

The physiological effects of gravity range from subtle to substantial, and influence numerous molecular and cellular events in addition to those solely associated with gravitropism. Many of the early plant space biology experiments resulted in morphological and physiological changes, manifested as cellular and phenotypic abnormalities. These include chromosomal breakage [20], failure to produce seed [21], altered or nonviable embryos [9], alterations in cell wall composition and properties [22], increased breakdown of xyloglucans [23], changes in polar auxin transport [24], or other morphological abnormalities [25]. Indeed, spaceflight appears to initiate both molecular and cellular remodeling throughout the plant. For example, spaceflight can induce significant genomic and epigenomic mutations [26]. In the absence of gravity plants rely on other environmental cues to initiate the morphological responses essential to successful growth and development, and the basis for that engagement lies in the differential expression of genes in an organ-specific manner [11, 16, 27], which is followed by a microgravity-driven remodeling of the proteome [28].

Reflecting on the early spaceflight experiments, we now know that a number of the early obtained results were more likely due to the rigors of the microgravity environment than to the lack of gravity itself. For example, altered starch content has been reported for different species of space grown plants: pepper [29], lepidium [30], maize [31], and Arabidopsis [32]. However, just improving plant ventilation during space flight was found to eliminate carbohydrate differences [33]. In addition, ethylene, a plant stress hormone, is a common problem in microgravity experiments. Plant ethylene production increases in space [34]. Elevated ethylene levels (1100–1600 ppb on a Shuttle) caused anomalous seedling growth of Arabidopsis in spaceflight studies, although they had no effect on relative graviresponsiveness [35]. Furthermore, ethylene levels on the MIR space station were very high (800–1200 ppb) during a Brassica spaceflight growth study [36]. While Brassica plants were capable of producing seed at this ethylene level, the same environment stopped a wheat crop from producing seed on board MIR [37]. Novel plant growth spaceflight hardware uses ethylene scrubbers to mitigate the negative effects of elevated ethylene levels in spacecraft [10, 11].

The absence of natural convection in space makes it easy for plants to become oxygen starved [38]. Hypoxia symptoms in seed include reduction in size of the protein bodies, failure of the protein bodies to fill, free floating lipid droplets in the cytoplasm, abnormally vacuolated cells, and degeneration of portions of the embryo. In a full life-cycle spaceflight experiment with Brassica, the protein bodies that were found to be 44% smaller, starch grains were aberrantly deposited in the seed, and the cotyledon cell number was reduced by 80% [36]. This study concluded that alterations in the oxygen and ethylene concentrations within developing siliques were problematic in the experiment [36]. While the Svet greenhouses used to grow Brassica on MIR used a fan to circulate air, the circulation rate was insufficient (below 0.5 m/s) to prevent hypoxia [38]. Control of the gaseous environment appears to be a key factor for plant reproduction in microgravity [39].

3.1. Plants for bio-regenerative life support systems

The logistical challenges of long-term human space exploration missions require a life support system capable of regenerating all the essentials for survival. The life support systems on the

http://dx.doi.org/10.5772/intechopen.74230

ISS provide oxygen via water electrolysis, absorb and remove carbon dioxide, and manage vaporous emissions (e.g., ammonia, acetone) from the astronauts themselves; water is recycled.

Central to the concept of bio-regenerative life support systems is the use of photosynthetic organisms and light to generate oxygen and food. Learning to grow plants in space is thus an essential goal for long duration space missions since crop growth in space will aid with air regeneration, food production and water recycling for astronauts during long-term space missions [2, 40]. Research on plants in space, in addition to producing key scientific insights into specific plant gravitropic and abiotic stress responses, fosters the overall development of bio-regenerative life support systems for the production of oxygen, food, and nutrients [41].

The cultivation of higher plants occupies an essential role within bioregenerative life support systems (BLSS), which are designed to provide a habitation environment similar to the Earth's biosphere for space missions with extended durations and in deep space. It contributes to all key functional aspects by closing the different loops in a habitat like oxygen production, carbon dioxide reduction, food production, water management, and metabolic waste recycling. Fresh crops are also expected to have a positive impact on crew psychological health.

Different designs and technological solutions have been implemented in higher plant flight experiments. Continuous subsystem improvements and increasing knowledge of plant response to the spaceflight environment has led to the design of current plant growth systems, the latest being the Vegetable Production System (Veggie) [42, 43] and the Advanced Plant Habitat [44]. Plants can adapt to extreme environments on Earth, and model plants have been shown to grow and develop through a full life cycle in microgravity. Adequate environmental control, including forced ventilation, trace gas control, and a well-functioning system for water and nutrient delivery are required for long-term plant growth in space [3, 45].

To put this issue in perspective, the planned early Martian missions (around 500 days overall duration) will primarily focus on water recycling, atmosphere regeneration, and stockpiling of food. Due to the different orbits between Earth and Mars, the launch/return window for the trip is limited either to 30 days, or longer than 2 years (about 780 days). These relatively long space missions can only be sustained with a bioregenerative life support system. Due to the long permanence of the crew and the difficulty to transport and store a large quantity of food, it is estimated that bioregenerative life support system should provide around 80–90% of the nourishment and oxygen needed, which translates to about 40–50 m^2 of plant growing area needed per crew member [45]. To satisfy this requirement, permanent greenhouses and/or sizeable agricultural modules for space would need to be developed.

3.2. Plants for food in space

Growing plants in space helps solve one of the biggest issues in space travel: the supply and the price of food. Space food has evolved since 1961, when the cosmonaut German Titov became the first human to eat in space. The first foods were highly engineered, thermo-stabilized and packaged, capable of meeting the rigid requirements imposed by spacecraft design. However, both the Apollo and the Shuttle missions demonstrated that astronauts did not consume sufficient nutrients, and determined that adequate nutrition begins with appropriate food presented to the consumer in a familiar form [46]. Accordingly, much progress has been made from the first tubed food. Today, the food for the astronauts on the ISS includes a variety of individually packaged, thermostabilized, irradiated, intermediate moisture, and natural form foods [46].

The ultimate goal of growing plants for food in space is to create a self-sustainable regenerative growth system, so plants for food could be continually grown in orbit, in Moon colonies, or on other planets. Challenges to growing plants in space are primarily in the areas of nutrient delivery, lighting, and ventilation (gas exchange). With adapted growth chambers, plant growth in space is similar to plant growth on Earth, except for some morphological traits. However, only small-scale experiments on plant growth have been performed in Earth orbit. These have not provided sufficient data on crop yield for space environments. [47]. Microgravity can reduce cell growth, alter gene expression and protein synthesis, and influence plant morphology – all aspects which critically affect plant cultivation in space. Seeds produced in space also seem to have different composition compared to seeds grown on Earth. As well as affecting the performance and nutritional content of space seeds, this could influence the flavor of plants produced in space, which might become a problem for crews reliant on plant-based diets during long space missions [47].

The theme of agriculture for space has contributed to, and benefited from, terrestrial, controlled environment agriculture; and will continue to do so into the future. For a comprehensive historical review of agricultural systems that have been developed for outer space see [4]. What started with studies on algal production in controlled environment agriculture in the 1950s in the USA and in the USSR has undergone significant improvements via NASA's Controlled Ecological Life Support Systems (CELSS) Program, Japan's Controlled Ecological Experiment Facility (CEEF), the European Space Agency's MELiSSA Project, and most recently, the Chinese Lunar Palace 1 plant factory [4, 48].

The innovative studies for space agriculture have resulted in the development of novel technologies, for both space and Earth applications. These include the use of light emitting diodes for growing crops, the demonstrations of vertical agriculture, use of hydroponic approaches for subterranean crops, crop yields that surpassed reported record field yields, the ability to quantify volatile organic compound production from whole crop stands, innovative approaches for controlling water delivery, and approaches for processing and recycling wastes back to crop production systems [4]. In addition, application of the space environment for mutagenesis and crop breeding has been suggested [26].

Recent research has focused on the possible growth of plants on the Moon and on Mars [49, 50]. In principle, it is possible to grow crops and other plant species in Martian and Lunar soil simulants, even without addition of nutrients. For the record, the Mars simulant can be obtained from a volcanic cone in Hawaii, and has a chemical composition similar to the Mars dirt that the Viking 1 lander analyzed; the Moon simulant comes from volcanic ash deposits near Flagstaff, Arizona. Beyond a Hollywood movie, experiments with 14 plant species in soils that simulate the Martian and lunar regolith suggest that future space colonizers may be able to farm their own food using local dirt [50]. Additional research is needed to improve our understanding of the water holding capacity and other physical characteristics of the extraterrestrial soils, the availability of reactive nitrogen and other (essential) nutrients, further combined with the addition of nutrients and creating a balanced nutrient availability, and the influence of gravity, light and other conditions [50]. Further efforts should include mechanistic modeling of plant growth for better understanding of the intricate and combined physical, biochemical, and morphological phenomena involved, necessary to accurately control and predict plant growth in space.

3.3. Plants for the mental well-being of major tom

Plant in space also provide a substantial non-nutritive value; they are not just for eating or producing oxygen. Plants generally act as a form of emotional sustenance sometimes called horticultural therapy, and can mitigate the negative psychological consequences of space travel. Humans have a preference for nature scenery. Humans (especially humans in a confined space) positively react to plants, and they derive a variety of physiological benefits from exposure to plants. These include human well-being, sense of mastery of the environment, social development, health support, overcoming boredom and mental fatigue, and stress reduction and recovery [51, 52]. Studies of the potential psychological consequences of long-term exposure to conditions common to long-term isolated environments indicated that humans are less stressed and perform better in conditions that include plants and natural settings [51, 52].

The spaceflight environment induces a host of physiological, biomedical, and environmental stressors to flight crews. Long duration spaceflight has revealed a group of stressors that impact crew performance and health: hypochondria, diminished motivation and performance, impaired cognitive ability, withdrawal, impulsive behavior, hallucinations, mood swings, helplessness, depression, and anger [53]. These have spurred the emergence of areas of specialty within the behavioral sciences, including space psychology, space human factors, space habitability, space performance, and space sociology [53]. In that context, the benefit of plants as a countermeasure for difficulties experienced by humans living in isolated or extreme environments, including space travel [54]. A symbiotic relationship between plants and space travelers, including a plant garden for Major Tom, is probably a very good idea.

3.4. Case study: growing *Arabidopsis thaliana* on the International Space Station

As the International Space Station was being assembled, we designed and custom-built a novel advanced plant growth chamber for microgravity experiments [55]. The ADVanced AStroCulture (ADVASC) was the first plant growth chamber flown on the ISS [56]. We used this chamber to grow *Arabidopsis thaliana* from seed-to-seed-to-seed (i.e., two consecutive full cycles of ontogenesis) wholly in microgravity, on the ISS. Arabidopsis plants were germinated, grown and maintained on the ISS prior to returning to Earth [10]. Some of these seeds were used in a subsequent experiment, to successfully produce a second (back-to-back) generation of microgravity-grown Arabidopsis [11].

The ADVASC plant growth unit was designed to control environment parameters including temperature, relative humidity, lighting, fluid nutrient and water delivery, and CO_2 and ethylene concentrations. Advanced control software provided control of each environmental parameter in the plant chamber, creating environmental conditions suitable for growing a wide variety of plant species. Auto-prime technologies eliminated the need for power during Space Shuttle ascent/descent, greatly relieving the shortage of Shuttle resources and the ISS crew time. Fault tolerance and recovery algorithm significantly increased overall system robustness and efficiency. Tele-science features allowed engineers and scientists to receive telemetry data, to send remote commands, and to monitor plant development status via the video images and other data (**Figure 1**).

The first flight of ADVASC provided an opportunity to study the patterns of plant growth and development, as well as seed and plant morphology in microgravity (first seed-to-seed Arabidopsis experiment on the ISS) [10]. The subsequent flight of ADVASC was used to obtain a second generation of microgravity-grown Arabidopsis plants (second seed-to-seed Arabidopsis experiment on the ISS), and to obtain fresh plant tissue for DNA microarray analysis (gene expression profiling) [11]. Since previous investigators found abnormalities in seed produced on long duration missions, we wanted to see if ADVASC's improvements in remote plant care had translated into improved seed quality. We were also interested to learn if microgravity would alter plant form and cause biochemical, cellular, and molecular changes.

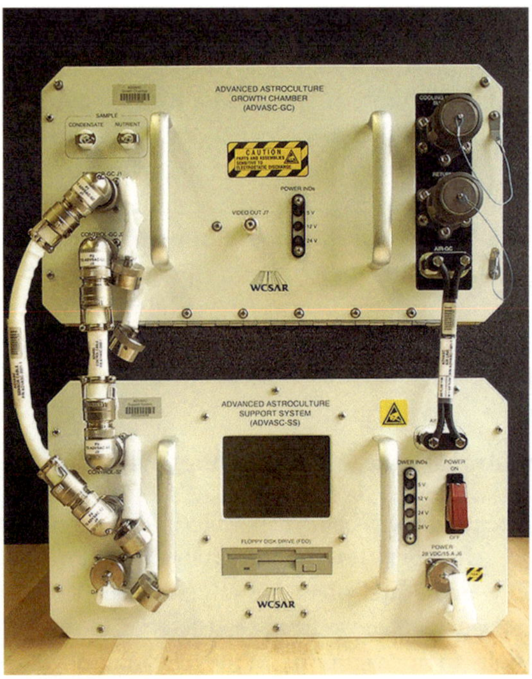

Figure 1. Advanced Astroculture (ADVASC) environmentally controlled plant growth chamber, designed for experiments on the ISS, able to support plant research for a maximum of 6 months in microgravity environment.

The first ADVASC payload with 91 *Arabidopsis thaliana* seeds planted in the root module was launched on STS-100 (ISS-6A), and returned to Earth on STS-104 (ISS-7A). During approximately 70 days in space, the experiment went through seed germination, plant development, seed formation, and seed maturity, which formed a complete life cycle. The experiment was designed to perform autonomously through the entire life cycle [10]. Post-mission analysis data shows that fully 90% of seeds germinated in space, which was similar to the 1 g–grown control plants (grown in a separate ADVASC growth chamber on Earth). Approximately 70% of seeds grew to produce siliques which contained mature seeds in space; An average of 24 siliques per plant were produced, each one containing an average of 36 seeds per silique; plants were healthy and growing normally with the exceptions of the roots and the inflorescent branches from the main stem of flowers. The directions that these organs grew were different in comparison to ground-controlled experiment, and were consistent with an apparent microgravity impact [10].

Plant growth and development in microgravity proceeded similarly to the ground controls that were grown under 1 *g* in an identical chamber [10, 11]. Morphologically, the most striking feature of space-grown Arabidopsis was that the secondary inflorescence branches and siliques formed nearly perpendicular angles to the inflorescence stems. The branches grew out perpendicularly to the main inflorescence stem, indicating that gravity is the key determinant of branch and silique angle, and that light has either no role or a secondary role in branch and silique orientation [10, 11]. Seed protein bodies were 55% smaller in space seed than in controls, but protein assays showed only a 9% reduction in seed protein content. Germination rates for space-produced seed were 92% indicating that the seed developed in microgravity were healthy and viable. We determined that gravity is not necessary for seed-to-seed growth of plants, though it plays a direct role in plant form, and may influence seed reserves [10, 11]. Indeed, it appears that plants undergo somewhat different growth and morphogenesis under space conditions; plant organs show automorphogenesis in space, which may be masked by gravimorphogenesis on earth, except when growing on a clinostat (**Figures 2** and **3**) [57].

Upon return of the plants to Earth, we conducted biochemical, cellular and molecular analyses. We observed a 55% reduction in protein body size; however, since the protein bodies in space-developed seed were filled and we did not observe any other signs of hypoxia such as degeneration of the embryos, deposition of starch grains or alterations in cell structures or cell numbers, we conclude that the aerial portions of the plant were not starved for oxygen. The high forced airflow rates (2–3 m/s) and accompanying ethylene removal provided by the growth chamber improved growing conditions for the aerial part of the plants when compared to the previous studies [9, 33, 39].

Root zone hypoxia could explain the reduced seed protein content. ADVASC uses passive airflow to move air through the root tray. Root zone hypoxia has been prevalent in space flight experiments [58, 59]. We used our own mix of porous arcillite matrix that is one of the favored rooting systems for space [60]. Arcillite reduces root zone hypoxia by allowing air to penetrate between the arcillite grains. Nonetheless, air movement through arcillite is restricted, especially if the spaces between arcillite grains are filled with roots, water, or both. If passive

Figure 2. First plant life cycle experiment on-board the International Space Station, showing *Arabidopsis thaliana* retrieved from the ISS; these were grown in the period between ISS 6A–ISS 7A missions.

Figure 3. Expedite the Processing of Experiments to the Space Station (EXPRESS) rack 1 on the ISS is pictured on-orbit May 14, 2001, with astronaut James Voss checking ADVASC functioning. Image credit: NASA/JSC. The EXPRESS rack is a multipurpose rack system that houses and supports research aboard the space station.

airflow through the arcillite is cut off then oxygen can only reach the roots by diffusion from the air above the soil, or by the arrival of oxygenated water. Diffusion rates are negligible when the diffusion distances are more than a few millimeters [60].

Approximately 80% of the roots formed a dense mat in the top 13 mm of arcillite, while the roots of the ground control plants penetrated deeply throughout the root tray. Evapotranspiration data showed that the porous tubes in the growth chamber delivered an average of 110 mL/d of aerated water during the major growth portion of the experiment. There was not enough oxygen in this amount of water to meet the physiological needs of the roots [58]. In the absence of moisture sensor in the root tray, we had no way of knowing the relative moisture level in the root tray. An anoxic root zone in space resembles an environment similar to flooded soil on earth. Anoxia reduces nitrogen uptake by the roots therefore seed protein content is reduced. On Earth, applying fertilizer to flooded plants improves seed protein content. Because our growth chamber used an artificial soil with no native nutrient value, the plants were fertilized four times during the experiment. This may explain how the plants achieved only 82% of the normal protein content in the seed [10, 11].

This was the first report of altered branch and silique angles for space-grown plants. The reduced branch angles and perpendicular growth of the siliques in space appear to be true microgravity phenotypes. The branching pattern seen in the first spaceflight experiment [10] was replicated during the second spaceflight experiment [11], indicating that this phenotype is persistent in Arabidopsis development on long duration space flights. Light plays a principle role in the "upright" or light-seeking growth habit of the primary axis of many plants, and is responsible for houseplants curving towards the nearest window. On Earth, this response interacts with negative gravitropism in the shoot and requires that shoot gravitropism experiments be conducted in the dark [61]. In our spaceflight experiments the primary axis of Arabidopsis always grew towards the light source, supporting a central role for light in the orientation of the primary axis. The reduced branch angles and tendency of the branches to ignore or curve away from the light source in space shows that gravity plays the key role in signaling branches to curve upwards on Earth. The reduced angles that the siliques made with the stems also show that gravity has a direct role in determining the silique angles. Since Arabidopsis branches do not naturally curve towards the light in microgravity, light plays either a negative or a secondary role in the branch form. Spaceflight appears to initiate cellular remodeling throughout the plant, yet specific strategies of the response are distinct among specific organs of the plant. In the absence of gravity plants rely on other environmental cues to initiate the morphological responses essential to successful growth and development; the basis for that engagement lies in the differential expression of genes in an organ-specific manner [27].

We also conducted the first ever transcriptional profiling of higher plants fully grown in microgravity [11]. The gene expression data were suggestive of the presence of an abiotic stress response. However, we cautioned with respect to deriving conclusions from our gene expression profiling study, because the observed expression patterns may be at least in part induced by other interacting suboptimal environmental conditions, e.g., an anoxic root zone in space. During the second seed-to-seed experiment on the ISS (that provided plants used for transcriptional profiling), technical issues interfered with the priming of the growth chamber and its transition into steady state [11]. These may have contributed to the observed gene expression patterns.

While Arabidopsis plants grown in microgravity may have shown some signs of root zone hypoxia, the ADVASC growth chamber in general provided a very good environment for growing plants on the ISS, and successfully eliminated most of the problems seen in previous plant spaceflight experiments, allowing us to discover alterations in plant form and architecture. We were thus able to successfully grow two consecutive generations of *Arabidopsis thaliana* in space, i.e., seed-to-seed-to-seed. Future experiments should be conducted to see if these alterations can be generalized across different species of plants. As well, future designs of space growth chambers (e.g., the Vegetable Production System [43] and the Advanced Plant Habitat [45]) should consider improving the root zone aeration to prevent root zone hypoxia.

4. Prospects

This is a very exciting time for space science, as the search for extraterrestrial life is one of the great intellectual enterprises of our species. At the same time, better understanding of the profound biodiversity and adaptability of life on Earth is part of the same continuum. Results from the performed space experiments were previously plagues by inconclusiveness due to the small number of experiments, small number of replicates, use of diverse flight hardware, growth conditions, limited possibilities for tissue preservation and subsequent analysis, etc. Future space experiments should therefore have standardized conditions for plant growth [3, 62]. Plus, it is the one area of space science in which you get to eat your experiment.

The theme of agriculture for space has contributed to, and benefited from, terrestrial, controlled environment agriculture; it will continue to do so into the future. The ISS ability to provide an opportunity for direct comparison of microgravity vs. 1 *g* (in on-board centrifuge) conditions, and for on-the-spot modification to the experiment conditions, create unprecedented advantages for plant space biology investigators. This is particularly helpful when investigators are surprised after taking a well-understood experiment on Earth and attempting to reproduce it on the ISS.

Understanding gene and protein expression is the key to unlocking the mechanisms behind microgravity-induced problems, and to finding effective countermeasures to spaceflight-induced phenotype alterations. Even though large-scale tests on growing crops for food production in microgravity are lacking, the body of acquired knowledge that there is little impediment to growing plants in microgravity, in outer space, and on other planets; even if the plants do experience some level of genotoxic stress and anatomic changes [49]. As human space exploration continues to advance, we should feel confident about our ability to grow plants on the Moon, on other planets, and on board spacecraft during long-term space missions. We still need to investigate how plants deal long-term with cosmic radiation and with the soils of other planets. We do, however, know that plants require specialized environments for growth and development in microgravity, including efficient watering and nutrient-delivery systems, precise environmental controls for temperature, humidity and air composition, and low-energy lighting. We already known how to produce such specialized growth chambers and greenhouses; we could design light absorption systems that take advantage of sunlight on the surface of planets and moons, to help us more efficiently grow plants in them.

Finally, it is not far beyond the realms of possibility that selected plant species can be genetically engineered and remotely controlled to provide food, clean air, and potable water, while at the same time acting as a source of raw materials and as small pharmaceutical factories, many miles away from Earth. Such "programmable plants" could uniquely support human missions in space by receiving and responding to remote signals for the synthesis of compounds needed yet unavailable off-the-shelf in deep space [6].

Acknowledgements

I would like to pay special thanks to those who risk their lives to advance our understanding of both Earth and space. In particular, thanks to the several astronaut crews for taking care of our experiments while onboard the ISS and on the Space Shuttle. In addition, I would like to thank the team of dedicated engineers and scientists at the Wisconsin Center for Space Automation and Robotics, University of Wisconsin-Madison, who devoted their time, energy, and creativity to spaceflight experiments on the Space Shuttle and the ISS. Funding was generously provided by NASA.

Conflict of interest

The author declares no conflict of interest.

Author details

Bratislav Stankovic†

Address all correspondence to: bratislav.stankovic@fulbrightmail.org

United States Patent and Trademark Office, Alexandria, VA, USA

†Bratislav Stankovic contributed to this article in his personal capacity; the views expressed are his own and do not necessarily represent the views of the USPTO or the United States Government

References

[1] Darrin A, O'Leary BL, editors. Handbook of Space Engineering, Archaeology, and Heritage. Boca Raton: CRC Press; 2009. 1035 p

[2] Stankovic B. 2001: A plant space odyssey. Trends in Plant Science. 2002;**6**:591-593. DOI: http://dx.doi.org/10.1016/S1360-1385(01)02158-6

[3] Musgrave M. Growing plants in space. CAB Reviews: Perspectives in Agriculture, Veterinary Science, Nutrition and Natural Resources. 2007;**2**(065):9. DOI: http://dx.doi.org/10.1079/PAVSNNR20072065

[4] Wolff SA, Coelho LH, Karoliussen I, Jost A-IK. Effects of the extraterrestrial environment on plants: Recommendations for future space experiments for the MELiSSA higher plant compartment. Life. 2014;**4**:189-204. DOI: 10.3390/life4020189

[5] Wheeler RM. Agriculture for space: People and places paving the way. Open Agriculture. 2017;**2**:14-32

[6] Brown CS, Winter-Sederoff H, Davies E, Stankovic, B. Plan(t)s for space exploration. In: Gilroy S, Masson PH, editors. Plant Tropisms. Oxford, UK: Wiley-Blackwell Publishing; 2008. pp. 183-195

[7] Odeh R, Guy C. Gardening for therapeutic people-plant interactions during long-duration space missions. Open Agriculture. 2017;**2**:1-13. DOI: 10.1515/opag-2017-0001

[8] Beischer DE, Fregly AR. Animals and man in space. In: A Chronology and Annotated Bibliography Through the Year 1960. Pensacola, FL: US Naval School of Aviation Medicine; 1962. ONR REPORT ACR-64 (AD272581)

[9] Merkys AL, Laurinavicius RS. Complete cycle of individual development of *Arabidopsis thaliana* (L.) Heynh. Plants on board the Salyut-7 orbital station. Doklady Akademii Nauk SSSR. 1983;**271**:509-512

[10] Link BM, Durst SJ, Zhou W, Stankovic B. Seed-to-seed growth of *Arabidopsis thaliana* on the International Space Station. Advances in Space Research. 2003;**31**:2237-2243

[11] Link BM, Busse JS, Stankovic B. Seed-to-seed-to-seed growth and development of *Arabidopsis* in microgravity. Astrobiology. 2014;**14**:866-875. DOI: 10.1089/ast.2014.1184

[12] Halstead TW, Dutcher FR. Plants in space. Annual Review of Plant Physiology. 1987;**38**: 317-345. DOI: https://doi.org/10.1146/annurev.pp.38.060187.001533

[13] Wolverton C, Kiss J. An update on plant space biology. Gravitational Space Biology. 2009; **22**:13-20

[14] Chebli Y, Geitmann A. Gravity research on plants: Use of single-cell experimental models. Frontiers in Plant Science. 2011;**2**(56):1-10. DOI: 10.3389/fpls.2011.00056

[15] Morrow R. A brief history of growing plants in space. Engineering and Technology for Sustainable World. American Society of Agricultural and Biological Engineers. 2014:**213**; 17-19

[16] Paul A-L, Zupanska AK, Ostrow DT, Zhang Y, Sun Y, Li J-L, Shanker S, Farmerie WG, Amalfitano CE, Ferl RJ. Spaceflight transcriptomes: Unique responses to a novel environment. Astrobiology. 2012;**12**:40-56

[17] Paul A-L, Wheeler RM, Levine HG, Ferl RJ. Fundamental plant biology enabled by the space shuttle. American Journal of Botany. 2013;**100**:226-234

[18] Vandenbrink JP, Kiss J. Space, the final frontier: A critical review of recent experiments performed in microgravity. Plant Science. 2016;**243**:115-119

[19] Kordyum EL, Chapman DK. Plants and microgravity: Patterns of microgravity effects at the cellular and molecular levels. Cytology and Genetics. 2017;**51**(2):108-116

[20] Krikorian AD, O'Connor SA. Karyological observations. Annals of Botany. 1984;**54**:49-63

[21] Mashinsky A, Ivanova I, Derendyaeva T, Nechitailo G, Salisbury F. "From seed-to-seed" experiment with wheat plants under space-flight conditions. Advances in Space Research. 1994;**14**:13-19

[22] Hoson T, Soga K, Wakabayahis K, Kamisaka S, Tanimoto E. Growth and cell wall changes in rice roots during spaceflight. Plant and Soil. 2003;**255**:19-26

[23] Soga K, Wakabayashi K. Stimulation of elongation growth and xyloglucan breakdown in Arabidopsis hypocotyls under microgravity conditions in space. Planta. 2002;**215**:1040-1046

[24] Ueda J, Miyamoto K, Yuda T, Hoshino T, Sato K, Fujii S, Kamigaichi S, Izumi R, Isioka N, Aizawa S, Yoshizaki I, Shimazu T, Fukui K. STS-95 space experiment for plant growth and development, and auxin polar transport. Biological Sciences in Space. 2000;**14**:47-57

[25] Link BM, Cosgrove DJ. Analysis of peg formation in cucumber seedlings grown on clinostats and in a microgravity (space) environment. Journal of Plant Research. 2000; **112**:507-516

[26] Shi J, Lu W, Sun Y. Comparison of space flight and heavy ion radiation induced genomic/ epigenomic mutations in rice (*Oryza sativa*). Life Sciences in Space Research. 2014;**1**:74-79. DOI: https://doi.org/10.1016/j.lssr.2014.02.007

[27] Paul A-L, Zupanska AK, Schultz ER, Ferl RJ. Organ-specific remodeling of the Arabidopsis transcriptome in response to spaceflight. BMC Plant Biology. 2013;**13**:112

[28] Rea G, Cristofaro F, Pani G, Pascucci B, Ghuge SA, Corsetto PA, Imbriani M, Visai L, Rizzo A. Microgravity-driven remodeling of the proteome reveals insights into molecular mechanisms and signal networks involved in response to the space flight environment. Journal of Proteomics. 2016;**137**:3-18

[29] Johnson SP, Tibbitts TW. The liminal angle of a plagiogeotropic organ under weightlessness. Bioscience. 1968;**18**:655-661

[30] Volkmann D, Behrens HM, Sievers A. Development and gravity sensing of cress roots under microgravity. Die Naturwissenschaften. 1986;**73**:438-441

[31] Moore R, McClelen CE, Fondren WM, Wang CL. Influence of microgravity on root-cap regeneration and the structure of columella cells in Zea mays. American Journal of Botany. 1987;**74**:218-223

[32] Brown CS, Tipathy BC, Stutte GW. Photosynthesis and carbohydrate metabolism in microgravity. In: Suge H, Takahashi H, editors. Plants in Space Biology. Sendai, Japan: Tohoku University Press; 1996. pp. 127-134

[33] Musgrave ME, Kuang A, Brown CS, Mathews SW. Changes in Arabidopsis leaf ultrastructure, chlorophyll and carbohydrate content during spaceflight depend on ventilation. Annals of Botany. 1998;**81**:503-512

[34] Klymchuk DO, Vorobyova TV, Chapman DK, Brown CS. Changes in vacuolation in the root apex cells of soybean seedlings in microgravity. Advances in Space Research. 2003; **31**:2283-2288

[35] Kiss J, Edelmann RE, Wood PC. Gravitropism of hypocotyls of wild-type and starch-deficient Arabidopsis seedlings in spaceflight studies. Planta. 1999;**209**:96-103

[36] Kuang A, Xiao Y, McClure G, Musgrave ME. Influence of microgravity on ultrastructure and storage reserves in seeds of *Brassica rapa* L. Annals of Botany. 2000;**85**:851-859

[37] Campbell WF, Salisbury FB, Bugbee B, Klassen S, Naegle E, Strickland DT, Bingham GE, Levinskikh M, Iljina GM, Veselova TD, Sytchev VN, Podolsky G, McManus WR, Bubenheim DL, Stieber J, Jahns G. Comparative floral development of Mir-grown and ethylene-treated, earth-grown super dwarf wheat. Journal of Plant Physiology. 2001;**158**:1051-1060

[38] Porterfield DM. The biophysical limitations in physiological transport and exchange in plants grown in microgravity. Journal of Plant Growth Regulation. 2002;**21**:177-190

[39] Musgrave ME, Kuang A, Matthews SW. Plant reproduction during spaceflight: Importance of the gaseous environment. Planta. 1997;**203**:S177-S184

[40] Sager JC, Drysdale AE. Concepts, components and controls for a CELSS. In: Goto E, Kurata K, Hayashi M, Sase S, editors. Plant Production in Closed Ecosystems: The International Symposium on Plant Production in Closed Ecosystems Dordrecht, The Netherlands: Kluwer Academic; 1996. pp. 205-223

[41] Zabel P, Bamsey M, Schubert D, Tajmar M. Review and analysis of over 40 years of space plant growth systems. Life Sciences in Space Research. 2016;**10**:1-16

[42] NASA Vegetable Production System (Veggie) [Internet]. 2017. Available from: https://www.nasa.gov/mission_pages/station/research/experiments/383.html [Accessed: Nov 15, 2017]

[43] Stutte G, Wheeler R, Morrow R, Newsham G. Operational evaluation of VEGGIE food production system in the habitat demonstration unit. In: Proceedings of the 41st International Conference on Environmental Systems, Portland, Oregon. Reston, Virginia: American Institute of Aeronautics and Astronautics; 2011. p. 3357

[44] NASA Advanced Plant Habitat [Internet]. Available from: https://www.nasa.gov/mission_pages/station/research/experiments/2302.html [Accessed: Nov 15, 2017]

[45] De Micco V, Aronne G, Colla G, Fortezza R, De Pascale S. Agro-biology for bioregenerative life support systems in long-term space missions: General constraints and the Italian efforts. Journal of Plant Interactions. 2009;**4**:241-252. DOI: 10.1080/17429140903161348

[46] Perchonok M, Bourland C. NASA food systems: Past, present, and future. Nutrition. 2002;**18**:913-920

[47] Poulet L, Fontaine J-P, Dussap C-G. Plant's response to space environment: A comprehensive review including mechanistic modelling for future space gardeners. Botany Letters. 2016;**163**:337-347

[48] Dong C, Shao L, Fu Y, Wang M, Xie B, Yu J, Liu H. Evaluation of wheat growth, morphological characteristics, biomass yield and quality in lunar Palace-1, plant factory, green house and field systems. Acta Astronautica. 2015;**111**:102-109. DOI: https://doi.org/10.1016/j.actaastro.2015.02.021

[49] Karoliussen I, Brinckmann E, Kittang A-I. Will plants grow on moon or Mars? Current Biotechnology. 2013;**2**:235-243

[50] Wamelink GWW, Frissel JY, Krijnen WHJ, Verwoert MR, Goedhart PW. Can plants grow on Mars and the Moon: A growth experiment on Mars and Moon soil simulants. PLoS One. 2014;**9**(8):e103138

[51] Lewis CA. Human health and well-being: The psychological, physiological, and sociological effects of plants on people. Acta Horticulturae. 1995;(391):31-39

[52] Ulrich RS, Simons RF, Losito BD, Fiorito E, Miles MA, Zelson M. Stress recovery during exposure to natural and urban environments. Journal of Environmental Psychology. 1991; **11**:201-230

[53] Morphew ME. Psychological and human factors in long duration spaceflight. McGill Journal of Medicine. 2001;**6**:74-80

[54] Bates S, Gushin V, Bingham G, Vinokhodova A, Marquit J, Sychev, V. A review of the literature and application to habitation systems for humans living in isolated or extreme environments. Habitation. 2009;**12**:33-40. DOI: https://doi.org/10.3727/ 154296610X12686999887201

[55] NASA Advanced Astroculture™ (ADVASC) Fact Sheet [Internet]. 2001. Available from: http://www.nasa.gov/centers/marshall/news/background/facts/advasc.html [Accessed: Nov 15, 2017]

[56] Zhou W, Durst SJ, De Mars M, Stankovic B, Link BM, Tellez G, Meyers RA, Sandstrom PW, Abba JR. Performance of the Advanced ASTROCULTURE Plant Growth Unit During ISS-6A/7A Mission. SAE Technical Paper 2002-01-2280. 2002

[57] Hoson T. Plant growth and morphogenesis under different gravity conditions: Relevance to plant life in space. Life. 2014;**4**:205-216. DOI: 10.3390/life4020205

[58] Stout SC, Porterfield DM, Briarty LG, Kuang A, Musgrave ME. Evidence of root zone hypoxia in *Brassica rapa* L. grown in microgravity. International Journal of Plant Science. 2001;**162**:249-255

[59] Porterfield DM, Matthews SW, Daugherty CJ, Musgrave ME. Spaceflight exposure effects on transcription, activity and localization of alcohol dehydrogenase in the roots of *Arabidopsis thaliana*. Plant Physiology. 1997;**113**:685-693

[60] Porterfield DM, Dreschel TW, Musgrave MEA. Ground-based comparison of nutrient delivery technologies originally developed for growing plants in the spaceflight environment. Horticulture. Technology. 2000;**10**:179-185

[61] Weise SE, Kiss JZ. Gravitropism of inflorescence stems in starch-deficient mutants of Arabidopsis. International Journal of Plant Science. 1999;**160**:521-527

[62] Ruyters G, Braun M. Plant biology in space: Recent accomplishments and recommendations for future research. Plant Biology. 2014;**16**:4-11

Chapter 11

Approaches to Assess the Suitability of Zooplankton for Bioregenerative Life Support Systems

Miriam Knie, Bernard Wolfschoon Ribeiro, Jessica Fischer, Burkhard Schmitz, Kay Van Damme, Ruth Hemmersbach, Donat-P. Häder and Christian Laforsch

Additional information is available at the end of the chapter

http://dx.doi.org/10.5772/intechopen.74261

Abstract

Future manned space exploration will send humans farther away from Earth than ever before (e.g., to Mars), leading to extended mission durations and thus to a higher demand for essentials such as food, water and oxygen. As resupplying these items from Earth is nearly impossible, aquatic bioregenerative life support systems (BLSS) appear to be a promising solution. Due to its central role in aquatic ecosystems, zooplankton could act as a key player in aquatic BLSS, linking oxygen liberating, autotrophic producers and higher trophic levels. However, prior to the utilization of BLSS in space, organisms proposed to inhabit these systems have to be studied thoroughly to evaluate any space-borne adverse traits, which may impede a proper function of the system. To investigate the impact of microgravity (μ*g*), in particular, several platforms are available, providing μ*g* periods ranging from seconds (Bremen drop tower and parabolic flights), to minutes (sounding rockets), up to even days and months (space flights and the International Space Station (ISS)). Furthermore, ground-based facilities, such as clinostats, enable the of candidate organisms to variable periods of simulated/functional μg. In this book chapter, research on zooplankton utilizing these methods is summarized.

Keywords: zooplankton, bioregenerative life support systems, microgravity, drop tower, parabolic flight, sounding rocket, clinostat, space

IntechOpen

1. Introduction

At the beginning of the spaceflight era in the late 1940s and early 1950s, it was uncertain whether humans will be able to survive a journey into space as it poses a hostile environment to life. Therefore, animals have been used to test the survivability under space conditions before the first manned space missions were launched. Initially, fruit flies acted as test organisms aboard a German V-2 rocket [1]. Up to now, a variety of vertebrate (e.g., monkeys [2], fish [3] and gerbils [4]) and invertebrate animals (e.g., spiders [5], snails [6] and ants [7]) has been used to investigate the impact of microgravity (μg) and cosmic radiation on various biological processes ranging from behaviour [8] to embryology [9]. Fourteen years after the first animal experiment, a manned mission followed with the launch of the spacecraft Wostock I. Since then, the sojourn times of humans in space have significantly increased from nearly 2 hours up to several months aboard the International Space Station (ISS) (e.g., [10]). The present objective of space exploration (e.g., to Mars [11]) will send humans farther away from Earth than ever before, leading to even longer mission durations. One of the key issues to be solved in advance of those missions is the supply of essentials, such as food, water and oxygen, since regular supply is only feasible to low-orbit platforms. A solution to create independency from regular delivery is to rely on autochthonous production using bioregenerative life support systems (BLSS). Besides the production of food, future BLSS could fulfil functions such as the regeneration of atmosphere, purification of water and waste, as well as food processing [12]. As the reaction of aquatic organisms to μg [3, 13–15] given the increased viscosity of their habitat compared to terrestrial systems [16, 17], most BLSS are based on aquatic systems.

The first systems for housing aquatic animals in space that led to the current BLSS were designed quite simple and did not aim to fulfil the functions of a BLSS, but provided initial insights into what needs to be taken into account for future systems. An example of such a primary system is special plastic bags filled with water and oxygen to keep killifish aboard Skylab 3 [13]. A further advanced system, the so-called *STATEX* container (derived from *STAT*olith *EX*periment), was used to test tadpoles of the South African clawed toad in the German-D1 mission in 1985 [18] and in addition with cichlid fish larvae in the Spacelab-D2 mission in 1993 [19]. The *STATEX* container was equipped with a centrifuge that allowed to perform inflight reference experiments under the same conditions, but exposed the animals to the same acceleration as the gravitational force on Earth (1 g). The animals were housed in small, petri-dish like, mini-aquaria into which oxygen transfer was provided by a gas-permeable biofoil. In 1980, also the National Space Development Agency of Japan (NASDA) started to develop experimental hardware to support aquatic animals in space. All these facilities included an artificial lung for the supply with oxygen and a feeding system. One of the major aims from the technical point of view was to achieve a completely closed water circuit and an effective water purification system, requiring a limited amount of water to be used in the space shuttle [20]. The first of these systems was the *Vestibular Function Experiment Unit* (VFEU) that was used to study the behaviour of Japanese carp in 1992 [21]. The subsequent *Aquatic Animal Experiment Unit* (AAEU) was flown in 1994 and included four different experiments with fish and newts (for an overview of the experiments, see [20]). The original VFEU was later on improved to accommodate marine fish under low temperature for two shuttle missions in 1998 [22]. In parallel, two more facilities to conduct space research with aquatic animals have been introduced: the *Aquatic Research Facility* (ARF) by the Canadian Space Agency and the *Autonomous Biological System* (ABS) by the National Aeronautics and Space Administration (NASA). Both systems were employed in the space shuttle mission STS-77 in 1996 [23, 24]. In the ARF, a highly sophisticated facility to study the development of sea urchins in μ*g* [23], oxygen was transferred into the experimental units via a gas-permeable biofoil. In the ABS, however, the oxygen needed for the animals (water fleas, small snails and small shrimps) was produced within the facility itself by an aquatic plant [24]. With this, the ABS represents the first facility brought into space, which combined organisms from different trophic levels and mimicked natural ecosystems. The ABS was used in the space shuttle and aboard the Mir Space Station [15]. In 1992, another German setup, the *Closed Equilibrated Biological Aquatic System* (C.E.B.A.S.), was introduced as a possible precursor for long-term multi-generation experiments with aquatic organisms on Earth and in space missions [25]. For the application in space, C.E.B.A.S. was miniaturized (C.E.B.A.S. minimodule) to fit into a Spacelab middeck locker [26]. It consisted of four subcomponents (zoological, botanical, microbiological and electronic component) and was successfully flown onto two space shuttle missions in 1998 [27]. However, due to the limited space preconditioned by the Spacelab locker, it was impossible to establish a self-sustaining artificial ecosystem. Hence, food for animals inhabiting this system still had to be provided by an automated feeder, causing limited mission duration. The impossibility to harvest the oxygen-producing plant *Ceratophyllum demersum* inside the running setup was another obstacle to reach the goal of a self-sustaining system, as a rapid increase in biomass led to mutual shading and thereby to a reduced photosynthetic activity. This issue might be solved by using unicellular algae for oxygen production, as their automated harvesting is less complex, rendering phytoplankton a foundation of aquatic BLSS.

Several systems based on phytoplankton have already been successfully tested (AQUARACK [28], BIORAT [29], CAES [6], AQUACELLS [30] and SIMBOX mini-ecosystem [31]), and it was shown that the amount of oxygen produced by *Euglena gracilis* is sufficient to sustain fish (OMEGAHAB [32]). However, fish food still has to be provided by an automated feeder. A solution to this problem is to produce it within the BLSS itself by the introduction of herbivorous zooplankton. In aquatic food webs on Earth, these organisms link oxygen-producing phytoplankton (microalgae) and higher trophic levels, such as fish, and may thus play this very role also in BLSS [33]. However, zooplankton is well studied with regard to ecological and evolutionary aspects (e.g., [34, 35]), but so far little is known about its performance in μ*g* (e.g., [15, 36, 37]). Gaining knowledge on aspects such as behaviour, survival and reproduction of zooplankton under space conditions is thus of great importance with regard to the establishment of stable biomass production in BLSS. Therefore, potential effects of μ*g* on candidate organisms have to be evaluated. To this assignment, different methods are available to researchers in order to achieve μ*g* conditions, ranging from short-term μ*g* in drop towers and parabolic flights to prolonged μ*g* phases aboard sounding rockets and the ISS. Furthermore, ground-based facilities, providing simulated/functional long-term μ*g*, such as clinostats, are a valuable tool for the assessment of zooplankton used for biomass production in BLSS for space application. In the following, examples of gravitational research on zooplankton using these methods are explained in detail.

2. The Bremen Drop Tower

Since its inauguration in 1990, the Bremen Drop Tower has been widely used for a variety of experiments, predominantly in fluid mechanics, physics and space research. The drop tower facility in Bremen consists of three basic elements: the tower itself with a height of 146 m (110 m free fall in 120 m vacuum drop tube) (**Figure 1A**), the catapult and drop apparatus, each allowing a different method to obtain microgravity (μg), and the experiment integration area (**Figure 1B**) including the control room, laboratories and a mechanical as well as an electronic workshop. With a g-value of about 10^{-6} g, the quality of the μg in the Bremen Drop Tower is exceptional compared to other facilities (cf. other chapters). In addition to the excellent μg quality, further favourable features are the daily accessibility and unproblematic safety regulations in comparison with parabolic flights and ISS experiments. Also, air traffic is not affected and risk of fire or contact with harmful substances can be mitigated. Yet another key benefit in using the drop tower represents the possibility to test experimental hardware as well as obtaining biological data and optimizing the procedures in preparation for subsequent campaigns under μg conditions. Easy accessibility of the tower and hardware, constant consultation and evaluation between researcher and the *Zentrum für angewandte Raumfahrttechnologie und Mikrogravitation* (ZARM) technicians allow a productive environment for experimental design as well as theoretical and practical implementation of hardware and test objects. This is especially useful when adjustments on the experiment on short notice are necessary. Furthermore, multiple launches and drops per week are possible, so configuration and hardware errors become evident almost immediately after experiment recovery.

There are two ways of using the drop tower for microgravity research. First, there is the drop apparatus, through which the drop capsule is lifted to a height of 120 m at the top of the drop

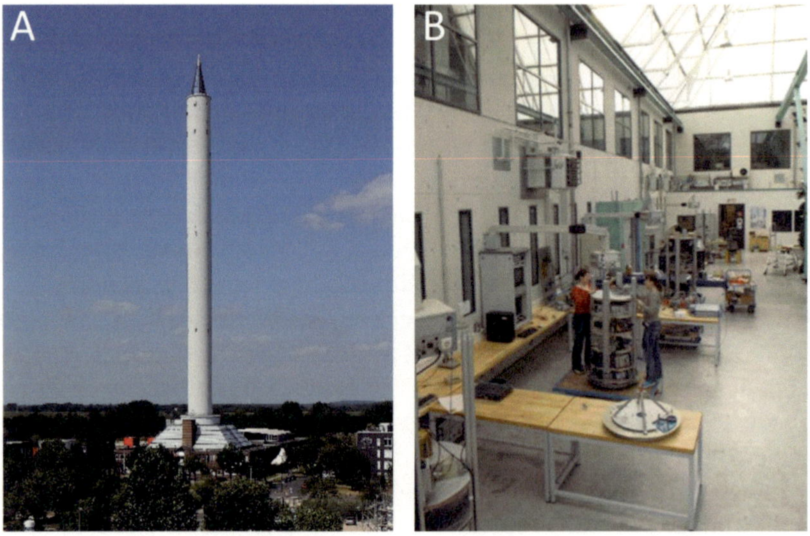

Figure 1. A: The Bremen Drop Tower (photo: ZARM). B: Experiment integration area (photo: ZARM Drop Tower User Manual 2012).

tube by a hoisting winch and finally released into free fall, resulting in about 4.7 seconds of μg. This method is the original one and was the only way to operate until the installation of the catapult system. The catapult system enables a second way of using the drop tower in shooting the drop capsule in a vertical parabola to the tip of the tower and back, resulting in about 9.3 seconds of real μg.

In order to withstand deceleration forces and to maintain an environment at constant atmospheric pressure, the experiments are integrated into one of two drop capsule types of different sizes. The drop capsule is a modular cylindrical container with a diameter of 800 mm and a length of 1.6 m in the first type or 2.4 m in the second type, while the space for experiments extends to a length of either 953 or 1718 mm according to the needs of the experimental hardware (**Figure 2**). The payload area is subdivided in experiment platforms on which the hardware can be placed. A maximum mass of up to 500 kg for the integrated capsule is possible. Subtracting the capsule net weight, a maximum payload mass between 161.5 and 264.4 kg is feasible, depending on the capsule type in use. The drop capsule is assembled with the experimental setup in the integration hall of the drop tower facility. The whole experiment integration process is monitored and assisted by specially assigned technicians of the drop tower staff (**Figure 3**). The integration process usually starts 10 days to 1 week before the first drop or

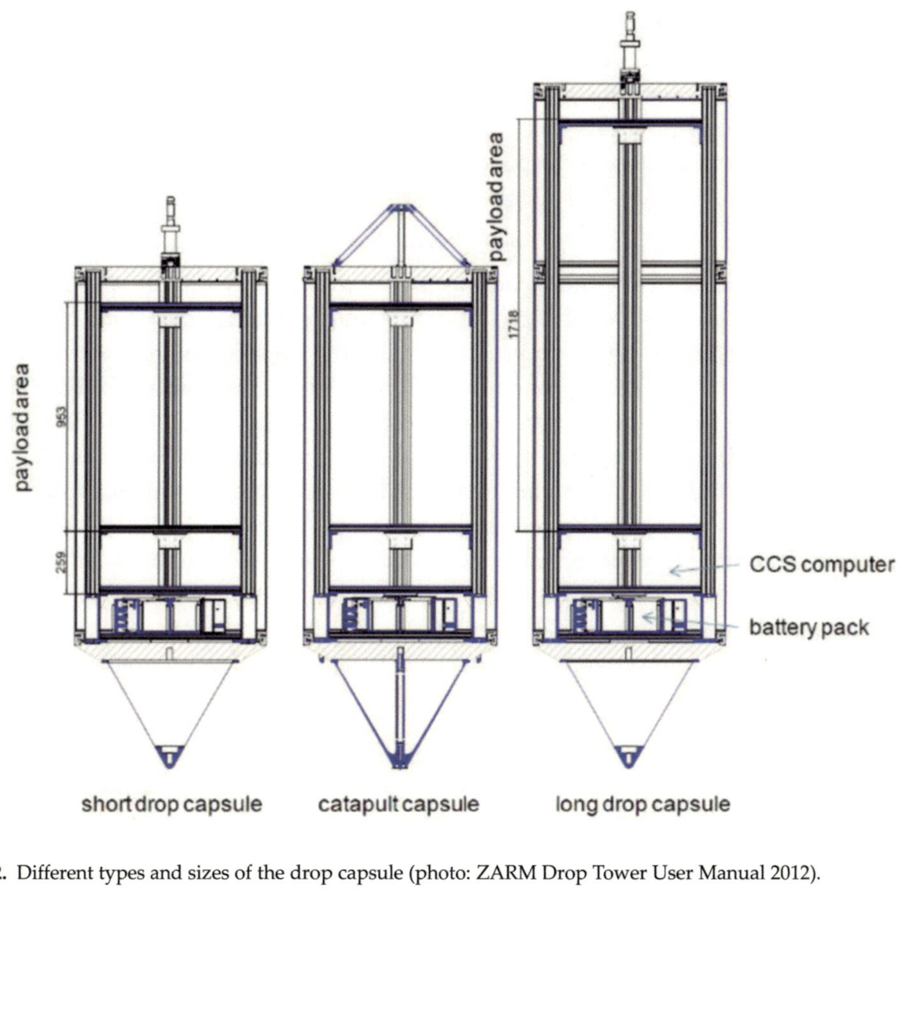

Figure 2. Different types and sizes of the drop capsule (photo: ZARM Drop Tower User Manual 2012).

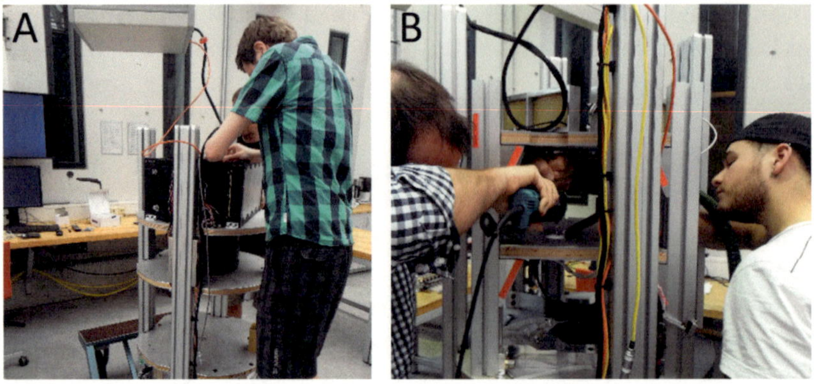

Figure 3. Different stages of the assisted integration process. A: Early stage of capsule assembly. B: Preparation of experiment platform by technicians of the drop tower staff.

catapult launch. It is usually possible to get extended preparation time as needed, e.g., in some biological experiment setups, when organisms require to reach a certain developmental stage and size. Acclimatized laboratories and storing areas for all chemical compounds are provided by the ZARM facility and installed in close range to the drop tower integration area. All elements essential to the experiment setup in the capsule, such as electronic devices, experiment containers and controllers, are mounted on the experiment platform and wired accordingly to controllers and power supply units. Sensitive samples are transferred to the corresponding containers shortly before finalization of the integration. Also, during this time, all preparations and integration details are organized in close correspondence with the drop tower staff. This procedure is important as it ensures a flawless and a highly adjusted operation of the experiment.

The basic setup of the capsule contains the base structure, the four-stringer-rack for experiment accommodation and a lid plate with interfaces and a release bolt (**Figure 4A, B**). The base structure always consists of a switchable power supply unit, a radio telemetry and telecommand system, a WLAN unit on top of the capsule and also the capsule control system (CCS) for experiment control [38]. The CCS is programmable and controls the units of the base structure and experiment sequences. It also controls any electronic device attached to the experiment platforms, such as high-speed cameras, servomotors and illuminating devices like LED arrays. It is therefore of highest importance to adjust the CCS programme individually to the present experiment hardware. This is done by the staff of the drop tower, i.e., the ZARM FAB mbH, in correspondence with the experimenters. For example, in experiments on the water flea *Daphnia* using RNAlater as preservative to determine the effect of μg on gene expression at specific moments in time, the exact time of RNAlater release from hydraulic-driven syringes for each fixation unit could be set and programmed in the CCS (**Figure 5**). The programming enabled a fixation at four different time points on each shot [39].

When operating on catapult mode, once the integration is completed, the drop capsule has to be balanced in order to ensure a safe launch procedure, as an unbalanced setting could lead to deviations from the flight vertical vector and in turn cause fatal damage. After balancing, the drop capsule is covered by a pressure-sealing aluminium sheath (**Figure 6**), before it is

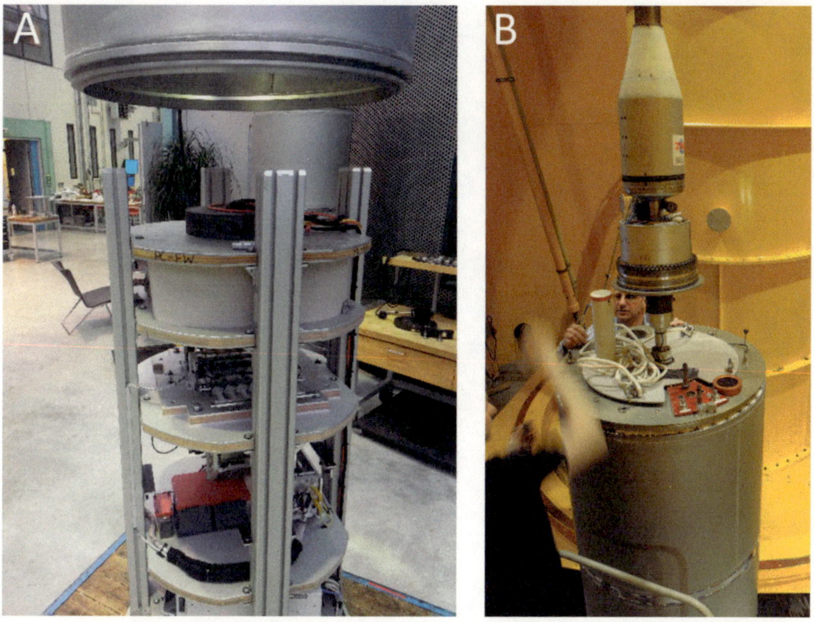

Figure 4. A: Four-stringer-rack for experiment accommodation. B: Drop capsule lid plate with interfaces and release bolt (photo: ZARM).

Figure 5. Hydraulic-driven syringes controlled by the capsule control system for fixation experiments with zooplankton.

transferred to the catapult or drop system of the tower. Before launching the experiment, the tower has to be evacuated for approximately 1.5 hours to create a vacuum in the drop tube in order to reduce the air resistance to a minimum.

The catapult has been installed 11 m below the base of the tower. It is based on a combined hydraulic-pneumatic system and consists of 12 pressure tanks located around the prolonged drop tube in the chamber. The catapult utilizes a pneumatic piston that accelerates the drop

Figure 6. A: Sealing process of the capsule with the aluminium sheath for pressurisation. B: Sealed capsule in the drop tower.

capsule to speeds up to 48 m/s within 0.28 s leading to a capsule velocity of around 175 km/h during the parabola [40]. The launch is started at the control room by the technical staff. While both methods achieve the same high quality of μg, the catapult operation involves a very short time frame (around 280 ms) of hypergravity at start acceleration. This may need to be taken into account depending on experiment design and requirements. In any case, the drop capsule is received at ground by a deceleration container filled with small polystyrene pellets, which ensure a safe landing of the capsule. The container is positioned below the drop tube during the catapult launch or the drop. After the shot or drop, the drop tube is ventilated again and the experiment can be recovered from the drop capsule for analysis. All experiment and monitoring data can be retrieved from the CCS. Following this procedure, up to three launches or drops per day are possible.

In the past, a variety of biological experiments were performed at the drop tower in Bremen. Research in plant biology focused on auxin transport, stress reactions in roots already present in the first few seconds in μg, and could show fluxes of nitric oxide, reactive oxygen species and oxygen in the apex zone of seedlings of *Zea mays* under μg conditions [41]. Already in the 1990s, the gravitactic orientation and its respective thresholds in the unicellular green alga *Euglena gracilis* were analysed using real-time image analysis [42, 43]. Present drop tower experiments are now being focused on gravity-related signalling pathways and adaptation mechanisms in *Euglena* as well as their helical swimming patterns via a three-dimensional motion analysis system. Experiments in the drop tower facility have been used not only to study swimming patterns, but also to investigate the effects of gravity on animal behaviour and postural control mechanisms under μg conditions. For instance, the catfish *Synodontis nigriventris* shows a ventral substrate response (VSR) behaviour. This behaviour is characterised by a turning of the fish's ventral side towards a respective surface, when a suitable substrate is close by. Without a substrate present, it swims upside down. In order to elucidate if the VSR is affected by a lack of gravity or coupled to a gravitational stimulus, some specimens of *S. nigriventris* were exposed to μ*g* in the drop tower. The experiments showed that the VSR can override the vestibular input in this particular species [44]. In order to analyse kinetotic ("motion sickness") behaviour in fish over a short period of time under μ*g*, Anken and Hilbig [45] exposed *Oreochromis mossambicus* to various gravity environments, ranging from 0.1 to 0.9 *g*. Along with establishing suitable thresholds for kinetosis levels, the drop experiments revealed that the short time frame of 4.7 s in specific gravity conditions is sufficient to induce kinetosis and also confirmed the feasibility of the procedures with the used hardware.

In zooplankton research, the search for suitable organisms for bioregenerative life support systems (BLSS) has risen in importance. Knowledge on graviperception and the involved organs in zooplankton organisms give an insight into how these animals cope with gravity. Future experiments in the drop tower with different planktonic species can help to find answers to these questions. By using drops under different μ*g* levels, thresholds for gravity perception of the selected species can be determined, as it has already been shown for some protists in clinostat experiments [46]. In order to establish such BLSS, the comprehension on how food webs function and how zooplankton organisms are affected by altered gravity conditions is of utmost importance. To test whether predator-prey interactions are affected by μ*g* is a first step to investigate the functioning of a BLSS based on multi-trophic levels. If the food chain is interrupted because the predators do not feed on their prey, energy transfers to higher trophic levels and hence biomass production is not possible, thus preventing food production for humans in space missions. Experiments focusing on the first trophic level at the Bremen Drop Tower showed that foraging and feeding of *Daphnia magna* are not significantly altered in μ*g* [37]. So far, little is known on how μ*g* acts on a molecular level in zooplankton. In human cells, studies in parabolic flights have revealed that already short-time exposure to μ*g* affects gene expression patterns [47], and also various studies in simulated microgravity report disturbances and changes of the cytoskeleton (e.g., [48, 49]). Also in *Daphnia magna,* a structural disruption of the cytoskeleton and an upregulation of energy metabolism-related proteins during clinorotation could be observed [50]. In further drop tower experiments, preservation of *Daphnia* with RNAlater in different gravity conditions can help to elucidate the first response as well as adaptation processes in altered gravity on the cellular basis [39].

In conclusion, the Bremen Drop Tower represents an extraordinary facility for various applications in space and microgravity research. A milestone in the drop tower history was undoubtedly the construction and inauguration of the catapult system in December 2004, which enabled μ*g* conditions up to almost twice the time achieved at free fall mode [40]. The engaging working environment coupled with an outstanding quality of μ*g* ensures a high quality of data and inspiration for future experiments. Easy handling and relatively low costs by using commercial hardware allow changes on short notice and step by step refining of experimental hardware and procedures during the integration phase and to some extent even during the actual experiment campaign. With up to three possible catapult launches or drops per day, not only good data quality, but also a sufficient amount of replicates are provided. The opportunities at the drop tower facility will continue to foster further research on biological systems related to varying gravity conditions and therefore secure its key position for short-term tests and as preparation environment for campaigns in more complex gravity-related research environments, like parabolic flights with aircrafts, sounding rocket launches and suborbital flights with new commercial vehicles. Thus, drop tower experiments are an indispensable tool for research on plankton as a key element for bioregenerative life support systems for future crewed long-term space missions.

3. Parabolic flights

The term parabolic flight describes a special flight manoeuvre where an aircraft follows a free-fall ballistic Keplerian trajectory [51]. Thereby, the resultant of all forces acting on the occupants of the aircraft other than gravity is nulled. This manoeuvre is started by accelerating the aircraft to gain velocity before pulling up to convert horizontal velocity into vertical velocity. During this climb, the gravity level increases. Upon reaching a sufficient upward velocity, the pilots reduce the thrust, compensating the effect of air drag and the aircraft starts to "fall" uphill (parabolic free fall) and the microgravity phase starts. The aircraft then passes the apogee of the parabola and starts to dive downwards. At the end of the parabola, the pilots pull up to stop the dive and the g-level increases again [52, 53] (**Figure 7**).

The first-ever parabolic flights were performed by National Aeronautics and Space Administration (NASA) pilot Scott Crossfield and Air Force pilot Charles E. Yeager at Edwards Air Force Base in California and at Wright Field in Ohio in 1951 [54]. From then on, parabolic flights became a valuable and frequently used tool for training astronauts (e.g., [55]), medical and physiological experiments on human subjects (e.g., [56]), space technologies (e.g., [57]), physics and material sciences (e.g., [58]), medical engineering and biotechnology (e.g., [59]) and life sciences (e.g., [60]).

The United States Space Agency NASA operated its own parabolic aircrafts until 2014. Since 2015, the *Zero Gravity Cooperation* (ZERO-G) and their aircraft, the G-FORCE ONE, a modified Boeing 727-200 are used for research [61, 62]. Further, the Canadian National Research Council (NRC) in association with the Canadian Space Agency (CSA) irregularly utilise parabolic flights for research aboard their Falcon-20, which has been modified for parabolic flights [63].

The European Space Agency (ESA) launched its own parabolic flight programme in 1984 [53], and since then, on average six scientific parabolic flight campaigns are carried out each year, reflecting the great interest in this platform. Since 2015, parabolic flights for research are performed using the Airbus A310 ZERO-G. It is the largest airplane for parabolic flights worldwide [64], owned and operated by Novespace, a subsidiary of the French National Space Center (CNES). A parabolic flight campaign usually consists of 3 consecutive flights conducting 31 parabolas each. The μg phase of each parabola has a duration of approximately 22 s and the residual acceleration acting on experimental set-ups is in the order of 10^{-2} g. Furthermore, modified parabolas can be flown during which partial g-levels (including lunar (0.16 g) and Martian (0.38 g) gravity) can be achieved.

In comparison to experiments aboard research satellites or the International Space Station, parabolic flights have several advantages. They have a short turnaround time of approximately

http://dx.doi.org/10.5772/intechopen.74261

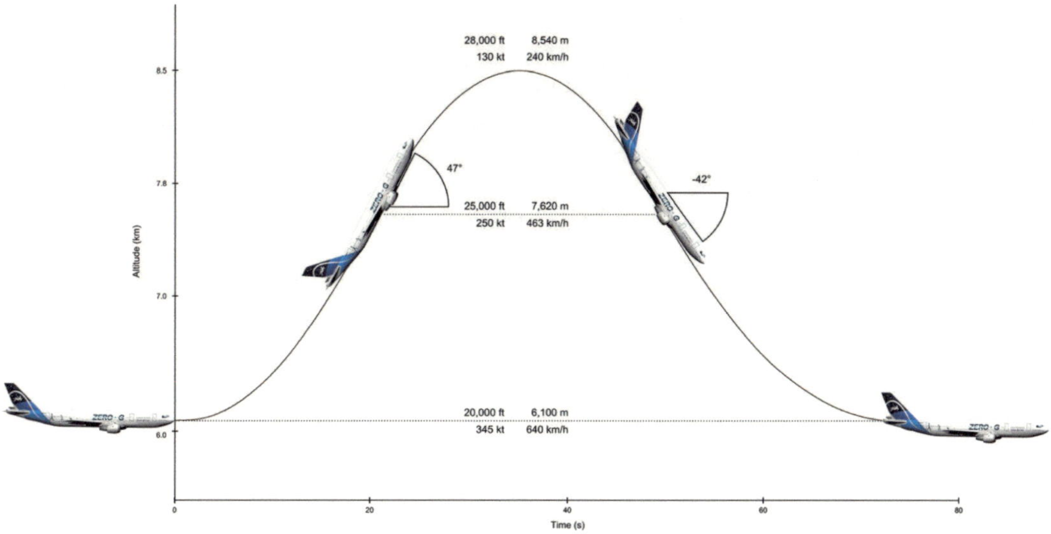

Figure 7. Schematic illustration of the parabolic flight manoeuvre. Courtesy of Novespace, France.

8 months between the experiment proposal and its performance. The scheduled campaign dates are reliable and the campaigns take place regularly. Furthermore, it is possible to use laboratory-type instrumentation, which also provides a high flexibility in the experimental approach. The major advantage is that investigators can directly interact with their experimental set-up or change experimental parameters during and in between parabolas. However, major disadvantages are the hypergravity phases interspaced between the phases of reduced gravity (microgravity or partial *g*), which can present a substantial prejudice to some research questions. The short duration of the reduced gravity phase is a further disadvantage since some issues may require a longer time period. Nevertheless, these periods are still sufficient to address a lot of questions in the area of life sciences. They range from the impact of μ*g* on the cellular level to shifts in physiological parameters and to the behaviour of whole organisms in reduced gravity conditions. Some of these variations/shifts are also likely to occur in zooplankton and are thus of interest with regard to the suitability of zooplankton for bioregenerative life support systems. A selection of those studies will be presented in the following paragraphs.

Parabolic flights are a well-suited tool to analyse the effects of μ*g* on the cellular and molecular level, such as the change of the electrophysiological properties of various cell types and the propagation velocity of action potentials [65]. Furthermore, bone cells are studied on a regular basis, as mechanical loading plays a critical role in their function and differentiation [66, 67]. Likewise, cytoskeleton experiments are frequently carried out, as it is redistributed and reorganized under reduced gravity [68, 69]. In order to find possible explanations for these phenomena, the impact of altered gravity on gene expression is increasingly being investigated in different organisms [47, 70]. Another research area in which parabolic flights are frequently utilized is physiology. A number of physiological changes are caused by μ*g*, such as an initial shift in the distribution of blood [71], which happens within a very short time frame. This might have an impact on the blood flow and could thus influence the pulmonary diffusing capacity [72]. Also isometric force production was examined during parabolic flights [73], since it is known that deviations from 1 *g* affect sensorimotor performance and thereby the ability to grasp objects or to operate the pedals, buttons and levers of machines. The behaviour of animals from different habitats under reduced gravity conditions was examined in a multitude of studies, showing that behavioural changes occur quite fast and that their responses are fairly diverse. In pigeons, flight movements are provoked at the transition from hypergravity to μ*g* at the beginning of the parabola and they display random head movements while free floating [74]. Likewise, the reaction of different species of amphibians and reptiles to diminished gravity was analysed in parabolic flights [75]. In terrestrial and semi-arboreal lizards, long-axis thrusting body motions and high-amplitude, high-frequency tail thrashing movements have been observed [75], whereas terrestrial and arboreal frogs take up a "sky diving" posture [76], to name just a few examples. Fish also show altered behaviour when subjected to parabolic flights, the so-called loop-swimming, which is exclusively exhibited in μ*g* [77].

In planktonic research, parabolic flights are mainly used to study spatial orientation, locomotion behaviour, physiological and cellular responses to reduced gravity conditions and to elucidate the gravireceptive organs. As an example, the removal of the graviceptor (rhopalia) of an asexually produced life stage of the jellyfish *Aurelia aurita* led to an inability to swim in 1 *g* conditions and a missing response to the *g*-force changes occurring during the parabolic flight, whereas unharmed control individuals swam loops or became immobilized [78]. This experiment showed the importance of intact rhopalia for orientation in *Aurelia aurita* at 1 *g* and during the *g*-force changes in parabolic flights. In the protist *Paramecium biaurelia* [79], the changes of graviorientation and gravikinesis were investigated at different accelerations. At first, *P. biaurelia* showed a significant preference for upward swimming (negative gravitaxis), but after 7 s of μ*g*, no significant swimming direction could be recorded. Another aim of this experiment was to test whether parabolic flights are suitable to examine the threshold of gravikinesis in these organisms. It was shown that parabolic flights are in principle suited for preliminary threshold studies on fast-reacting biological systems. The unicellular freshwater flagellate *Euglena gracilis* was used in a number of parabolic flights. These organisms are of special interest with regard to aquatic bioregenerative life support systems, as they could act as a foundation of such systems. *E. gracilis* shows a pronounced negative gravitaxis, which is most likely mediated by an active physiological mechanism involving changes of internal calcium concentrations and the membrane potential [80]. It was shown that the μ*g* phase leads to a pronounced loss in the precision of orientation in *E. gracilis* and to a decrease of the intracellular calcium concentration, which indicates that calcium signalling is involved in the graviperception and orientation. Similar experiments have been performed with *Astasia longa*, a close relative of *E. gracilis* [81]. They show a negative gravitaxis in the absence of other external stimuli apart from gravity. During μ*g*, however, a clear deterioration of gravitactic orientation was detected, which improved during the subsequent hypergravity phase. Also, the cytosolic calcium levels showed acceleration-dependent changes, with a transient increase upon increasing acceleration. Like in *E. gracilis*, these findings confirm the model of gravitaxis, which assumes the presence of mechanosensitive channels, activated upon deviation from the vertical swimming direction.

In addition, parabolic flights can be used to address responses of microcrustaceans to altered gravity. It was previously shown that the locomotion behaviour of the water flea *Daphnia magna* and of ostracods is modified in μg, due to a disturbance of spatial orientation [15, 36]. As this could impair food uptake of the animals, a parabolic flight experiment was performed to analyse if the ostracod species *Heterocypris incongruens* is still able to forage [37]. Since the direction of incident light is used as orientational cue by many crustaceans, the possible influence of illumination on the behavioural response was included in the experimental design (for technical details, see [37]; **Figures 8** and **9**). As feeding could not be directly observed in ostracods, due to the non-transparent carapace valves covering the mouthparts, the sojourn time on food was used as indirect proxy for the feeding duration. The fact that feeding behaviour was not significantly affected in μg renders *H. incongruens* as suitable candidate for future BLSS.

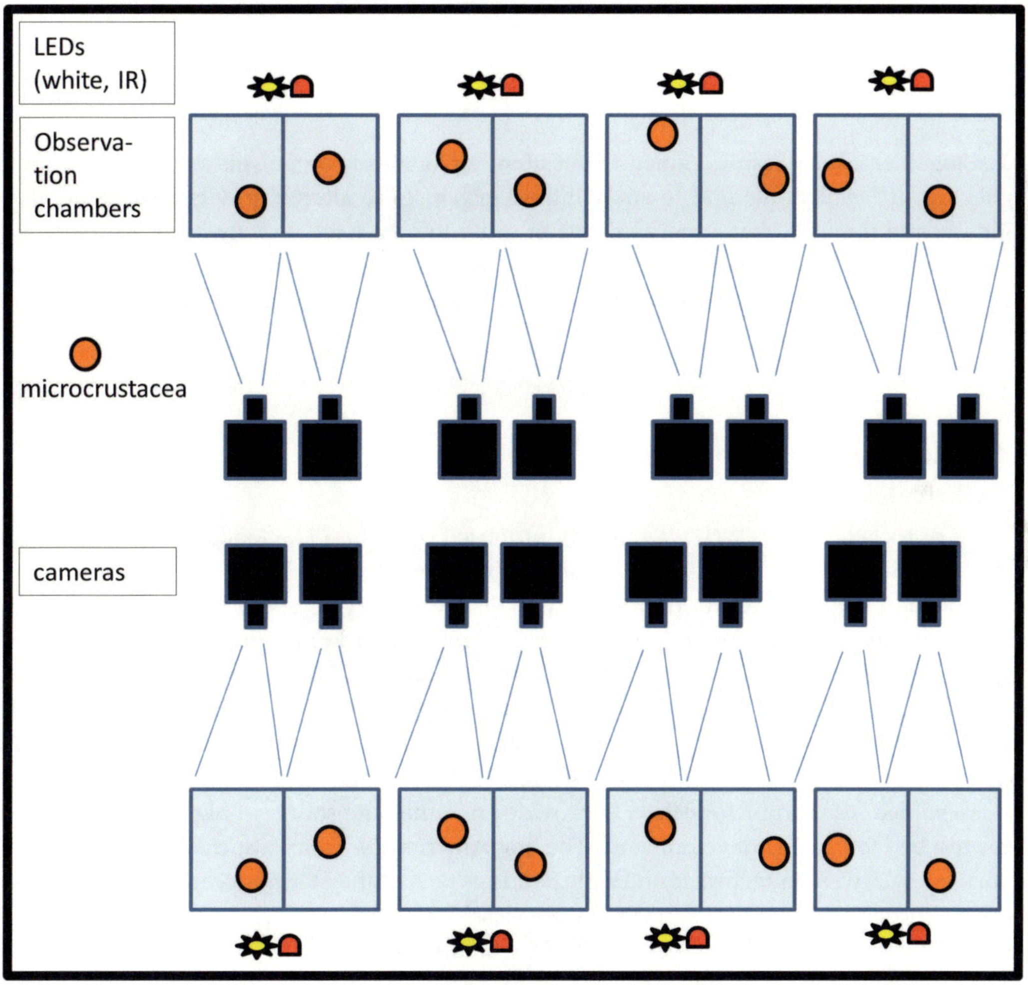

Figure 8. Schematic drawing of experimental set-up for the analysis of microcrustaceans in altered gravity conditions during parabolic flights.

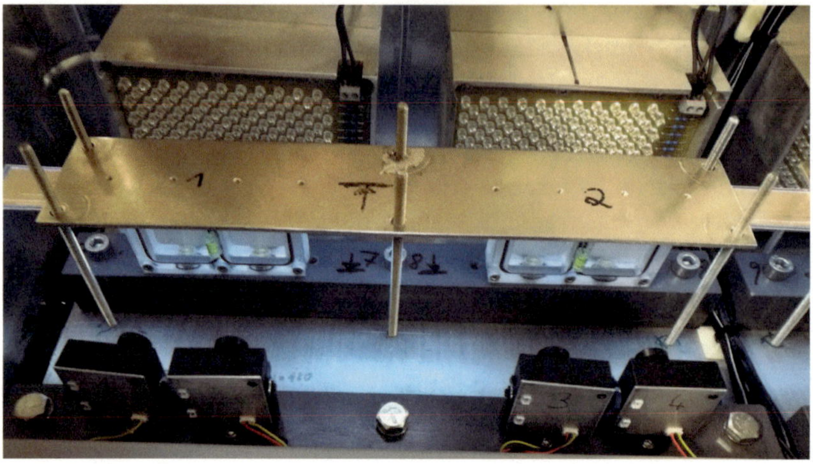

Figure 9. Picture of the experimental set-up used with microcrustaceans in parabolic flights.

Taken together, all the results gained with different planktonic organisms show that parabolic flights are an excellent method to analyse their responses to altered gravity conditions, especially μ*g*, and the underlying mechanisms. In order to obtain a comprehensive assessment of the suitability of planktonic organisms for bioregenerative life support systems, it is thus necessary to include data from other platforms such as clinostats, the drop tower and sounding rockets.

4. Sounding rocket experiments

The use of rockets for research was already proposed by Robert H. Goddard at the beginning of the twentieth century [82]. In 1933, the first instrumented, liquid-fuelled sounding rocket was launched by the Russian Tikhonravov [83]. The history of sounding rockets was profoundly affected by the beginning of the German military rocket programme in 1930 and the beginning of rocket research at the Guggenheim Aeronautical Laboratory, California Institute of Technology (GALCIT), in 1936 [84]. The German work culminated in the development of the V-2 rocket. After World War II, the American army was the first to inherit an underground V-2 factory. The seized rocket parts were transferred to White Sands Missile Range in New Mexico and assembled into complete rockets to provide an immediate source of high-altitude vehicles and a test bed for further developments. The first American V-2 was launched in 1946 [85], and in total, 67 V-2s were launched from White Sands as part of the Hermes programme [86]. Even though almost half of the V-2 s launched were classified as failures, the experience gained with this rocket provided the knowledge to build sounding rockets tailored specifically to space research. A comprehensive overview on the first sounding rockets developed in the United States is given in [84]. A breakthrough in the use of rocket technology for civilian purposes was initiated by the International Geophysical Year (IGY) from July 1957 to the end of 1958, which was devoted to global atmospheric research and prompted the launch of some 200 sounding rockets worldwide [87]. The impetus given by IGY to sounding rocket activities led to the establishment of national sounding rocket programmes in many countries. An elaborate overview on the European developments is given in [87]. Up to now, sounding rockets are frequently used in microgravity and space research all over the world embedded in well-structured programmes (e.g., NASA Sounding Rocket Program [88], Australian Space Research Institute Small Sounding Rocket Program [89] and European Sounding Rocket Programs [90]).

The TEXUS Sounding Rocket Programme (Technologische EXperimente Unter Schwerelosigkeit) was first funded in 1976 by the German Ministry for Research and Technology, as a preparatory programme for the first Spacelab mission in 1983, and from the end of 2005 by the Federal Ministry of Economics and Technology, both acting through the DLR Space Agency in Bonn. The European Space Agency (ESA) joined in that project from 1981 and the first experiment flew on the German TEXUS 6 mission in 1982.

Skylark VII two-stage solid fuel launchers (first stage: Goldfinch IID; second stage: Raven XI) manufactured by British Aerospace were usually employed in the TEXUS programme. The mission-related tasks, such as the provision of the rocket motor, the service systems and the launch service, are covered by an industrial consortium led by Airbus Defence and Space (Bremen, Germany). Since TEXUS 42, launched in December 2005, the Brazilian two-stage solid propellant VSB30 rocket motor has been used for TEXUS missions.

The European long-duration sounding rocket programme MAXUS started in 1990 and finally the MiniTEXUS was added in 1993 to the family of sounding rockets. The TEXUS/MiniTEXUS and MAXUS rockets consist of two major sections: the motor and the payload, which is mounted on top of it. The modular payload comprises the Recovery System with the parachute, the Service Module and the Experiment Modules. On a MAXUS rocket, additionally a Guide Control System (GCS) and a Telemetry and Tracking Unit (TTU) complement the payload (see **Figure 10**). The MiniTEXUS flight offers 3 1/2 minutes of microgravity flight, the TEXUS more than 6 minutes and the MAXUS flight about 13 minutes of microgravity [91, 92].

All missions are launched from the European rocket launch site ESRANGE near Kiruna in the north of Sweden. On its ballistic flight, microgravity (μ*g*) conditions (10^{-4} *g*) prevail for 3 1/2 minutes on a MiniTEXUS flight, for more than 6 minutes on a TEXUS flight and for about 13 minutes on a MAXUS flight [91, 92]. The payload of the rocket, meaning the tip that contains the Experiment Modules as well as the Recovery and Service System, comes down on a parachute and is transported back to the launch site by a helicopter. Scientific experiments are housed in modules stacked one atop the other within the rocket. Each experiment is directly monitored and controlled by researchers on the ground through telecommanding and TV systems. Scientific data are either directly transmitted during the flight by telemetry or saved after the payload has been recovered. The TEXUS/MAXUS missions offer a unique environment especially for biological research due to the availability of well-equipped laboratories close to the launcher for the preparation and post-flight analysis of biological specimen and due to the possibility of late access to the rocket and early retrieval of samples after the flight. Furthermore, safety requirements for this unmanned programme are less stringent than those of manned missions and experiments are less expensive.

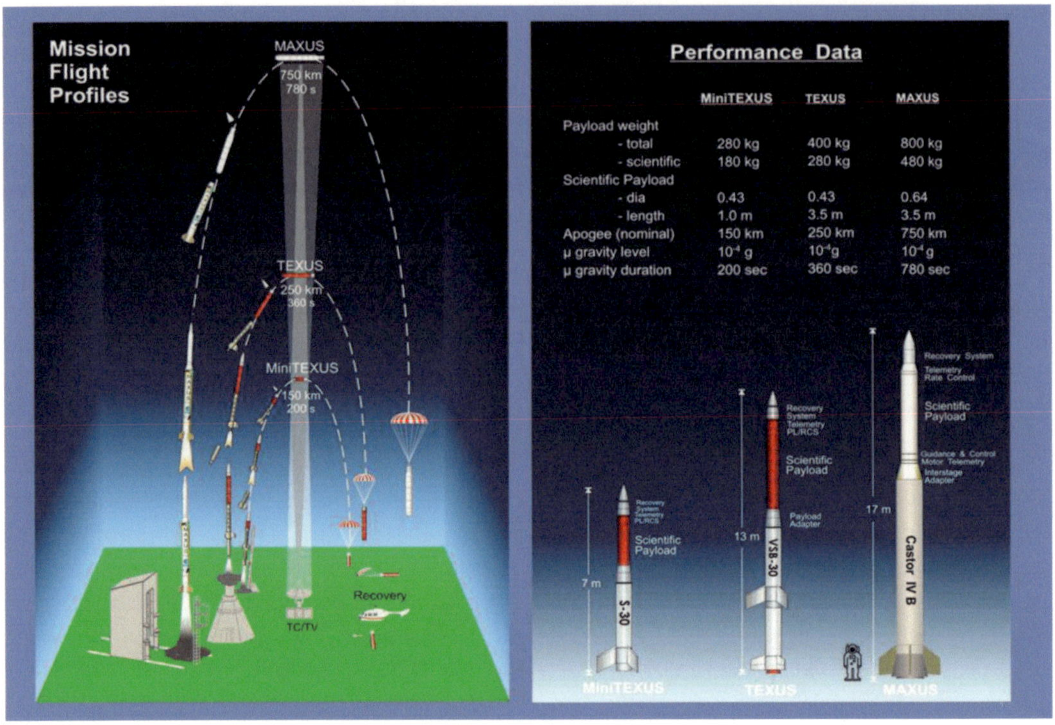

Figure 10. TEXUS/MAXUS launcher and flight profile. Courtesy of Airbus Defense & Space, Germany.

The major difference to other microgravity platforms such as parabolic flights and drop tower tests is the extended duration of the μ*g* phase and the high microgravity quality compared to parabolic flights.

Hence, a plethora of basic research was performed with different organisms during the various TEXUS and MAXUS missions. The gravitaxis and phototaxis in the flagellate *Euglena gracilis* were studied during the TEXUS 23 and TEXUS 28-30 missions [93], and the scientists were able to show that in the absence of light, the Earth's gravitational field is responsible for orientation in *E. gracilis,* as the directed upwards swimming was replaced by swimming in random distributions in μ*g*. Through the application of light during the flight, it was demonstrated that the cells showed both positive and negative phototaxis in μ*g* and the precision of orientation was higher than under terrestrial conditions. In the TEXUS 35 mission, the threshold for gravitaxis in *E. gracilis* was determined and it was shown that reorientation of the cells started at 0.12 *g* and the precision of gravitaxis increased with the increase of applied acceleration [94]. A further experiment in the TEXUS 36 mission dealt with the physiological mechanism of gravitaxis in *E. gracilis* and it was demonstrated that cAMP is involved in gravitaxis and the increase in this second messenger is triggered by mechanical stimuli [95]. A close relative of *E. gracilis,* the flagellate *Astasia longa,* was studied in the MAXUS 3 mission with respect to intracellular calcium levels at different accelerations, and a clear change due to changes from μ*g* to accelerated conditions could be observed [80]. During the TEXUS 27 and 28 missions, the swimming behaviour of the ciliate *Paramecium biaurelia* was analysed under μ*g* conditions. A random distribution of swimming directions was observed after 80 s of μ*g*, showing that gravity is the stimulus for the directed upwards swimming at 1 *g* [96]. Another organism whose locomotion behaviour was studied with a TEXUS mission (TX 48) is the cichlid fish *Oreochromis mossambicus*. This experiment gave evidence that fish are able to adapt to extreme gravitational habitat, as about 40% of the animals immediately started to swim normal after the launch and about 14% were able to regain normal swimming during the μ*g* phase [97].

Just recently, also the locomotion behaviour of zooplankton (*Daphnia magna, Daphnia cucullata* and *Heterocypris incongruens*) and predator-prey interactions between different trophic levels (predators: *Triops cancriformis*, larvae of the phantom midge; prey: *D. cucullata*) were studied in the TEXUS 52 mission. The long phase of μ*g* achieved with TEXUS facilitated to analyse which adaptation strategies towards weightlessness may occur in zooplankton and, in addition, a molecular study was carried out to investigate the influence of μ*g* on the expression of different genes, for example, stress markers.

For the observation of the organisms during the flight and preparation for post-flight analysis, specific experiment modules were developed that provide specific features like cameras with different magnifications, manipulators or centrifuges to apply a well-defined gravitation stimulus and fixation systems for the in-flight preservation of the specimen.

A typical experiment module design configuration is the TEXUS experiment module 06-31 (TEM 06-31). This TEM was accommodated on several TEXUS flights, namely TEXUS 45, 48 and 52, in the years 2008–2015 [98]. The design of the experiment module was always adapted to the scientific objectives of each dedicated flight campaign. For the last flight on TEXUS 52, the experiment module was equipped with a combination of any of the previous design features for observation and fixation. Thus, a description of this configuration covers all of the previous flight configurations.

The experiment module TEM 06-31 is presented in **Figure 11**. It consists of three different experiment platforms dedicated to a specific research objective and the Experiment Service System, which houses the electronic system for the automatic operation and control of the experiment module during the flight preparation phase and the flight phase.

Each of the experiment platforms is equipped with so-called late access units (LAU). These units house the aquatic systems in water-filled containers. The containers are sealed and connected via a short capillary tube to a small gas volume to compensate volume variations caused by temperature variations. The pressure inside the sealed containers is maintained at ambient pressure during the flight of the sounding rocket when the LAUs are exposed to a vacuum environment.

For the launch preparation, the LAUs are equipped with the organisms a few hours prior to the nominal rocket lift-off, checked for leak tightness and finally integrated into the rocket via late access hatches about 1 1/2 hours prior lift-off.

The three experiment platforms provide the following technical features for the observation or fixation of the specimen.

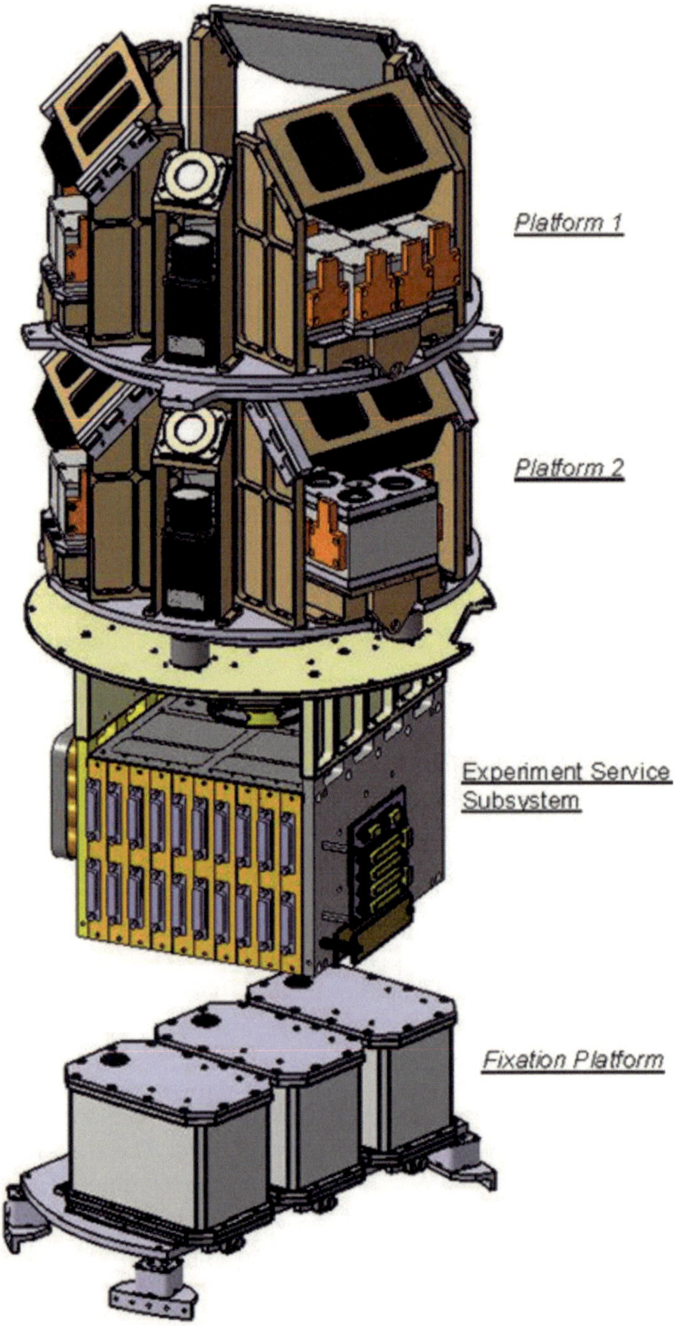

Figure 11. TEXUS experiment module 06-31. Courtesy of Airbus Defense & Space, Germany.

Platform 1: This platform is dedicated to the observation of the aquatic organisms' swimming behaviour under microgravity. Three LAUs with each accommodating eight sealed and water-filled containers are installed on this platform. The observation is performed from the top side of the containers via a set of mirrors and three CCD cameras equipped with an on-board video recording system for post-flight analysis of the video data. The illumination of the containers is performed via a backlight infrared illumination system, which is located beneath the late access units. The wavelength of the infrared light is not visible by the organisms.

Platform 2: The platform 2 design is based on the concept of platform 1 with the following main modification to study predator-prey interactions under microgravity conditions during microgravity: The sealed containers are equipped with additional mechanism for removing a barrier to a second small volume (see **Figure 12**). In the small volume, the prey is housed during the launch and released by operating the mechanism in the μg phase. In the main volume of the container, a predator is housed. The observation system is identical to platform 1.

Fixation Platform: This platform accommodates three LAUs. Each of these LAUs contains two sealed containers housing the aquatic specimen in a water volume of approximately 15 ccm. These LAUs are equipped with a mechanism containing a fixative volume of up to 20 ccm. This mechanism can be activated at any time during the mission. When activated, the water in the container will be completely substituted by the fixative solution within seconds.

Sounding rockets represent a highly suitable tool not only to analyse the behaviour and behavioural adaptability of zooplankton but also to study interactions between organisms from different trophic levels, which are a basic prerequisite for the establishment of food chains in bioregenerative life support systems (BLSS).

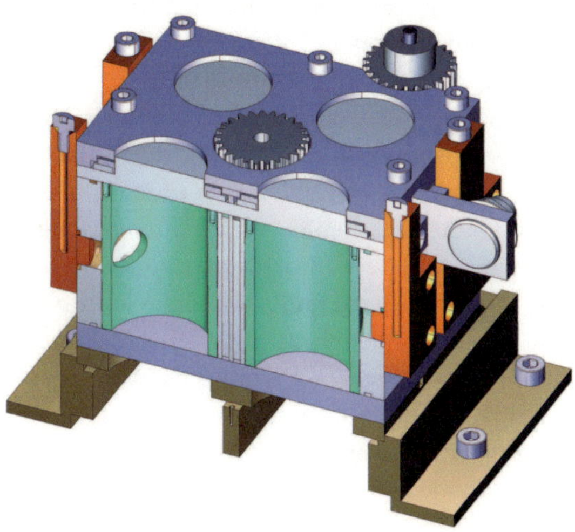

Figure 12. Observation unit with separation mechanism. Courtesy of Airbus Defense & Space, Germany.

5. Zooplankton experiments on space flights

Experiments that have directly assessed the survival and behaviour of zooplankton under microgravity conditions in actual space flights are few and basic. The first experiments more specifically observing the behaviour of microcrustaceans (cladocerans and ostracods) on board a spacecraft were carried out in basic ABS units (containing hornwort, amphipods and gastropods; [99]) on an Endeavour and Atlantis/Mir mission in 1996–1997, the latter for 4 months [15, 36]. Besides rudimentary observations on survival, which was more successful for ostracods than for cladocerans, and on aberrant swimming behaviour, no additional data were collected on zooplankton [36]. The most recent experiment was launched on the SpaceX Mission-8 in April, 2016. The experiment, developed by NanoRacks LCC and King's College London (UK) and sponsored by NASA, was carried out on the ISS as part of the educational International Space School Educational Trust (ISSET) Mission Discovery [100]. Behaviour of *Daphnia* was observed weekly by the astronauts (for the duration of a month) in small closed cell culture flasks fed with phytoplankton (*Chlorella* sp.) through a syringe connected to the flask, fitting into the dimensions of the ISS NanoRacks Platform, where light and temperature can be controlled (see [101]). The short trial illustrated the challenges of designing and conducting strongly size- and time-constrained experiments with zooplankton in the ISS and showed the limitations of what can be deduced from short observations during allocated crew time, indicating the importance of (and the need for) aimed ground-based work to improve space experiments on microgravity in zooplankton and expected outcomes.

From the few space experiments, we know that survival of live zooplankton up to several months is possible. Experiments are also possible after the animals are revived from dormancy. Besides being crucial components as intermediate consumers in the miniature artificial freshwater ecosystems, an additional important feature of zooplankton, in particular cladocerans, is their ability to produce drought- and cold-resistant dormant eggs. Exposure of *Daphnia* and *Eucypris ornate* (Ostracoda) resting stages (dormant eggs, which are encased in a chitinous casing, called *ephippium,* in *Daphnia*) to outer space and subsequent hatching on earth have shown that dormant eggs remain viable after having been transferred to outer walls of space platforms during missions (Biorisk experiment; [102]). Although undefined, effects of cosmic radiation and of microgravity on the viability of the dormant eggs are present. In the EXPOSE-R project, researchers showed that after 18 months in space, 11–35% of *Daphnia* ephippia and 7% of ostracod resting eggs hatched in comparison to Earth controls [103], still comparatively higher than dormant eggs of killifish under the same conditions. In general, animals with an anhydrobiotic stage or state show a higher tolerance to gamma radiation when desiccated than when hydrated due to the presence of high levels of protective molecules (e.g., tardigrades; [104]) and efficient DNA repair systems. As a by-product of adaptations against desiccation and freezing, such organisms show a high tolerance to a wide range of extreme conditions, and the study of dormancy in zooplankton is useful for space exploration [105].

http://dx.doi.org/10.5772/intechopen.74261

6. Simulation of weightlessness

Space flight experiments offer long periods of microgravity, but research in the near-Earth orbit is expensive and limited by the small amount of flight opportunities [106]. Other facilities such as drop towers, parabolic flights and sounding rockets provide only limited periods of microgravity ranging from few seconds to several minutes. In addition, application of the latter methods includes phases of hypergravity (reaching up to 30 *g* catapult acceleration at the Bremen Drop Tower [40]) before the onset of microgravity, which can present a considerable prejudice to some research questions.

The history of clinostats started at the end of the eighteenth century when Sachs and Pfeffer exposed plants on a device, which enables rotation of an object around an axis perpendicular to the direction of the gravity vector. By means of this so-called clinostat, the role of gravity for plant gravitropism could be visualized.

To some extent, weightlessness (microgravity) can be simulated on ground. Though gravity is a unique natural force - permanently present and acting - experimenters try to create a condition in which the organism "looses" its orientation with respect to gravity [107]. Gravity cannot be switched off on Earth, but its direction can be randomized. This situation is achieved when a test system is rotated on a horizontal axis, perpendicular to the direction of the gravity vector. The aim of this 2D clinostat principle is that the exposed organism can no longer detect the gravity vector. Consequently, several parameters have to be considered. From physical principles, it is obvious that speed of rotation and diameter determine the quality of the simulation. Furthermore, physiological parameters such as reaction time and thresholds for stimulus perception of the respective organism are of relevance, which are, however, in most cases not known. Many organisms have been exposed to simulated microgravity on clinostats, which have been adapted to several experimental demands (for review, see [108]) (**Figure 13**): clinostats for adherent or suspended organisms, aquatic systems and in combination with online analysis such as photomultipliers or microscopy. In some cases, a direct comparison with results obtained in real microgravity was possible and validated 2D clinorotation as appropriate method to simulate microgravity. An example is the random swimming displayed by previously gravitactic organisms [109].

Clinostats are essential parts to provide a comprehensive view on the impact of altered gravity on biological systems. Otolith growth in cichlid fish is slowed down by hypergravity, while microgravity during space flight increases their growth. Long-term clinorotation on a 2D fast-rotating clinostat confirmed these results in late-stage zebrafish [110]. Callus cells of *Arabidopsis thaliana* were exposed for 8 hours on a horizontal or a vertical clinostat [111]. The amount of glucose and fructose decreased while the starch content increased. In order to understand the physiological mechanism, the proteome was analysed after clinorotation. Eighty proteins were found to show qualitative and quantitative differences after horizontal rotation. Eighteen proteins showed a significant expression alteration under horizontal rotation in contrast to a vertical rotation. Rat alveolar macrophages (NR8383) were exposed on a

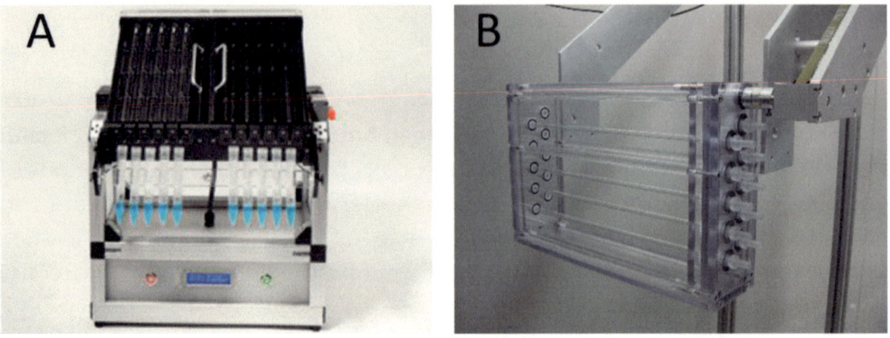

Figure 13. Clinostat with one rotation axis (2D clinostat), constantly running to simulate microgravity conditions for planktonic organisms and in general for cells in suspension. A: Clinostat with up to 10 pipettes, which can be emptied during clinorotation in prepared tubes. B: Clinostat constructed for aquatic systems to be placed into an aquarium.

2D clinostat to study the effect of gravity on the oxidative burst reaction, which is a key element in the immune response and cellular signalling [112]. The results indicate that reactive oxygen species (ROS) release is reduced in microgravity and enhanced in hypergravity and the adaptation to altered gravity occurs within seconds. Flagellates and ciliates were also observed under 2D clinorotation as well as on the random positioning machine (RPM) [49]. 2D clinorotation (60 rpm) was found to mimic the results obtained under real microgravity: the swimming direction was random, swimming paths kept their high linearity and the velocity was not impaired. In contrast, the protists showed erratic movements and frequent spontaneous changes in swimming direction on the RPM.

Exposure of the microcrustacean *Daphnia cucullata* in a 2D clinostat demonstrated that this method does not generate small scale turbulences, which resemble those generated by predators during their movement in the water column [113]. Otherwise, these organisms would express predator-induced defences such as peaked and elongated helmets as well as longer tail spines [114, 115]. However, the narrow tubes in a clinostat - a prerequisite to limit residual accelerations - limit long-term (embryonal) developmental studies as oxygen and food supply are severely restricted. Using a proteomic approach, key proteins and pathways involved in the response of *Daphnia* to simulated microgravity on a 2D clinostat (60 rpm with a residual gravity of ~0.008 *g*) were identified. Assuming that clinorotation is an appropriate simulation for *Daphnia*, the data indicate that microgravity will have an impact on the actin cytoskeleton disruption and breakdown of protein structures in general as well as an increase of energy demands. Interestingly, most of the proteins found to be affected are well-conserved throughout taxa and suggest that a lack of gravity affects similar molecular processes in a variety of organisms [50].

The studies with microcrustaceans and other organisms support the assumption that experiments in space can partly be replaced by studies using clinostats but also underline the need for experiments in real microgravity with respect to long-term exposure and adaptation processes [37].

A further technical approach to simulate weightlessness is the principle of 3D clinorotation. In contrast to a 2D clinostat, a second rotation axis perpendicular to the other axis characterizes a 3D clinostat. Furthermore, the operation is performed in a constant or randomly changing mode with respect to speed and rotation direction; the latter configuration has been named "random positioning machine (RPM)". So far, no evidence for an advantage of a 3D clinostat and a RPM over a 2D clinostat has been presented. Even more, induced side effects have to be critically considered. By using dinoflagellates (*Pyrocystis noctiluca*) as fast and sensitive reporter system, shear forces were made visible as they provoked bioluminescence [116]. The results show that the mechanical stress is higher on a RPM than during constant clinorotation, thereby proofing fast and constant 2D clinorotation as simulation method with negligible small side effects in contrast to random operation modes tested so far.

Magnetic levitation has been shown to be unsuitable to simulate microgravity in unicellular organisms, as magnet forces have a severe influence on the behaviour of the exposed organisms, as demonstrated in the cases of *Euglena* and *Paramecium* [117].

To sum up, ground-based studies in combination with long-term space flights are valuable tools to provide a comprehensive view on the role of gravity on the behaviour, physiology and genetics of motile microorganisms, which promise further insights into the complex molecular machinery of graviperception, signal transduction and responses. 2D clinostat running fast and with a restricted effective radius has been approved as the optimal simulation approach. However, due to the size limitations on clinostats, only small objects can be exposed under optimal simulation conditions on these instruments in order to avoid effects of radial accelerations. All results obtained under simulation conditions should be verified in selected space experiments in real microgravity.

7. Synopsis

Gravity plays a dominant role for spatial orientation of planktonic organisms and probably all eukaryotic organisms are capable of perceiving gravity [118]. Typical representatives of planktonic unicellular protists are *Euglena* and *Paramecium*. Both show a precise gravitactic orientation. The flagellate *Euglena*, which can grow photoautotrophically or heterotrophically, changes its swimming direction age-dependently ranging from preferentially downward swimming in the water column (positive gravitaxis) in young cells to pronounced upward swimming (negative gravitaxis) in older cells [119]. The underlying reason remains unknown so far, but a pure physical reason, i.e., changes in shape, could be excluded. The ciliate *Paramecium* shows negative gravitaxis, the precision of which is modulated by the oxygen concentration in the water [120, 121]. *Euglena* and *Paramecium* respond to a variety of other environmental stimuli such as oxygen gradients and light, assuming a complex interaction of underlying signalling pathways. Here, microgravity and hypergravity have become important analytical tools.

By using a plethora of experimental devices and designs on ground and in space, such as 2D and 3D clinostats, random positioning machines, sounding rocket flights, drop towers, satellites and shuttles, the mechanism of gravitactic orientation in unicellular organisms, which is summarized here as an example, was elucidated [30, 109, 122–124]. The first proof that gravitactic orientation in *Euglena* and *Paramecium* is due to the detection of the Earth's gravity field (1 *g*) and not due to the magnetic field lines or a chemical gradient was obtained during a

TEXUS (technical experiments under microgravity) sounding rocket flight [125, 126]. In microgravity, the cells swam randomly, while the 1 *g* controls displayed precise negative gravitaxis. In order to determine the threshold of gravitaxis, the slow rotating centrifuge microscope NiZeMi was installed in the Space Shuttle Columbia during the second international microgravity mission (IML-2). In orbit, *Euglena* and *Paramecium* cells were subjected to increasing accelerations via a centrifuge. The threshold for gravitactic orientation was found at $\leq$0.16 *g* and saturation at 0.64 *g* for *Euglena* [127] and in the range of 0.16–0.32 *g* for *Paramecium* [128]. This was confirmed in a subsequent sounding rocket flight [43]. Using the same instrument on ground showed that the *Euglena* and *Paramecium* cells even increase their gravitactic orientation >1 *g* (hypergravity) and can orient with respect to a centrifugal acceleration up to 9 *g* [129, 130].

One early hypothesis that *Euglena* cells orient themselves using a passive buoy effect [131] could be clearly falsified [132, 133]. Rather they use mechanosensitive ion channels as gravisensors of the large transient receptor potential protein (TRP) family [134], which is found in many organisms serving diverse functions such as photoperception, nociperception, thermosensation, taste, osmolar sensation and mechanosensing [135, 136]. Using RNA interference (RNAi) [137], the specific gravireceptor in *Euglena* could be identified as TRPC7 [134]. This Ca^{2+} gating channel can be efficiently blocked by gadolinium [138, 139]. Also, in *Paramecium*, mechano(gravi-)sensitive ion channels have been identified, which modulate the membrane potential and the ciliary activity [140]. Several calmodulins are found in *Euglena*. During gravistimulation, Ca^{2+} ions are gated into the cell where they bind specifically to one of the calmodulins (CaM.2), which is involved in the gravitaxis signal transduction chain. This was confirmed by RNAi [141]: after blocking the synthesis of CAM.2, gravitaxis was impaired. The changing Ca^{2+} concentrations during gravistimulation could be recorded using the fluorescence of Calcium Crimson induced by microgravity and hypergravity phases during parabolic flight manoeuvres [80] as well as on a centrifuge during a sounding rocket flight (MAXUS 3) [142]. After gravitactic stimulation, calmodulin activates an adenylyl cyclase, which converts ATP to cAMP, which was confirmed on a sounding rocket flight (TEXUS 36) [95]. The gravitactic sensory transduction chain in *Paramecium* also involves cAMP [143, 144]. A phosphodiesterase quenches the cAMP signal and decreases gravitactic orientation in *Euglena*, while inhibition of the enzyme or the application of the analogue 8-Bromo-cAMP enhances the precision of gravitaxis [145, 146]. In the final step, cAMP activates one (PK.4) of the five protein kinases A found in *Euglena* [147], which is thought to finally control the flagellar reorientation [118].

Euglena is an excellent candidate for bioregenerative life support systems. Being photosynthetic, it can absorb carbon dioxide exhaled by astronauts during long space flights, e.g., to Mars, and emit oxygen, which can be used by the astronauts. In addition, it is able to utilize ammonia, which is toxic to many other organisms. As a proof of principle, a closed system was constructed and run for more than 600 days. The system was composed of an 11-L tank populated by *Euglena* and a zoological compartment, which contained 15 snails (*Biomphalaria glabrata*) and 4 adult swordtail fish (*Xiphophorus hellerii*) [148]. Subsequently, several closed environmental life support systems (Aquacells, Omegahab) have been developed for space flights on Russian Foton satellites. The *Euglena* suspension was located in a 1450-ml cylindrical aquarium (in the Aquacells experiment on the Foton M2 mission) connected to a fish tank (35

http://dx.doi.org/10.5772/intechopen.74261

larval cichlids (*Oreochromis mossambicus*)), via membrane tubes, which allowed exchange of oxygen, carbon dioxide and ammonium excreted by the fish [30]. This assembly was successful in sustaining the fish during the space flight. In addition, the swimming behaviour and cell shape of the flagellates were video recorded. Another closed aquatic ecosystem, containing the chlorophyte *Chlorella*, *Euglena* and the snail *Bulinus*, was flown 17.5 days in orbit in cooperation with Chinese scientists on board the Shenzhou 8 spacecraft [31, 149]. After return, transcription of genes involved in signal transduction, oxidative stress defence, cell cycle regulation and heat shock responses were analysed using quantitative PCR. The analysis showed that *Euglena* suffered stress upon short-term exposure to microgravity since of the 32 tested genes, 18 stress-induced genes were upregulated. These results confirm the suitability of *Euglena* within a biological life support system.

Each of the approaches presented in this chapter is suitable to gain knowledge on how organisms respond to altered gravity conditions, especially microgravity, and can be regarded as stand alone. However, as the example of *Euglena* impressively shows, it is advisable to combine the different approaches in order to achieve a comprehensive understanding of how organisms deal with the absence of gravity and what the underlying physiological and genetic mechanisms are. These studies play a crucial role in the selection of suitable zooplankton organisms for bioregenerative life support systems.

Acknowledgements

The authors gratefully acknowledge the staff of ZARM FAB mbH in Bremen and Novespace in Bordeaux for assistance and support in creating this chapter. Special thanks go to Thorben Könemann from ZARM in Bremen and Ralf Anken from DLR in Cologne for reviewing the manuscript and providing valuable input and advice.

Author details

Miriam Knie[1], Bernard Wolfschoon Ribeiro[1], Jessica Fischer[1], Burkhard Schmitz[2], Kay Van Damme[3], Ruth Hemmersbach[4], Donat-P. Häder[5] and Christian Laforsch[1]*

*Address all correspondence to: christian.laforsch@uni-bayreuth.de

1 Animal Ecology I, University of Bayreuth, Bayreuth, Germany

2 Airbus Defence and Space GmbH, Bremen, Germany

3 Senckenberg Research Institute and Museum of Nature, Frankfurt am Main, Germany

4 German Aerospace Center (DLR), Institute of Aerospace Medicine, Gravitational Biology, Cologne, Germany

5 Department of Biology, Friedrich-Alexander University Erlangen-Nürnberg, Germany

References

[1] Campbell MR, Garbino A. History of suborbital spaceflight: Medical and performance issues. Aviation, Space, and Environmental Medicine. 2011;**82**:469-474. DOI: 10.3357/ASEM.2921.2011

[2] Mack PB. Bone density changes in a *Macaca nemestrina* monkey during the biosatellite 3 project. Aerospace Medicine. 1971;**42**:828-833

[3] Ijiri K. Fish mating experiment in space-what it aimed at and how it was prepared. Biological Sciences in Space. 1995;**9**:3-16. DOI: 10.2187/bss.9.3

[4] Okuneva AD, Vikhlyantsev IM, Shpagina MD, Rogachevskii VV, Khutzyan SS, Podlubnaya ZA, Grigoriev AI. Changes in titin and myosin heavy chain isoform composition in skeletal muscles of Mongolian gerbil (*Meriones unguiculatus*) after 12-day spaceflight. Biophysics. 2012;**57**:581-586. DOI: 10.1134/S0006350912050144

[5] Witt PN, Scarboro MB, Daniels R, Peakall DB, Gause RL. Spider web-building in outer space: Evaluation of records from the skylab spider experiment. The Journal of Arachnology. 1976;**4**:115-124. DOI: 10.2307/3705083

[6] Wang GH, Li GB, Hu CX, Liu YD, Song LR, Tong GH, Liu XM, Cheng ET. Performance of a simple closed aquatic ecosystem (CAES) in space. Advances in Space Research. 2004;**34**:1455-1460. DOI: 10.1016/J.ASR.2004.06.005

[7] Countryman SM, Stumpe MC, Crow SP, Adler FR, Greene MJ, Vonshak M, Gordon DM. Collective search by ants in microgravity. Frontiers in Ecology and Evolution. 2015;**3**:25. DOI: 10.3389/fevo.2015.00025

[8] Benguría A, Grande E, De Juan E, Ugalde C, Miquel J, Garesse R, Marco R. Microgravity effects on *Drosophila melanogaster* behavior and aging. Implications of the IML-2 experiment. Journal of Biotechnology. 1996;**47**:191-201. DOI: 10.1016/0168-1656(96)01407-1

[9] Ma BH, Cao YJ, Zheng WB, Lu JR, Kuang HB, Lei XH, Lv YH, Zhang T, Duan EK. Real-time micrography of mouse preimplantation embryos in an orbit module on SJ-8 satellite. Microgravity Science and Technology. 2008;**20**:127-136. DOI: 10.1007/s12217-008-9013-8

[10] Trappe S, Costill D, Gallagher P, Creer A, Peters JR, Evans H, Riley DA, Fitts RH. Exercise in space: Human skeletal muscle after 6 months aboard the International Space Station. Journal of Applied Physiology (Bethesda, Md: 1985). 2009;**106**:1159-1168. DOI: 10.1152/japplphysiol.91578.2008

[11] National Research Council. Pathways to Exploration: Rationales and Approaches for a U.S. Program of Human Space Exploration. Washington, DC: National Academies Press; 2014. DOI: 10.17226/18801

[12] Schwartzkopf SH. Design of a controlled ecological life support system. Bioscience. 1992;**42**:526-535. DOI: 10.2307/1311883

[13] von Baumgarten RJ, Simmonds RC, Boyd JF, Garriott OK. Effects of prolonged weightlessness on the swimming pattern of fish aboard Skylab 3. Aviation, Space, and Environmental Medicine. 1975;**46**:902-906

[14] Black S, Larkin K, Jacqmotte N, Wassersug R, Pronych S, Souza K. Regulative development of *Xenopus laevis* in microgravity. Advances in Space Research. 1996;**17**:209-217. DOI: 10.1016/0273-1177(95)00637-T

[15] Ishikawa Y, Kobayashi K, Mizutani H, Kawasaki Y, Koike J, Ijiri K, Yamashita M, Sugiura K, Poynter J, MacCallum T, Anderson G. Concluding remarks of autonomous biological systems (ABS) experiments. Biological Sciences in Space. 1998;**12**:394-399. DOI: 10.2187/bss.12.394

[16] Bluem V, Paris F. Aquatic modules for bioregenerative life support systems based on the C.E.B.A.S. Biotechnology. Acta Astronautica. 2001;**48**:287-297. DOI: 10.1016/S0094-5765(01)00025-X

[17] Slenzka K. Life support for aquatic species - past; present; future. Advances in Space Research. 2002;**30**:789-795. DOI: 10.1016/S0273-1177(02)00397-6

[18] Neubert J, Briegleb W, Schatz A. The frog-statolith-experiment (STATEX) of the German Spacelab mission D 1- Scientific background and technical description. In: International Astronautical Federation, 34th International Astronautical Congress (AF-83-184); 10–15 Oct 1983; Budapest, Hungary. pp. 1-6

[19] Neubert J, Briegleb W, Schatz A, Bromeis B, Link-Hommes A, Rahmann H, Slenzka K, Horn E, Eßeling K, Sebastian C. Spacelab mission D-2 experiment Statex "gravity perception and neuroronal plasticity" Comparison of near weightlessness effects on two waterliving vertebrates. In: Proceedings of the 5th European Symposium on 'Life Sciences Research in Space (ESA SP-366); Sep 26–Oct 1. France: Arcachon; 1994. pp. 77-81

[20] Sakimura T, Suzuki T, Matsubara S, Uchida S, Kato M, Tanemura R, Honda S. NASDA aquatic animal experiment facilities for space shuttle. Biological Sciences in Space. 1999;**13**:314-320. DOI: 10.2187/bss.13.314

[21] Mori S. Carp experiment in space microgravity – A visual-vestibular sensory conflict model. Biological Sciences in Space. 1997;**11**:327-333

[22] Uchida S, Matsubara S, Kato M, Sakimura T, Nakamura HK, Ogawa N, Nagaoka S. VFEU water quality control in STS-95 mission. Biological Sciences in Space. 1999;**13**:321-326. DOI: 10.2307/3556657

[23] Schatten H, Chakrabarti A, Levine HG, Anderson K. Utilization of the aquatic research facility and fertilization syringe unit to study sea urchin development in space. Journal of Gravitational Physiology. 1999;**6**:43-53

[24] Ishikawa Y, Anderson G, Poynter J, MacCallum T, Frye R, Kawasaki Y, Koike J, Kobayashi K, Mizutani H, Sugiura K, Ijiri K, Ishikawa Y, Saito T, Shiraishi A. Mini

ecosystem in space – preliminary experiment on board STS-77. Biological Sciences in Space. 1996;**10**:105-111

[25] Bluem V. C.E.B.A.S. a closed equilibrated biological aquatic system as a possible precursor for a long-term life support system? Advances in Space Research. 1992;**12**:193-204. DOI: 10.1016/0273-1177(92)90024-R

[26] Bluem V, Stretzke E, Kreuzberg K. C.E.B.A.S.-AQUARACK project: The mini-module as tool in artificial ecosystem research. Acta Astronautica. 1994;**33**:167-177. DOI: 10.1016/0094-5765(94)90122-8

[27] Bluem V, Andriske M, Paris F, Voeste D. The C.E.B.A.S.-Minimodule: Behaviour of an artificial aquatic ecological system during spaceflight. Advances in Space Research. 2000;**26**:253-262. DOI: 10.1016/S0273-1177(99)00569-4

[28] Lebert M, Porst M, Häder D-P. Long-term culture of *Euglena gracilis*: An AQUARACK progress report. In: Bräucker R, editor. Proceedings of the CEBAS Workshops Annual Issue 11. Bochum, Germany: Ruhr-University of Bochum; 1995. pp. 223-231

[29] Demey D, Hermans V, Cornet J-F, Leclercq J-J, Lasseur C, Delahaye A. BIORAT: Preliminary Evaluation of Biological Life Support in Space Environment (No. 2000-01-2384). SAE Technical Paper 2000. DOI: 10.4271/2000-01-2384

[30] Häder D-P, Richter PR, Strauch SM, Schuster M. Aquacells - flagellates under long-term microgravity and potential usage for life support systems. Microgravity Science and Technology. 2006;**18**:210-214. DOI: 10.1007/BF02870411

[31] Preu P, Braun M. German SIMBOX on Chinese mission Shenzhou-8: Europe's first bilateral cooperation utilizing China's Shenzhou programme. Acta Astronautica. 2014; **94**:584-591. DOI: 10.1016/j.actaastro.2013.08.022

[32] Strauch SM, Schuster M, Lebert M, Richter P, Schmittnagel M, Häder D-P. A closed ecological system in a space experiment. In: Proceedings of Life in Space for Life on Earth (ESA SP-663). Angers, France, June 22–27, 2007; pp. 22-27

[33] Tollrian R, Dodson SI. Inducible defences in cladocera: Constraints, costs, and multipredator environments. In: Tollrian R, Harvell CD, editors. The Ecology and Evolution of Inducible Defenses. Princeton: Princeton University Press; 1999. pp. 177-202

[34] Pace ML, Cole JJ, Carpenter SR, Kitchell JF. Trophic cascades revealed in diverse ecosystems. Trends in Ecology and Evolution. 1999;**14**:483-488. DOI: 10.1016/S0169-5347(99)01723-1

[35] Meester LD, Dawidowicz P, Loose C, van Gool E. Ecology and evolution of predator-induced behavior of zooplankton: Depth selection behavior and diel vertical migration. In: Tollrian R, Harvell CD, editors. The Ecology and Evolution of Inducible Defenses. Princeton: Princeton University Press; 1999. pp. 160-176

[36] Ijiri K, Mizuno R, Narita T, Ohmura T, Ishikawa Y, Yamashita M, Anderson G, Poynter J, MacCallum T. Behavior and reproduction of invertebrate animals during and after a long-term microgravity: Space experiments using an autonomous biological system (ABS). Biological Sciences in Space. 1998;**12**:377-388. DOI: 10.2187/bss.12.377

[37] Fischer J, Schoppmann K, Laforsch C. Life history responses and feeding behavior of Microcrustacea in altered gravity – Applicability in bioregenerative life support systems (BLSS). Microgravity Science and Technology. 2017;**29**:241-249. DOI: 10.1007/s12217-017-9545-x

[38] ZARM Drop Tower Bremen User Manual. 2012. Available from: https://www.zarm.uni-bremen.de/fileadmin/user_upload/drop_tower/Users_Manual_0412.pdf [Accessed: Oct 11, 2017]

[39] Knie M, Schoppmann K, Eck H, Wolfschoon RB, Laforsch C. The "*Daphnia*" lynx mark I suborbital flight experiment: Hardware qualification at the drop tower Bremen. Microgravity Science and Technology. 2016;**28**:345-349. DOI: 10.1007/s12217-015-9453-x

[40] von Kampen P, Kaczmarczik U, Rath HJ. The new drop tower catapult system. Acta Astronautica. 2006;**59**:278-283. DOI: 10.1016/j.actaastro.2006.02.041

[41] Mugnai S, Pandolfi C, Masi E, Azarello E, Monetti E, Comparini D, Voigt B, Volkmann D, Mancuso S. Oxidative stress and NO signalling in the root apex as an early response to changes in gravity conditions. BioMed Research International. 2014;**2014**:10. DOI: 10.1155/2014/834134

[42] Häder D-P, Rosumi A, Schäfer J, Hemmersbach R. Gravitaxis in the flagellate *Euglena gracilisis* controlled by an active gravireceptor. Journal of Plant Physiology. 1995;**146**:474-480. DOI: 10.1016/S0176-1617(11)82011-0

[43] Häder D-P, Porst M, Tahedl H, Richter P, Lebert M. Gravitactic orientation in the flagellate *Euglena gracilis*. Microgravity Science and Technology. 1997;**10**:53-57

[44] Anken R, Hilbig R. Swimming behaviour of the upside-down swimming catfish (*Synodontis nigriventris*) at high-quality microgravity – A drop-tower experiment. Advances in Space Research. 2009;**44**:217-220. DOI: 10.1016/j.asr.2009.02.009

[45] Anken RH, Hilbig R. A drop-tower experiment to determine the threshold of gravity for inducing motion sickness in fish. Advances in Space Research. 2004;**34**:1592-1597. DOI: 10.1016/j.asr.2004.01.023

[46] Hemmersbach R, Häder D-P. Graviresponses of certain ciliates and flagellates. FASEB Journal. 1999;**13**(Suppl):S69-S75

[47] Grosse J, Wehland M, Pietsch J, Ma X, Ulbrich C, Schulz H, Saar K, Hubner N, Hauslage J, Hemmersbach R, Braun M, van Loon J, Vagt N, Infanger M, Eilles C, Richter P, Baltz T, Einspanier R, Sharbati S, Grimm D. Short-term weightlessness produced by parabolic flight maneuvers altered gene expression patterns in human endothelial cells. FASEB Journal. 2012;**26**:639-655. DOI: 10.1096/fj.11-194886

[48] Vorselen D, Roos WH, MacKintosh FC, Wuite GJL, Van Loon JJWA. The role of the cytoskeleton in sensing changes in gravity by nonspecialized cells. FASEB Journal. 2014;**28**:536-547. DOI: 10.1096/fj.13-236356

[49] Häder D-P, Braun M, Grimm D, Hemmersbach R. Gravireceptors in eukaryotes – A comparison of case studies on the cellular level. NPJ Microgravity. 2017;**3**:13. DOI: 10.1038/s41526-017-0018-8

[50] Trotter B, Otte KA, Schoppmann K, Hemmersbach R, Fröhlich T, Arnold GJ, Laforsch C. The influence of simulated microgravity on the proteome of *Daphnia magna*. NPJ Microgravity. 2015;**1**:15016. DOI: 10.1038/npjmgrav.2015.16

[51] Studer M, Bradacs G, Hilliger A, Hürlimann E, Engeli S, Thiel CS, Zeitner P, Denier B, Binggeli M, Syburra T, Egli M, Engelmann F, Ullrich O. Parabolic maneuvers of the Swiss air force fighter jet F-5E as a research platform for cell culture experiments in microgravity. Acta Astronautica. 2011;**68**:1729-1741. DOI: 10.1016/j.actaastro.2010.11.005

[52] Karmali F, Shelhamer M. The dynamics of parabolic flight: Flight characteristics and passenger percepts. Acta Astronautica. 2008;**63**:594-602. DOI: 10.1016/j.actaastro.2008.04.009

[53] Pletser V. Short duration microgravity experiments in physical and life sciences during parabolic flights: The first 30 ESA campaigns. Acta Astronautica. 2004;**55**:829-854. DOI: 10.1016/j.actaastro.2004.04.006

[54] Fisher AC. Aviation medicine on the threshold of space: General considerations. Annals of International Medicine. 1959;**50**:1542-1549

[55] Voas RB. Project mercury astronaut training program. Annals of International Medicine. 1961;**4**:22-40

[56] Videbaek R, Norsk P. Atrial distension in humans during microgravity induced by parabolic flights. Journal of Applied Physiology. 1997;**83**:1862-1866

[57] Block J, Bäger A, Behrens J, Delovski T, Hauer LC, Schütze M, Schütz R, Spröwitz T. A self-deploying and self-stabilizing helical antenna for small satellites. Acta Astronautica. 2013;**86**:88-94. DOI: 10.1016/j.actaastro.2011.10.012

[58] Brillo J, Lohofer G, Hohagen FS, Schneider S, Egry I. Thermophysical property measurements of liquid metals by electromagnetic levitation. International Journal of Materials and Product Technology. 2006;**26**:247. DOI: 10.1504/IJMPT.2006.009469

[59] Frerichs I, Dudykevych T, Hinz J, Bodenstein M, Hahn G, Hellige G, Zhang Y, Passmore PJ, Bayford R, Herrmann P, Weisser G, Frerichs Z. Gravity effects on regional lung ventilation determined by functional EIT during parabolic flights. Journal of Applied Physiology. 2001;**91**:39-50

[60] Paul AL, Manak MS, Mayfield JD, Reyes MF, Gurley WB, Ferl RJ. Parabolic flight induces changes in gene expression patterns in *Arabidopsis thaliana*. Astrobiology. 2011; **11**:743-758. DOI: 10.1089/ast.2011.0659

[61] NASA. Flight Opportunities [Internet]. 2017. Available from: https://www.nasa.gov/directorates/spacetech/flightopportunities/index.html [Accessed: Nov 13, 2017]

[62] Zero Gravity Corporation. ZERO-G Research Programs [Internet]. 2017. Available from: http://gozerog.com/index.cfm?fuseaction=Research_Programs.welcome [Accessed: Nov 13, 2017]

[63] University of Toronto, Earth, Atmospheric and Planetary Physics [Internet]. 2010. Available from: http://www.atmosp.physics.utoronto.ca/~workshop/presentations_2010/Parabolic%20-%20C%20Swail.pdf [Accessed: Nov 11, 2017]

[64] Pletser V, Rouquette S, Friedrich U, Clervoy JF, Gharib T, Gai F, Mora C. The first European parabolic flight campaign with the Airbus A310 ZERO-G. Microgravity Science and Technology. 2016;**28**:587-601. DOI: 10.1007/s12217-016-9515-8

[65] Kohn FPM, Ritzmann R. Gravity and neuronal adaptation, in vitro and in vivo – From neuronal cells up to neuromuscular responses: A first model. European Biophysics Journal. 2017:1-11. DOI: 10.1007/s00249-017-1233-7

[66] Rodan GA. Perspectives: Mechanical loading, estrogen deficiency, and the coupling of bone formation to bone resorption. Journal of Bone and Mineral Research. 2009;**6**:527-530. DOI: 10.1002/jbmr.5650060602

[67] Guignandon A, Vico L, Alexandre C, Lafage-Proust MH. Shape changes of osteoblastic cells under gravitational variations during parabolic flight – relationship with PGE2 synthesis. Cell Structure and Function. 1995;**20**:369-375. DOI: 10.1247/csf.20.369

[68] Crawford-Young SJ. Effects of microgravity on cell cytoskeleton and embryogenesis. International Journal of Developmental Biology. 2006;**50**:183-191. DOI: 10.1387/ijdb.052077sc

[69] Yang F, Dai Z, Tan Y, Li Y. Effects of altered gravity on the cytoskeleton of neonatal rat cardiocytes. Microgravity Science and Technology. 2010;**22**:45-52. DOI: 10.1007/s12217-008-9103-7

[70] Hausmann N, Fengler S, Hennig A, Franz-Wachtel M, Hampp R, Neef M. Cytosolic calcium, hydrogen peroxide and related gene expression and protein modulation in *Arabidopsis thaliana* cell cultures respond immediately to altered gravitation: Parabolic flight data. Plant Biology. 2014;**16**:120-128. DOI: 10.1111/plb.12051

[71] Norsk P, Foldager N, Bonde-Petersen F, Elmann-Larsen B, Johansen TS. Central venous pressure in humans during short periods of weightlessness. Journal of Applied Physiology. 1987;**63**:2433-2437

[72] Väida P, Kays C, Rivière D, Téchoueyres P, Lachaud JL. Pulmonary diffusing capacity and pulmonary capillary blood volume during parabolic flights. Journal of Applied Physiology. 1997;**82**:1091-1097

[73] Mierau A, Girgenrath M, Bock O. Isometric force production during changed-Gz episodes of parabolic flight. European Journal of Applied Physiology. 2008;**102**:313-318. DOI: 10.1007/s00421-007-0591-8

[74] Oosterveld WJ, Greven AJ. Flight behaviour of pigeons during weightlessness. Acta Oto-Laryngologica. 1975;**79**:233-241. DOI: 10.3109/00016487509124679

[75] Wassersug RJ, Roberts L, Gimian J, Hughes E, Saunders R, Devison D, Woodbury J, O'Reilly JC. The behavioral responses of amphibians and reptiles to microgravity on parabolic flights. Zoology. 2005;**108**:107-120. DOI: 10.1016/j.zool.2005.03.001

[76] Wassersug RJ. Vertebrate biology in microgravity: Basic questions about how complex organisms respond to spaceflight and microgravity can only be answered by long-term study. American Scientist. 2001;**89**:46-53. DOI: 10.2307/27857399

[77] Hilbig R, Anken RH, Sonntag G, et al. Effects of altered gravity on the swimming behaviour of fish. Advances in Space Research. 2002;**30**:835-841. DOI: 10.1016/S0273-1177(01)00641-X

[78] Spangenberg D, Phillips R, Kostas A. Effect of graviceptor excision in *Aurelia ephyra* swimming/pulsing behavior at 1 G and during parabolic flight. American Society for Gravitational and Space Biology. 1989;**2**:24

[79] Krause M, Bräucker R, Hemmersbach R. Graviresponses of *Paramecium biaurelia* during parabolic flights. Protoplasma. 2006;**229**:109-116. DOI: 10.1007/s00709-006-0207-x

[80] Richter PR, Schuster M, Wagner H, Lebert M, Häder D-P. Physiological parameters of gravitaxis in the flagellate *Euglena gracilis* obtained during a parabolic flight campaign. Journal of Plant Physiology. 2002;**159**:181-190. DOI: 10.1078/0176-1617-00605

[81] Richter PR, Schuster M, Lebert M, Häder D-P. Gravitactic signal transduction elements in *Astasia longa* investigated during parabolic flights. Microgravity Science and Technology. 2003;**14**:17-24

[82] Goddard RH. The possibilities of the rocket in weather forecasting. Proceedings of the National Academy of Sciences. 1920;**6**:493-495

[83] Ordway FI. International Missile and Spacecraft Guide. New York City: McGraw-Hill; 1960

[84] Corliss WR. NASA Sounding Rockets, 1958–1968: A Historical Summary. The NASA Historical Report Series. 1st ed. Washington DC: Scientific and Technical Information Office, National Aeronautics and Space Administration; 1971

[85] Ley W. Rockets, Missiles, and Space Travel. New York City: Viking Press; 1961

[86] White LD. Final Report, Project Hermes V-2 Missile Program. Schenectady: General Electric; 1952

[87] Seibert G. The history of sounding rockets and their contribution to European space research. ESA-HSR-38, History Study Reports, No. 38. Noordwijk: ESA Publications Division; 2006

[88] Welcome to the NASA Sounding Rocket Science Website. [Internet]. 2017. Available from: https://rscience.gsfc.nasa.gov/index.html [Accessed: Oct 28, 2017]

[89] SSRP. Australian Space Research Institute. [Internet]. 2017. Available from: http://asri.org.au/project/ssrp/ [Accessed: Oct 28, 2017]

[90] European Sounding Rocket Programmes. [Internet]. 2017. Available from: https://web.archive.org/web/20070323162515/http://spaceflight.esa.int/users/index.cfm?act=default.page&level=11&page=1792 [Accessed: Oct 28, 2017]

[91] Preu P, Joop O, Ruyters G, Kuhl R, Braun M, Forke R. TEXUS Sounding Rockets for Microgravity Research. 2014:1-47. Available from: http://www.dlr.de/rd/Portaldata/28/Resources/dokumente/rw/Broschuere_TEXUS_lowres.pdf [Accessed: Nov 27, 2017]

[92] Franke B. Forschungsraketen. Lemwerder: Stedinger Verlag; 2007

[93] Häder D-P. Gravitaxis and phototaxis in the flagellate *Euglena studied* on TEXUS missions. Life Sciences Experiments Performed on Sounding Rockets (1985–1994): TEXUS 11-32, MASER 3-6 MAXUS 1. 1997;**1206**:77-79

[94] Richter PR, Lebert M, Tahedl H, Häder D-P. Physiological characterization of gravitaxis in *Euglena gracilis* and *Astasia longa* studied on sounding rocket flights. Advances in Space Research. 2001;**27**:983-988. DOI: 10.1016/S0273-1177(01)00163-6

[95] Tahedl H, Richter P, Lebert M, Häder D-P. cAMP is involved in gravitaxis signal transduction of *Euglena gracilis*. Microgravity Science and Technology. 1998;**11**:173-178

[96] Hemmersbach R: Swimming behaviour of the ciliate *Paramecium* in weightlessness. Life Sciences Experiments Performed on Sounding Rockets (1985–1994): TEXUS 11-32, MASER 3-6 MAXUS 1. 1997;**1206**:93-98

[97] Knie M, Shcherbakov D, Hilbig R. Individual behavioural adaptability to diminished g-forces and calcium uptake of inner ear otoliths in fish. A sounding rocket experiment (TX 48). In: Proceedings of Life in Space for Life on Earth (ESA SP-706); Aberdeen, United Kingdom. 18-22 June 2012; 6 p

[98] Schütte A. TEXUS and MAXUS an overview of the latest developments for the future. In: Proceedings of the 19th ESA Symposium on European Rocket and Balloon Programmes and Related Research (ESA SP-671); Bad Reichenhall, Germany. 7-11 June 2009. 373–377

[99] MacCallum TK, Anderson GA, Poynter JE, Stodieck LS, Klaus DM. Autonomous biological system (ABS) experiments. Biological Sciences in Space. 1998;**12**:363-365

[100] Barber C. NASA - NanoRacks–Mission Discovery Biomedical Experiments 2 [Internet]. 2016. Available from: https://www.nasa.gov/mission_pages/station/research/experiments/2337.html [Accessed: Dec 1, 2017]

[101] Lumpp JE, Erb DM, Clements TS, Rexroat JT, Johnson MD. The CubeLab standard for improved access to the international space station. In: 2011 Aerospace Conference; Mar 5–12, 2011; Big Sky, Montana, USA. pp. 1–6. DOI: 10.1109/AERO.2011.5747232

[102] Novikova N, Gusev O, Polikarpov N, Deshevaya E, Levinskikh M, Alekseec V, Okuda T, Sugimoto M, Sychev V, Grigoriev A. Survival of dormant organisms after long-term exposure to the space environment. Acta Astronautica. 2011;**68**:1574-1580. DOI: 10.1016/j.actaastro.2010.05.019

[103] Rabbow E, Rettberg P, Barczyk S, Bohmeier M, Parpart A, Panitz C, Horneck G, Burfeindt J, Molter F, Jaramillo E, Pereira C, Weiß P, Willnecker R, Demets R, Dettmann J, Reitz G. The astrobiological mission EXPOSE-R on board of the international Space Station. International Journal of Astrobiology. 2015;**14**:3-16. DOI: 10.1017/S1473550414000202

[104] Jönsson KI. Tardigrades as a potential model organism in space research. Astrobiology. 2007;**7**:757-766. DOI: 10.1089/ast.2006.0088

[105] Alekseev VR, Sychev VN, Novikova ND. Studying the phenomenon of dormancy: Why it is important for space exploration. In: Diapause in Aquatic Invertebrates Theory and

Human Use. Dordrecht, The Netherlands: Springer; 2007. pp. 207-214. DOI: 10.1007/978-1-4020-5680-2_14

[106] Ruyters G, Friedrich U. From the Bremen Drop Tower to the international space station ISS – Ways to weightlessness in the German space life sciences program. Signal Transduction. 2006;**6**:397-405. DOI: 10.1002/sita.200600108

[107] Sievers A, Hejnowicz Z. How well does the clinostat mimic the effect of microgravity on plant cells and organs? ASGSB Bulletin. 1992;**5**:69-75

[108] Brungs S, Egli M, Wuest SL, Christianen PCM, van Loon JJWA, Ngo Anh TJ, Hemmersbach R. Facilities for simulation of microgravity in the ESA ground-based facility programme. Microgravity Science and Technology. 2016;**28**:191-203. DOI: 10.1007/s12217-015-9471-8

[109] Herranz R, Anken R, Boonstra J, Braun M, Christianen PCM, de Geest M, Hauslage J, Hilbig R, Hill RJA, Lebert M, Medina FJ, Vagt N, Ullrich O, van Loon JJWA, Hemmersbach R. Ground-based facilities for simulation of microgravity: Organism-specific recommendations for their use, and recommended terminology. Astrobiology. 2013;**13**:1-17. DOI: 10.1089/ast.2012.0876

[110] Brungs S, Hauslage J, Hilbig R, Hemmersbach R, Anken R. Effects of simulated weightlessness on fish otolith growth: Clinostat versus rotating-wall vessel. Advances in Space Research. 2011;**48**:792-798. DOI: 10.1016/j.asr.2011.04.014

[111] Wang H, Hui QZ, Sha W, Zeng R, Qi CX. A proteomic approach to analysing responses of *Arabidopsis thaliana* callus cells to clinostat rotation. Journal of Experimental Botany. 2006;**57**:827-835. DOI: 10.1093/jxb/erj066

[112] Adrian A, Schoppmann K, Sromicki J, Brungs S, von der Wiesche M, Hock B, Kolanus W, Hemmersbach R, Ullrich O. The oxidative burst reaction in mammalian cells depends on gravity. Cell Communication and Signaling: CCS. 2013;**11**:98. DOI: 10.1186/1478-811X-11-98

[113] Fischer J, Schoppmann K, Knie M, Laforsch C. Responses of microcrustaceans to simulated microgravity (2D-Clinorotation) – Preliminary assessments for the development of bioregenerative life support systems (BLSS). Microgravity Science and Technology. 2016; **28**:337-344. DOI: 10.1007/s12217-015-9470-9

[114] Laforsch C, Tollrian R. Extreme helmet formation in *Daphnia cucullata* induced by small-scale turbulence. Journal of Plankton Research. 2004;**26**:81-87. DOI: 10.1093/plankt/fbg114

[115] Tollrian R, Laforsch C. Linking predator kairomones and turbulence: Synergistic effects and ultimate reasons for phenotypic plasticity in *Daphnia cucullata*. Archiv für Hydrobiologie. 2006;**167**:135-146. DOI: 10.1127/0003-9136/2006/0167-0135

[116] Hauslage J, Cevik V, Hemmersbach R. *Pyrocystis noctiluca* represents an excellent bioassay for shear forces induced in ground-based microgravity simulators (clinostat and random positioning machine). NPJ Microgravity. 2017;**3**:12. DOI: 10.1038/s41526-017-0016-x

[117] Hemmersbach R, Simon A, Waßer K, Hauslage J, Christianen PCM, Albers PW, Lebert M, Richter P, Alt, W, Anken R. Impact of a high magnetic field on the orientation of

gravitactic unicellular organisms – A critical consideration about the application of magnetic fields to mimic functional weightlessness. Astrobiology 2014;**14**:205-215. DOI: 10.1089/ast.2013.1085

[118] Häder D-P, Hemmersbach R. Gravitaxis in *Euglena*. In: Shigeoka S, Schwartzbach S, editors. *Euglena*: Biochemistry, Cell and Molecular Biology. Cham: Springer; 2017. pp. 237-266. DOI: 10.1007/978-3-319-54910-1_12

[119] Häder D-P, Hemmersbach R, Lebert M. Gravity and the Behavior of Unicellular Organisms, Vol. 40. Cambridge: Cambridge University Press; 2005

[120] Hemmersbach-Krause R, Häder D-P. Negative gravitaxis (geotaxis) of *Paramecium*-demonstrated by image analysis. Applied Microgravity Technology. 1990;**4**:221-223

[121] Hemmersbach-Krause R, Briegleb W, Häder D-P. Dependence of gravitaxis in *Paramecium* on oxygen. European Journal of Protistology. 1991;**27**:278-282. DOI: 10.1016/S0932-4739(11)80065-9

[122] Vogel K, Hemmersbach-Krause R, Kühnel C, Häder D-P. Swimming behavior of the unicellular flagellate, *Euglena gracilis*, in simulated and real microgravity. Micrograv. Science and Technology. 1993;**4**:232-237

[123] Häder D-P, Lebert M. Graviperception and gravitaxis in algae. Advances in Space Research. 2001;**27**:861-870. DOI: 10.1016/S0273-1177(01)00149-1

[124] Hemmersbach R, Strauch SM, Seibt D, Schuber M. Comparative studies on gravisensitive protists on ground (2D and 3D clinostats) and in microgravity. Microgravity – Science and Technology. 2006;**18**:257-259. DOI: 10.1007/BF02870423

[125] Hemmersbach-Krause R, Briegleb W, Häder D-P, Vogel K, Grothe D, Meyer I. Orientation of *Paramecium* under the conditions of weightlessness. Journal of Eukaryotic Microbiology. 1993;**40**:439-446. DOI: 10.1111/j.1550-7408.1993.tb04937.x

[126] Häder D-P, Vogel K, Schäfer J. Responses of the photosynthetic flagellate, *Euglena gracilis*, to microgravity. Applied Microgravity Technology. 1990;**3**(2):110-116

[127] Häder D-P, Rosum A, Schäfer J, Hemmersbach R. Graviperception in the flagellate *Euglena gracilis* during a shuttle space flight. Journal of Biotechnology. 1996;**47**:261-269. DOI: 10.1016/0168-1656(96)01374-0

[128] Hemmersbach R, Voormanns R, Briegleb W, Rieder N, Häder D-P. Influence of accelerations on the spatial orientation of *Loxodes* and *Paramecium*. Journal of Biotechnology. 1996;**47**:271-278. DOI: 10.1016/0168-1656(96)01337-5

[129] Hemmersbach R, Voormanns R, Häder D-P. Graviresponses in *Paramecium biaurelia* under different accelerations: Studies on the ground and in space. The Journal of Experimental Biology. 1996;**199**:2199-2205

[130] Häder D-P, Reinecke E, Vogel K, Kreuzberg K. Responses of the photosynthetic flagellate, *Euglena gracilis*, to hypergravity. European Biophysics Journal. 1991;**20**:101-107. DOI: 10.1007/BF00186258

[131] Fukui K, Asai H. Negative geotactic behavior of *Paramecium caudatum* is completely described by the mechanism of buoyancy-oriented upward swimming. Biophysical Journal. 1985;**47**:479-482. DOI: 10.1016/S0006-3495(85)83940-0

[132] Richter P, Ntefidou M, Streb C, Lebert M, Häder D-P. Cellular perception and transduction mechanisms of gravity in unicellular organisms. Current Topics in Plant Biology. 2002;**3**:143-154

[133] Lebert M, Häder D-P. Negative gravitactic behavior of *Euglena gracilis* can not be described by the mechanism of buoyancy-oriented upward swimming. Advances in Space Research. 1999;**24**:851-860. DOI: 10.1016/S0273-1177(99)00966-7

[134] Häder D-P, Richter PR, Schuster M, Daiker V, Lebert M. Molecular analysis of the graviperception signal transduction in the flagellate *Euglena gracilis*: Involvement of a transient receptor potential-like channel and a calmodulin. Advances in Space Research. 2009;**43**:1179-1184. DOI: 10.1016/j.asr.2009.01.029

[135] Nilius B, Appendino G, Owsianik G. The transient receptor potential channel TRPA1: From gene to pathophysiology. Pflugers Archiv European Journal of Physiology. 2012; **464**:425-458. DOI: 10.1007/s00424-012-1158-z

[136] Nilius B, Owsianik G. The transient receptor potential family of ion channels. Genome Biology. 2011;**12**:218. DOI: 10.1186/gb-2011-12-3-218

[137] Tuschl T, Elbashir SM, Lendeckel W. RNA Interference Mediating Small RNA Molecules. 2014. Google Patents

[138] Häder D-P, Hemmersbach R. Graviperception and graviorientation in flagellates. Planta. 1997;**203**:S7-S10. DOI: 10.1007/PL00008118

[139] Lebert M, Häder D-P. How *Euglena* tells up from down. Nature. 1996;**379**:590

[140] Ogura A, Machemer H. Distribution of mechanoreceptor channels in the paramecium surface membrane. Journal of Comparative Physiology A: Neuroethology, Sensory, Neural, and Behavioral Physiology. 1980;**135**:233-242. DOI: 10.1007/BF00657251

[141] Daiker V, Lebert M, Richter P, Häder DP. Molecular characterization of a calmodulin involved in the signal transduction chain of gravitaxis in *Euglena gracilis*. Planta. 2010; **231**:1229-1236. DOI: 10.1007/s00425-010-1126-9

[142] Häder D-P, Lebert M. Mechanism of gravitactic signal perception and signal transduction of *Euglena gracilis*. Microgravity News. 1998;**11**:14

[143] Hemmersbach R, Braun M. Gravity-sensing and gravity-related signaling pathways in unicellular model systems of protists and plants. Signal Transduction. 2006;**6**:432-442. DOI: 10.1002/sita.200600106

[144] Bräucker R, Cogoli A, Hemmersbach R. Graviperception and graviresponse at the cellular level. In: Horneck G, Baumstark-Khan C, editors. Astrobiology: The Quest for the Conditions of Life. Heidelberg: Springer Verlag; 2001. pp. 284-297

[145] Lebert M, Richter P, Häder D-P. Signal perception and transduction of gravitaxis in the flagellate *Euglena gracilis*. Journal of Plant Physiology. 1997;**150**:685-690. DOI: 10.1016/S0176-1617(97)80284-2

[146] Streb C, Richter P, Ntefidou M, Lebert M, Häder D-P. Sensory transduction of gravitaxis in *Euglena gracilis*. Journal of Plant Physiology. 2002;**159**:855-862. DOI: 10.1078/0176-1617-00769

[147] Daiker V, Häder D-P, Richter PR, Lebert M. The involvement of a protein kinase in phototaxis and gravitaxis of *Euglena gracilis*. Planta. 2011;**233**:1055-1062. DOI: 10.1007/s00425-011-1364-5

[148] Porst M, Lebert M, Häder D-P. Long-term cultivation of the flagellate Euglena Gracilis. Microgravity Science and Technology. 1997;**10**:166-169

[149] Nasir A, Strauch SM, Becker I, Sperling A, Schuster M, Richter PR, Weißkopf M, Ntefidou M, Daiker V, An YA, Li XY, Liu YD, Lebert M. The influence of microgravity on *Euglena gracilis* as studied on Shenzhou 8. Plant Biology. 2014;**16**:113-119. DOI: 10.1111/plb.12067

Chapter 12

Are We Alone? The Search for Life on Mars and Other Planetary Bodies

Stephanie A. Smith, Andrzej Paszczynski and
Susan E. Childers

Additional information is available at the end of the chapter

http://dx.doi.org/10.5772/intechopen.75437

Abstract

Extensive research has shown the potential for microorganisms to survive in some of the most extreme environments on Earth. Our current understanding of diverse life on Earth implies that, even though the surface of Mars is very inhospitable to life, it is possible that there may be indigenous microorganisms on Mars, especially in the protective subsurface. Ultimately, a better understanding of microbial diversity on Earth is needed to determine the limits of life to help determine the potential for life on Mars and other exoplanets.

Keywords: microorganisms, extreme environments, life on Mars, exoplanets

1. Introduction

The search for extraterrestrial life is bolstered by our long-standing quest to determine if we are alone in the Universe. Mars and Europa are two likely candidates to target in the search for extraterrestrial life, since both have carbon, potential energy sources, and water in some form [1–4]. The current focus to search for life on Mars is supported by the fact that although Mars is quite cold and dry, current conditions are thought to be analogous to conditions on early Earth when single-celled life was gaining a foothold [5]. Furthermore, because there is a diversity of microorganisms known to thrive in the most inhospitable habitats on Earth, it is not unreasonable to think that microorganisms could live on Mars.

While continuing Mars explorations confirm that all of the basic necessities for microbial life are present, it remains unclear whether microorganisms that are metabolically capable of living on Mars can actually survive in the Mars environment. The Mars surface presents a very inhospitable habitat for life because of the intense radiation, highly oxidizing conditions, concentrated evaporative salts, and extremely low water activity. Determining if microbes can survive those surface conditions, including tolerance to radiation (both ionizing and non-ionizing), desiccation, and oxidizing environments of microorganisms that utilize the carbon and energy resources available on Mars is vital. Some information is available on the survival of spore-forming microbes in a Mars-like environment, but much more information is needed regarding survival potential of different types of microorganisms. This chapter will focus on the search for life on planetary bodies.

2. The Martian environment

Although Mars is considered to be at the outer edge of the habitable zone of our solar system, the idea that there could potentially be life on Mars, especially in the subsurface, is not unfathomable. Although it can be expected that different areas of Mars would have somewhat different environments dependent on location, overall the Martian environment is quite inhospitable to most life as we know it. Average temperatures on Mars can range from −10 to −76°C with an average surface temperature of −65°C although temperatures can fluctuate from as high as 25°C to −123°C [6–8].

Mars is considered to be quite dry, but recent information suggests otherwise. Studies of the Gale Crater by the rover Curiosity found hydration of soils to be as much as 2.25 wt%. This finding was consistent with findings by both Viking 1 and 2 and the Mars Odyssey [9]. What is unknown is if there is an underground source of water. Geophysical and geochemical features on Mars indicate that there may have been water on the surface at some time in the past but it is unknown to what extent surface water would have existed. Features include alluvial fans in craters, dendritic valley networks, and the presence of specific minerals thought to only form in the presence of water. One hypothesis is that hydrothermal environments associated with craters from impacts and volcanism could have easily provided a source of liquid water on Mars [10].

The Martian atmosphere is much different from that on Earth. Mars has primarily a CO_2 atmosphere (95.3%) compared to the CO_2 content in Earth's atmosphere (0.04%). Earth's atmosphere consists mainly of N_2 (78.1%) while there is only 2.7% N_2 in the Martian atmosphere. The O_2 concentration on Earth is 20.9% whereas Mars' atmosphere contains only about 0.1% O_2 [7]. Studies by Mumma et al. [11] showed the presence of methane in extended plumes that appeared to be released from discrete regions on Mars. One of the principal plumes contained as much as 19,000 metric tons of methane, an amount comparable to that of a massive hydrocarbon seep in Santa Barbara, California. However, analyses by the Mars rover Curiosity found no detectable atmospheric methane. Although results are contradictory, it is possible that the location of the rover was too far from the methane seeps and prevented the detection of methane in the atmosphere.

The surface of Mars is subjected to both cosmic ionizing radiation and solar UV radiation. Ionizing radiation on Mars is believed to be 100X higher than on Earth, ranging from 100 to

200 mSv/a compared to Earth 1–2 mSv/a [7]. The UV-B and UV-C fluxes on Mars are nearly 5× higher than they are on Earth with fluences of 361 kJ/m² and 78 kJ/m², respectively [12]. On Mars, the high atmospheric concentration of CO_2 neutralizes incoming UV radiation <200 nm, however wavelengths >200 nm still reach the Martian surface [7]. Of note is that some of the UV radiation may be attenuated at times by the presence of dust storms in a particular region.

Data collected during the Viking missions showed that the surface of Mars was highly oxidized compared to the Mars atmosphere [13]. Mapping of hydrogen peroxide (H_2O_2) on Mars using infrared high-resolution imaging spectroscopy indicated H_2O_2 abundance on Mars is 15 ± 10 ppb although prior mapping showed concentrations as high as 40 ppb [14]. The formation of peroxides could occur in the presence of hematite, trace amounts of water, and UV radiation [7]. A more likely scenario is that radiolysis of ice or water would create a larger amount of peroxide formation. It has been reported that the surface ice of Europa contains as much as 1,300,000 ppb H_2O_2 which is generated from radiolysis of ice [15]. Additionally, perchlorate, a strong oxidizing agent, was found by the Phoenix Lander to be present in Martian soils in concentrations of 2.1–2.6 mM [16].

Martian soils contain few nutrients to support life as we know it, and the soils themselves pose a harsh environment. Martian soils were expected to be acidic, but the Phoenix Lander showed that the soils at its landing site were mildly basic with a pH of 7.7 ± 0.5 [17]. Salt tolerance would be required for life to survive and grow on Mars due to the high salt concentrations found in Martian soils in the form of NaCl, $MgSO_4$, $CaSO_4$, $FeSO_4$, MgCl and $CaCl_2$ [17]. The lack of water, the intense radiation and oxidative conditions make the Martian surface quite inhospitable to life.

3. Special regions on Mars

3.1. Introduction

Mars Special Regions are regions where organisms are likely to survive. NASA Procedural Requirement (NPR) 8020.12D [18] defines these areas as regions that have a high potential for the existence of extant Martian life forms, have sufficient water activity (0.5–1.0 aw) and have sufficiently warm temperatures (−25° C lower limit) to permit replication of Earth organisms. Areas that have observed features that may be associated with the presence of water must also be classified as Special Regions. It is noted that these parameters may need to be changed as our understanding of Mars and life on Earth evolve and as our technological capabilities improve [19, 20].

3.2. Formulating special regions

The COSPAR colloquium on special regions stated that "Preventing terrestrial biological contamination from becoming established and widespread on Mars is essential to our ability to protect high-priority science goals on Mars" [20]. The current standards are based solely on protecting science goals and not on protecting Mars in and of itself. The NRC study takes a precautionary principle approach by stating that there is insufficient data to determine which regions of Mars should be considered "special" and that all of Mars should be considered "special" until it can be proven otherwise [19, 20]. The COSPAR disagreed and concluded that there was sufficient data to arrive at a conclusion as to which areas of Mars would be defined as "Special Regions" [20].

The COSPAR colloquium concluded with the enactment of the standards that are currently in NPR 8020.12D [18]. Two main standards, water activity and temperature, are the basis for determining which regions should be taken into consideration. One area in need of additional research is that of microbial growth and reproduction at low temperatures. It was noted that most of the work in this area has been performed on laboratory isolates and more environmental data is needed to begin to define the lower temperature for life. It was concluded that investigations were needed to determine if microbial reproduction at water activities of lower than 0.6 is possible, that more studies are needed using Mars simulated environments, and that knowledge of reproduction of communities rather than isolates is essential to improve our understanding of life. COSPAR also noted that a larger phylogenetically diverse array of organisms needs to be studied and diurnal, seasonal and long-term variations in the Martian surface need to be better understood [20].

The NASA Planetary Protection Office made some initial suggestions to try to define special regions. The parameters were set as: (1) the existence of liquid water in "pure" form or in strong brines up to 5.5 M $CaCl_2$; (2) regions of current or active volcanism or enhanced heat flow which is yet unknown; (3) permafrost through 100% water ice, including segregated ground ice, ice-rich frozen ground, polar caps and subsurface ice; (4) subpermafrost groundwater and (5) any gully system that may be indicative of recent water activity within the last <50,000 years. The Special Regions Science Analysis Group (SRSAG) determined that regions should be defined as non-special if the temperature remains below −20°C or the water activity remains below 0.5 for a period of 100 years after spacecraft arrival [21]. Ultimately, the SRSAG developed a map of regions that are considered "significant" and of interest for determining special region boundaries. Of note is that the current definition of special regions mostly takes into account the known and sets the water activity and temperature parameters slightly below what is currently known.

4. The relationship between life on Earth, and the potential for life on Mars

4.1. Life in extreme environments

Despite our limited knowledge of microbes on Earth, everywhere we have looked for microbes and we have been able to find them. It appears that life inhabits all places on Earth including some of the most extreme environments imaginable. Microorganisms have been discovered surviving and reproducing in hot springs, at terrestrial depths exceeding 2 km, in the most arid of deserts, and in hydrothermal vents on the ocean floor. Microbial life has been found in extremely cold places such as in Antarctica and Greenland, and microorganisms have been described as reproducing and thriving at temperatures as low as −15°C. Many microorganisms can grow in salt at concentrations exceeding 20% NaCl, or 2 M $MgSO_4$, and others thrive in either very acidic or very alkaline environments. Microbes can conserve energy by respiring some of the most extreme compounds, such as U, Mn, Se, As, S and Cl-based molecules. Life at either high or low atmospheric pressures has been described, as well as organisms that are highly resistant to radiation and oxidative conditions. Most of the organisms surviving in these types of environments have a symbiotic relationship with other organisms in the same community. For example, methane-oxidizing archaea (MOA) are known to live in symbiosis with sulfate-reducing bacteria (SRB) in deep hydrothermal vents on the ocean floor. MOAs break down methane to CO_2 and H_2, and the H_2 is then utilized by the SRBs to reduce SO_4^{-2} to HS^- [22]. These types of relationships between organisms are far from uncommon.

Earth microorganisms have developed physiological and biochemical mechanisms to be able to survive in a variety of extreme niches. As previously stated, it would not be unreasonable to expect niches on Mars, although considered extreme, to support microbial life of some sort as well. The remainder of this chapter will discuss what is known about how microorganisms survive some of these extreme environmental conditions and how this information is relative to the potential for life on Mars. Although this portion of the chapter will focus on bacteria, it should always be kept in mind that many of the topics discussed apply to archaea and fungi as well.

4.2. Survival at low temperatures

Average temperatures on Mars can range from −10 to −76°C with an average surface temperature of −65°C although temperatures can fluctuate from as high as 25°C to as low as −123°C [6–8]. For an organism to be able to thrive on Mars it would need to be able to grow and reproduce in these frigid temperatures. An exception would be a subsurface environment that was geothermally heated though no such areas have been discovered on Mars.

A number of psychrophilic (cold-loving) organisms have been isolated from many regions of the Arctic and Antarctic where there are polar ice sheets, glaciers and permafrost. Additionally, microorganisms are known to inhabit the ocean floor where temperatures are ≤4°C [23]. These organisms are comprised of representatives from the Eukarya (algae, fungi and yeast), Bacteria and Archaea. Morozova et al. [24] identified several methanogenic archaea that were able to survive not only low temperatures ranging from −75 to 20°C, but could also simultaneously survive low humidity and a 95.3% CO_2 atmosphere. The methanogens that survived best under these conditions were isolated from permafrost. Six isolates from permafrost and nine known species of *Carnobacterium* were found to grow not only at 23°C, but also at 0°C, under low pressure and in a CO_2-enriched anoxic atmosphere [25]. A strain of *Serratia liquefaciens*, a common mesophilic organism often found as a contaminant in bathtubs, was shown to be capable of growth at 0°C as well as at low pressure and CO_2-enriched anoxic atmospheres [26]. Mykytczuk et al. [27] identified a *Planococcus* isolate that grows and divided at −15°C and is still metabolically active at −25°C.

Despite these organisms being interesting in themselves, what is even more interesting is the ability of these organisms to make both physiological and biochemical modifications to survive in such environments. *Psychrobacter arcticus* 273-4, a bacterium capable of growing at temperatures as low as −10°C, was found to downregulate genes related to energy metabolism and carbon incorporation, and upregulate genes required for maintenance of membranes, cell walls and nucleic acid motion. Furthermore, this organism turns on the expression of a cold-shock DEAD-box RNA helicase A, a protein that may be key for maintaining life in cold temperatures [28]. *Planococcus halocryophilus* Or1 grew at subzero temperatures by forming encrustations around the cell and increasing the ratio of saturated to branched fatty acids in the cytoplasmic membrane [27]. This is unique because often growth at lower temperatures results in a higher content of unsaturated, polyunsaturated and methyl-branched fatty acids to increase membrane fluidity at these temperatures. In many organisms, enzymes involved in transcription, translation, protein folding and stabilization of DNA and RNA show activity at very low temperatures and are adapted to life in cold environments. Antifreeze-like proteins have been seen in Antarctic lake microbes and trehalose and exopolysaccharides might also provide cryoprotection for psychrophiles [29]. Although scientists are far from having a full understanding at life in cold temperatures, studies like the ones above provide insights as to how these organisms adapt to their extreme environment. Additionally, the microbes are models to further our understanding of how organisms may survive on Mars, and can be useful as we continue the search for life on cold planets and moons.

4.3. Tolerance to high salt

Due to the high salt concentrations found in Martian soils in the form of NaCl, $MgSO_4$, $CaSO_4$, $FeSO_4$, MgCl and $CaCl_2$, salt tolerance would be required for life to survive and grow on Mars [17]. Salts can be chaotropic as they influence water activity, affect cell turgor, and are major stressors of cellular systems [30]. It is estimated that 1/4th of the Earth's land is covered by salt and salt water makes up the majority of Earth's water. On Mars, it is estimated that sulfurous salts are more common than chlorinated salts by a ratio of 3:1. On Earth the most common type of salt is NaCl but many brines also contain $MgCl_2$, $MgSO_4$ and other salts [17, 30]. Studying hypersaline environments from Earth increases our understanding of how organisms can adapt to these extreme environments.

Many *Bacillus* sp. are salt-tolerant and thus of special interest with regard to growth under high salt conditions. Previous studies in our laboratory have shown that many different species of *Bacillus*, including *pumilus*, *licheniformis*, *horti*, *mannailyticus* and *cellulosilyticus*, as well as species belonging to other genera including *Paenibacillus*, *Amphibacillus* and *Alkalibacterium*, could grow under salt concentrations as high as 10% NaCl. Several of these organisms also showed growth in media containing 20% NaCl. These isolates were collected from the Alvord Basin in Oregon where the soils are known to have elevated salt concentrations [31]. The ability of *Bacillus* sp. to grow under these conditions is not uncommon and many organisms which have been identified as non-spore formers can also grow in high NaCl concentrations.

A diversity of prokaryotes was discovered residing in deep hypersaline anoxic basins in the Mediterranean Sea; basins that are nearly saturated with $MgCl_2$ (5 M). In addition to growing in extremely high concentrations of $MgCl_2$, the microorganisms were involved in sulfate reduction and methanogenesis, and contributed to the cycling of carbon [32]. Furthermore, the overall microbial community was unique because the bacteria and archaea identified were not related to organisms normally found in seawater, and the archaea branched deeply within the Euryarchaeota indicating they comprised a new order.

It is estimated that the majority of salt on Mars would likely be $MgSO_4$, with lower concentrations of NaCl and $CaCl_2$. Studies by Crisler et al. [17] focused on the growth of microorganisms under high $MgSO_4$ concentrations using microorganisms collected from the Great Salt Plains in Oklahoma. Though the microbes were not identified, it was found that 35% of the organisms from the bacterial collection could grow in medium containing 2 M $MgSO_4$ and at least 80% could grow in the presence of 10% $MgSO_4$ [17]. Studies using isolates collected from the Mars Science Laboratory (MSL) pre-launch showed that a large percentage of the organisms from the MSL were able to grow in media containing 1 M or 2 M $MgSO_4$ (Smith, unpublished).

Although scientists are still learning more about how life survives in these extreme, high salt environments, we do know that the cells must have special physiological and biochemical properties to survive such environments. The primary factors for surviving these conditions are the amount of energy generated during dissimilatory metabolism and the mode of osmotic adaption utilized [33]. A review of studies from 1999 concluded that aerobic respiration, denitrification, and both oxygenic and anoxygenic photosynthesis can occur under the highest salt concentrations but autotrophic oxidation of ammonia and nitrate, some forms of methanogenesis and sulfate reduction were never found at salt concentrations >100–200 gl^{-1} [33]. Processes identified as occurring, albeit poorly, at salt concentrations >200 gl^{-1} included fermentation, aerobic autotrophic oxidation of sulfur compounds, sulfate reduction by incomplete oxidizers and some other forms of methanogenesis.

Oren hypothesized based on his findings that life at high salt concentrations is energetically expensive, and the upper salt concentration limit at which dissimilatory processes occur is determined partly by bioenergetics constraints. Given this the main factors that determine whether a certain type of organism can make a living at high salt concentrations are the amount of energy gained during its dissimilatory metabolism and the mode of osmotic adaptation used. Based on his review of halophiles, Oren stated that the energy cost associated with salt exclusion and pumping ions out was unfavorable and that the "salt-in" strategy was energetically favored. Given this the following types of metabolism are most likely to occur under high salt concentrations: (i) those that use light as the energy source, (ii) aerobic respiration, denitrification, and other highly exergonic dissimilatroy processes coupled with large production of ATP and (iii) types of metabolism performed by organisms that use the "salt-in" strategy even when the amount of ATP obtained in their dissimilatory processes is low [33]. Oren hypothesizes that the salt-in option would be energetically favorable to organisms, and it is clear that organisms have made adaptations to their molecules to thrive under high salt conditions and allow for the "salt-in" option. Studies by Tehei et al. [34] identified a malate dehydrogenase and tRNA molecules, from the archaeon *Haloarcula marismortui*, that are protected in the presence of high salt. The salt protected the tRNA molecules from thermal degradation while the malate dehydrogenase was protected from thermal denaturation. While studying the lipid composition of *Halobacillus halophilus*, Lopalco et al. [35] found that the organism increased the number of shorter chains and incorporated unsaturated chains in the lipid core structures.

It was believed that these changes compensated for an increase in phospholipid packing and rigidity, and sulfoglycolipid polar heads. It is believed that these changes allowed for homeostasis of membrane fluidity and permeability under high salt stress conditions.

Although many more studies need to be conducted to have a full understanding of how organisms survive these high salt environments, these studies do show that life under these conditions is possible and even, in some cases, protective. Given this, it would not be unreasonable to think that such microorganisms would be able to thrive on Mars in the salty Martian soils. Oren includes organisms using light as the energy source, however this would be unlikely on Mars since organisms living on this planet would also have to survive other conditions on the surface such as desiccation, and high radiation (to be discussed later). It is more likely that organisms on Mars would utilize exergonic dissimilatory processes or utilize types of metabolism which allowed for the "salt-in" strategy [33].

4.4. Tolerance to pH extremes

The ability of organisms to withstand alkaline pH is a factor to consider when discussing life on Mars. Initially, it was thought that the Martian soil was likely to be acidic but results by the Phoenix Lander showed that the soils at that site were mildly basic with a pH of 7.7 ± 0.5 [16]. Although the pH at the Phoenix Lander study site was only slightly basic, it is possible that other soils on Mars are more basic.

Alkaliphiles are organisms which grow above neutral pH whereas extreme alkaliphiles generally grow in the pH range of 10.0–14.0. Studies on alkaliphilic organisms have mostly focused on *Bacillus* sp. with the most extensive studies having been performed on *B. halodurans* and *B. pseudofirmus* [36]. The biggest hurdle facing alkaliphilic organisms is the ability to maintain homeostasis and maintain chemiosmosis. Alkaliphiles use transporters to help catalyze proton transport and these transporters include proton-pumping respiration chains, proton-coupled ATPases, and secondary active transporters. Often the uptake of protons is unequal where $2H^+$ are exchanged for one Na^+ ion. Studies have shown that even in extreme alkiliphiles, the pH remains relatively neutral to slightly alkaline in the cytoplasm even though the surrounding medium might be extremely alkaline. There is still much to be learned but it is clear that organisms have easily adapted to alkaline environments thus it would not be difficult for organisms to grow in Martian soils.

4.5. Surviving desiccation

Surviving desiccation is absolutely necessary if a microorganism is to survive on Mars as organisms must be able to survive the desiccating environment until they can come into contact with a water source suitable for growth. Only after finding suitable water activity, such as a polar ice cap or subsurface water sources, could the organisms then potentially become active.

As previously discussed, Mars is considered to be quite dry, and soils contain only 2.25 wt% water [9]. However, this analysis was performed on soils on the Mars surface so we do not know what the soil water content is at deeper depths. It is not known if there is a source of subsurface water, but geographical features of Mars indicate that there may have been water on the surface at some time in the past. It is not unreasonable to think that the water would have seeped into the subsurface and may still be present to some degree. Additionally, hydrothermal environments on Mars associated with craters from impacts and volcanism could have easily provided a source of liquid water, and crater impacts generating water are a potential concern today [11]. It may be possible for an organism to remain dormant for an extended period of time, then flourish after a wind storm has transferred the organism to a water source or water flows from a crater impact.

Several studies have shown that desiccation resistance in microorganisms is far from rare, and not only includes spore-forming microorganisms such as *Bacillus*, but non-spore-forming organisms such as *Moraxella* and *Staphylococcus* as well [37, 38]. Overall, dehydration of cells leads to severe cell damage by causing structural changes to lipid membranes and proteins, cross linking and polymerization of DNA molecules, inhibiting or altering enzyme activity, changing membrane permeability, and altering or mutating genetic information. DNA in the cell is at most risk to the desiccating environment since loss of water can lead to partial DNA denaturation [39]. Spore-forming organisms such as species belonging to the genera *Bacillus* and *Clostridium* are more likely to resist desiccation as the spore coat provides protection against a desiccating environment. The water content of spores is reduced to 25–45% of the cell's wet weight causing proteins to become immobile and ceasing enzymatic activity altogether [39]. However, the overall resistance of the spore to the desiccating environment is mostly due to protection of the dehydrated core by the cortex and spore coat layers while the DNA is protected by small DNA binding-acid soluble which protect the DNA from chemical and enzymatic reactivity [39].

Many non-spore-forming organisms have been shown to be resistant to desiccation. Studies by La Duc et al. [40] identified several isolates of *Pseudoaltermonas, Psychrobacte*r and *Acinetobacter* that survived a 7-day incubation at a Rh of 18 ± 3%. Several *Moraxella* sp. have been shown to survive a 30°C incubation for 35 days under dry conditions [37]. *Staphylococcus aureus* can survive on dry plastic surfaces for more than 1097 days [38]. The methanogens, *Methanobacterium wolfeii, Methanosarcina barkeri* and *Methanobacterium formicicum* survived desiccation for 90–120 day incubation periods [41]. Studies on Amazonian oxbow lake sediments showed that desiccation for 1 year at 4°C not only increased the overall abundance of *Methanocellales* and *Methanosarcinaceae,* but also increased the rates of CH_4 production after rewetting [42].

Although it is clear that the spore coat protects spore-forming organisms from a desiccating environment, it is relatively unclear how non-spore-formers survive similar environments. Studies by de Goffau et al. [45] have shown that cells can maintain intracellular water activity above that in their environment as long as the microbes can generate more water metabolically than is lost to the environment. However, this would require that the organisms were metabolically active which would be questionable under most desiccating environments such as the case of *Staphylococcus aureus* residing on a dry surface where there would be little to no nutrients [38]. Studies by Chaibenjawong and Foster [38] showed that the mutants *clpX, sigB* and *yjbH* were required for desiccation resistance in *Staphylococcus auerus*. *ClpX* and *yjbH* are both important for protein turnover while *sigB* plays a role in overall stress resistance [38]. It is likely that there are several factors involved in the desiccation resistance of non-spore-forming organisms but more studies on these unique organisms will need to be performed before we have a comprehensive understanding of these systems.

4.6. Exposure to an oxidative environment

Data from the Viking missions showed that the surface of Mars was highly oxidized compared to its atmosphere [13]. Additional studies of Mars have shown that H_2O_2 abundance can range from 15 ± 10 ppb to 40 ppb [14]. The formation of peroxides can occur in the presence of hematite, trace amounts of water, and UV radiation, and radiolysis of ice or water can create even larger amounts of peroxide formation approaching 0.13% as seen on Europa [7, 15]. For an organism to survive on Mars it would need to have mechanisms to protect itself from this oxidizing environment.

A number of microbes collected directly from spacecraft assembly facilities or pre-launch spacecraft are highly resistant to 5% H_2O_2. An isolate of *Acinetobacter radioresistens*, collected from the Mars Odyssey spacecraft, showed only a 2 log reduction after exposure to 100 mM H_2O_2. Even after exposure to 320 mM H_2O_2 there was still incomplete killing of all of the microbes [44]. Studies by Kempf et al. [43] have shown recurrent isolation of H_2O_2-resistant *Bacillus pumilus* from the JPL spacecraft assembly facility. Both vegetative cells and spores of these isolates survived exposure to 5% H_2O_2. Spores were less susceptible to killing showing only a 1–5 log reduction compared to vegetative cells which experienced a 5–8 log reduction. The examples just mentioned are far from a comprehensive list of organisms that have resistance to H_2O_2, but they demonstrate that organisms are able to withstand these types of exposures.

There have been numerous attempts to try to understand how microorganisms protect themselves from H_2O_2 exposure. Most of these studies have been performed in *Bacillus* species although there is some knowledge overall about how bacteria cope with this stress. Three well studied mechanisms are the peroxide responsive regulators OxyR, PerR and OhrR that also act as transcription regulators. OxyR and PerR are mainly involved in the detection of H_2O_2 whereas OhrR is involved in the sensing of organic peroxides and sodium hypochlorite. When exposed to peroxides, specific cysteine residues on OxyR and OhrR and histidine residues on PerR are oxidized by an Fe-catalyzed reaction. These transcriptional regulators are not only involved in H_2O_2 sensing, but also serve in the formation of biofilms, host immune response evasion, and antibiotic resistance [46].

Beyond general sensing of H_2O_2, genes involved in protein protection, such as groES, dnaK and clp tend to be upregulated thus also serving to protect the cell [47]. These proteins may be important for stabilizing the enzymes involved in the actual conversion of H_2O_2 to water and O_2, including catalases, peroxiredoxins, and peroxidases [48]. Studies in *Bacillus subtilus* have identified σ^B-dependent stress genes that are also involved in resistance to oxidative stress. Ultimately, the work performed by Reder et al. [49] identified 47 general stress response genes that were required for survival to superoxide, 6 genes required for protection from H_2O_2 stress and 9 genes that were required to protect against both.

Studies of the highly resistant strain, *Bacillus pumilus* SAFR-032, collected from JPL's spacecraft assembly facility, have identified many genes involved in H_2O_2 resistance overall [48]. Checinska et al. [50] looked further into the role of two manganese catalase proteins in the SAFR-032 spore coat, YjqC and BPUM_1305, which had been previously identified by others. It was concluded that the synergistic activity of YjqC and BPUM_1305, along with other coat oxidoreductases, contributes to the increased resistance of SAFR-032 to H_2O_2 over other *Bacillus pumilus* strains. This work has greatly improved our knowledge of the resistance of SAFR-032 to H_2O_2.

4.7. Exposure to radiation

The ability of an organism to survive radiation is paramount if the organism is to survive near the surface of Mars and pose a planetary protection threat. The radiation exposure on Mars is much more intense than it is on Earth because Mars lacks a magnetic field to deflect incoming charged particles and the atmosphere is <1% that of Earth [51]. There are 2 major types of radiation to be concerned with on Mars. The first type of radiation, Galactic Cosmic Rays (GCR), originates outside of our solar system and is formed from events such as supernovas. The second type of radiation, Solar Cosmic Radiation (SCR), originates from the sun and consists of both a constant flow of radiation as well as brief bursts [39, 51]. In the past, the overall radiation level on Mars has been based solely on calculations and modeling. New studies using data collected from the MSL found that the radiation in flight to Mars is approximately two times higher than the radiation on the surface of Mars (0.21 mGy/day vs. 0.48 mGy/day). The lower radiation level on the Mars surface is due in part to some atmospheric shielding by the Martian atmosphere, which is not provided to the spacecraft en route, and because radiation from GCR is modulated by SCR [51].

SCR can consist of both ionizing (e.g. gamma radiation) and non-ionizing radiation (e.g. UV radiation). This section will focus mostly on UV radiation since that has been the focus of the majority of previous studies. It is of note that ionizing radiation can be of more concern since it can penetrate through the Martian soils thus potentially making the first meter of soil inhabitable [51]. Solar UV radiation is divided into 3 spectral ranges; UV-A (315–400 nm), UV-B (280–315 nm) and UV-C (200–280 nm). UV-B and UV-C radiation are of the most concern since DNA has high absorption at those wavelengths and can be mutated leading to cellular inactivation [39]. Radiation of biological cells can cause breaks in molecular bonds including single and double strand breaks in DNA and photolysis of amino acids [52]. Calculations have suggested that DNA weighted irradiance on the Martian surface would be three orders of magnitude greater than that on Earth meaning that microbes would need to be resistant to much higher levels of UV radiation to sustain life on the surface of Mars [53].

Most of the research on radiation resistance and/or survival of microorganisms have been performed on spore-forming organisms since they are of the most interest to planetary protection and tend to be hardy due to their protective spore coat. Studies by Wassman et al. [54] exposed *Bacillus subtilis* spores to low Earth orbit and simulated Martian conditions for 559 days aboard the ESA's EXPOSE-E facility. Although results showed that there was 100% survival of *Bacillus subtilis* MW01 spores to simulated Martian conditions (UV $\lambda \geq 200$ nm), only a $\leq 8\%$ of spores survived low Earth orbit conditions (UV $\lambda \geq 110$ nm). Studies on *Bacillus pumilus* spores showed 10–40% viability on the EXPOSE facility versus a survival rate of 85–100% under dark simulated Martian atmospheric conditions. However, when the same studies were performed on the super tolerant *Bacillus pumilus* SAFR-032 strain, a 7 log reduction in viability was observed [55]. Overall SAFR-032 spores showing UVC resistance remain viable even after exposures up to 2000 J/m^2 [56]. Comparative proteomic studies showed that superoxide dismutase was present in higher concentrations in the space exposed isolates and exhibited higher UV-C resistance than the ground control counterparts [55]. Tauscher et al. [57] studied the effects of *Bacillus subtilis* spores exposed to simulated Mars solar radiation for an equivalent of 42 min of Mars solar radiation. Radiation exposure reduced spore viability by 3 logs but measure of germination metabolism was only reduced by <1 log. They concluded that the spores can retain the ability to initiate germination-associated metabolic processes and produce viable signature molecules despite being rendered nonviable.

It has been estimated that spores are 10–50 times more resistant than growing cells to UV radiation at 254 nm. This is due to a difference in the UV photochemistry of the DNA as well as error-free repair of any photoproducts formed by the UV light. Instead of forming thymine dimers as a photoproduct, spores tend to form thymine adducts instead; furthermore, small acid soluble proteins (SASPs) appear to suppress cyclobutane pyrimidine dimers [26]. Relative to gamma radiation, spores are significantly more resistant due to the decreased levels of water in the spore coat compared to vegetative cells which may reduce the amount of hydroxyl radicals formed overall [58]. SASPs do not appear to play a role in γ-radiation resistance [26].

Many non-spore-forming organisms have also been identified as being UV-resistant. Studies by Montero-Calasanz Mdel et al. [59] identified an isolate of *Geodermatophilus tzadiensis* that showed resistance to UV light at 254 nm. A highly radiation resistant isolate from the *Moraxella-Acinetobacter* group showed increased survival after a repeated exposure to UV light. Ultimately, this isolate was able to withstand a UV dose of 5940 J/m^2 with a 48% survival rate [60]. Antarctic Dry Valley bacteria closely related to *Brevundimonas*, *Rhodococcus*, and *Pseudomonas*, all showed resistance to γ-radiation. Surprisingly, these organisms, along with *Deinococcus radiodurans*, all showed increased resistance to γ-radiation when irradiated at −79°C [52].

Although the ability of non-spore-forming organisms to survive radiation appears to be poorly understood, there are some studies which have given clues to how these organisms survive. Keller et al. showed that the UV light resistance mechanism for survival was not associated with increased mutagenesis when the *Moraxella-Acinetobacter* isolate was repeatedly exposed to UV [60]. Studies on several strains of *Staphylococcus aureus* showed that UV-C resistance increased as the organisms entered into stationary growth phase, a characteristic that was attributed in part to the expression of σB during this phase [61]. Exposure of the lipids and proteins of *Acinetobacter* sp. PT511.2G and *Pseudomonas* sp. NT511.2B to ultraviolet radiation caused an increase of methyl groups that were associated with lipids, causing lipid oxidation, and alterations in lipid composition in addition to changes in propionylation, glycosylation, and/or phosphorylation of cell proteins [62]. The authors concluded that these changes may account for differences in UV sensitivity.

Ultimately, there are many microorganisms, both spore-forming and non-spore-forming, that are able to survive exposure to radiation and could potentially survive on Mars. For example, *Deinococcus radiodurans* would only be eradicated from the top several meters of Martian soil after a period of a few million years based on the radiation that currently reaches Mars. However, if the organism were to start growing again, then the clock would start over, and organisms could continue to stay dormant and survive up through today. This has implications for the potential for life to exist on Mars.

4.8. Conservation of energy

Unlike Earth, the Martian environment provides very little nutrients to sustain life. Any microbes that may already be on Mars would have to make a living using the limited nutrients that are available. As previously discussed, Mars has a mostly CO_2 atmosphere (95.3%) with low amounts of N_2 (2.7%) and O_2 (0.1%) [7]. However, studies by Mumma et al. [11] have shown the presence of methane in extended plumes that appeared to be released from discrete regions containing as much as 19,000 metric tons of methane. Additionally, previous studies have shown high amounts of salts including $MgSO_4$ and $FeSO_4$ [17]. Two of the most abundant compounds on Mars are Fe and S and there is evidence that there are large concentrations of sulfur in the Martian regolith [65]. Perchlorate, a strong oxidizing agent, was shown by the Phoenix Lander to be present in Martian soils in concentrations of 2.1–2.6 mM [16]. All of these compounds are potential chemical energy sources that can be used by microorganisms to survive.

The large methane plumes on Mars are of unknown origin. These plumes seasonally fluctuate but the amount of methane produced is on par with methane plumes on Earth that are known to be of biotic origin. Although the Mars rover Curiosity has found no detectable atmospheric methane, it is possible that the location of the rover prevented the detection of methane in the atmosphere since these methane plumes have been seen at polar regions rather than mid-latitude regions. Methanogenesis has become a well-known method for microorganisms to conserve energy. Many archaea, such as *Methanosarcina*, can use various carbon compounds to produce methane [63]. H_2 can readily be oxidized with the large amounts of CO_2 in the atmosphere to generate energy via methane production [64]. Once this methane is available, it could be oxidized by methanotrophicarchaea in the presence of sulfate-reducing bacteria to complete a methane cycle which would support at least 3 types of organisms [65]. An overview of the reaction might look something like this:

$$2H_2 + CO_2 \rightarrow CH_4 + O_2 \text{ (methanogenic archaea)}$$

$$2CH_4 + 2H_2O \rightarrow CH_3COOH + 4H_2 \text{ (methane - oxidizing archaea)}$$

$$4H_2 + SO_4^{2-} + H^+ \rightarrow HS^- + 4H_2O \text{ (sulfate - reducing bacteria)}$$

The electron donor H_2 could easily be generated by photochemical dissociation of water [66] and it has already been determined that there are large amounts of sulfate, especially in the form of $MgSO_4$ and $FeSO_4$ in the Martian soils [17, 67].

More likely energy sources fairly abundant in near surface soils on Mars are inorganics such as iron or sulfur [8]. An electron donor such as H_2 could be used to reduce Fe(III) or sulfate during respiration, with utilization of CO or CO_2 as a source of carbon. Sulfate and iron reduction by organisms on Earth have been very well studied. These organisms play very important roles in the biogeochemical cycling of carbon, nitrogen, sulfur, and other metals [68]. Studies by Karr et al. [69] identified a group of sulfate-reducing bacteria residing in the permanently frozen freshwater lake, Lake Fryxell, in Antarctica. These organisms are able to utilize the reduction of sulfate to conserve energy under very cold conditions (4°C). There have also been studies showing that Fe respiration under alkaline conditions is possible. Studies by Williamson et al. [70] identified organisms that could easily reduce Fe(III) at pH 10. These studies show that it is possible for these reactions to occur under cold or alkaline conditions. Once Fe or S has been reduced it is available for oxidation by other organisms.

Perchlorate, detected in soils by the Phoenix Mars Lander, is one of the more interesting potential electron acceptors recently discovered on Mars [16, 71]. More than 50 microorganisms on Earth are known to respire perchlorate coupled to the oxidation of H_2 or small organic acids, a metabolism that has been intensely studied over the past decade [72, 73]. This group of organisms is quite diverse and many have been found in environments that might seem, on the surface, to be inhospitable such as paper mill waste. Studies by Ju et al. [74] showed bacteria in sludge that were capable of oxidizing both Fe° and S° while reducing perchlorate. The enrichment culture was also able to oxidize S^{2-} and $S_2O_3^{2-}$ to support the reduction of perchlorate, and they also confirmed the disproportionation of S° to S^{2-} and SO_4^{2-}. Thus perchlorate reduction would tie in neatly to both the Fe and S cycles.

Although Mars seems inhospitable and lacks an abundant supply of nutrients, there are plenty of nutrients available to support anaerobic life on the red planet. The studies discussed above show that the organisms could work together to supply nutrients for one another within a complex ecosystem. Additionally, many of the organisms discussed above can survive in extreme environments on Earth while still making a living as evidenced by many of these processes still taking place at low temperatures or in alkaline environments.

5. Conclusions

Despite all that we know, there is still much to be learned with regard to the absolute limits for life. In order to answer these questions, we must have a better understanding of life on Earth. With regard to the potential for indigenous populations on other planets and moons, research has shown repeatedly that life can exist in the harshest of environments. Although this was not covered in depth in this chapter, life has been found in some of the most dry or frigid environments on Earth such as the Atacama Desert or Antarctica. It is not unreasonable to believe that microorganisms, similar to those found on Earth, could be thriving on locations such as Mars or Europa, especially in the subsurface where radiation would be lower and there would be a better chance for the existence of liquid water. While searching for life on other planets and moons, we look for the signs of life that are already known such as the presence of carbon and water. It may be possible that if we find life in these distant places that we may discover new limits to life in extremis.

Acknowledgements

This work was funded by NASA ROSES award #15-PPR15-0006 NASA EPSCoR award # NNX11AQ30A. We wish to thank members of the NASA and JPL Planetary Protection groups for the financial and collaborative support of our research work. The opinions expressed in this article are the authors' own and do not reflect the view of JPL or NASA.

Conflict of interest

The authors do not have any conflict of interest to report.

Author details

Stephanie A. Smith[1]*, Andrzej Paszczynski[2] and Susan E. Childers[3]

*Address all correspondence to: steph.smith@wsu.edu

1 Washington State University, Pullman, WA, USA

2 University of Idaho, Moscow, ID, USA

3 Colby College, Waterville, ME, USA

References

[1] Malin MC, Edgett KS. Evidence for recent groundwater seepage and surface runoff on Mars. Science. 2000;**288**:2330-2335

[2] Malin MC, Edgett KS. Evidence for persistent flow and aqueous sedimentation on early Mars. Science. 2003;**302**:1931-1934

[3] McKay CP. What is life-and how do we search for it in other worlds? PLoS Biology. 2004;**2**(9):1260-1263

[4] Roth L, Saur J, Retherford KD, Strobel DF, Feldman PD, McGrath MA, Nimmo F. Transient water vapor at Europa's south pole. Science. 2014;**343**(6167):171-174

[5] Rasmussen B, Blake TS, Fletcher IR, Kilburn MR. Evidence for microbial life in synsedimentary cavities from 2.75 Ga terrestrial environments. Geology. 2009;**37**:423-426

[6] Schofield JT, Barnes JR, Crisp D, Haberle RM, Larsen S, Magalhães JA, Murphy JR, Seiff A, Wilson G. The Mars pathfinder atmospheric structure investigation/meteorology (ASI/MET) experiment. Science. 1997;**278**(5344):1752-1758

[7] Horneck G. The microbial case for Mars and its implication for human expedition to Mars. Acta Astronautica. 2008;**63**:1015-1024

[8] Crawford RL, Newcombe DA. The potential for extant life in the soils of Mars. In: Dion P, Nautiyal CS, editors. Microbiology of Extreme Soils, Soil Biology. Vol. 13. NY: Springer-Verlag; 2008. pp. 225-243

[9] Meslin PY, Gasnault O, Forni O, Schröder S, Cousin A, Berger G, Clegg SM, Lasue J, Maurice S, Sautter V, Le Mouélic S, Wiens RC, Fabre C, Goetz W, Bish D, Mangold N, Ehlmann B, Lanza N, Harri A-M, Anderson R, Rampe E, McConnochie TH, Pinet P, Blaney D, Léveillé R, Archer D, Barraclough B, Bender S, Blake D, Blank JG, Bridges N,

Clark BC, DeFlores L, Delapp D, Dromart G, Dyar MD, Fisk M, Gondet B, Grotzinger J, Herkenhoff K, Johnson J, Lacour J-L, Langevin Y, Leshin L, Lewin E, Madsen MB, Melikechi N, Mezzacappa A, Mischna MA, Moores JE, Newsom H, Ollila A, Perez R, Renno N, Sirven J-B, Tokar R, de la Torre M, d'Uston L, Vaniman D, Yingst A, Science Team MSL. Soil diversity and hydration as observed by ChemCam at Gale Crater, Mars. Science. 2013;**341**:12386701-123867010

[10] Westall F, Loizeau D, Foucher F, Bost N, Betrand M, Vago J, Kminek G. Habitability on Mars from a microbial point of view. Astrobiology. 2013;**13**(9):887-897

[11] Mumma MJ, Villanueva GL, Novak RE, Hewagama T, Bonev BP, DiSanti MA, Mandell AM, Smith MD. Strong release of methane on Mars in northern summer. Science. 2009;**323**:1041-1045

[12] Cockell CS. The ultraviolet environment of Mars: Biological implications past, present and future. Icarus. 2000;**146**:343-359

[13] Slesak I, Slesak H, Kruk J. Oxygen and hydrogen peroxide in the early evolution of life on earth: In silico comparative analysis of biochemical pathways. Astrobiology. 2012; **12**(8):775-784

[14] Encrenaz T, Greathouse TK, Richter MJ, Bezard B, Fouchet T, Lefevre F, Montmessin F, Forget F, Lebonnois S, Atreya SK. Simultaneous mapping of H_2O and H_2O_2 on Mars from infrared high-resolution imaging spectroscopy. Icarus. 2008;**195**:547-556

[15] Johnson RE, Quickenden TI, Cooper PD, McKinley AJ, Freeman CG. Europa the production of oxidants in Europa's surface. Astrobiology. 2003;**3**(4):823-850

[16] Hecht MH, Kounaves SP, Quinn RC, West SJ, Young SMM, Ming DW, Catling DC, Clark BC, Boynton WV, Hoffman J, DeFlores LP, Gospodinova K, Kapit J, Smith PH. Detection of perchlorate and the soluble chemistry of Martian soil at the phoenix lander site. Science. 2009;**325**(5936):64-67

[17] Crisler JD, Newville TM, Chen F, Clark BC, Schneegurt MA. Bacterial growth at the high concentrations of magnesium sulfate found in Martian soils. Astrobiology. 2012; **12**(2):98-106

[18] NASA. Planetary Protection Provisions for Robotic Extraterrestrial Missions. NPR 8020.12. Washington, D.C: NASA; 2011

[19] National Research Council. Preventing the Forward Contamination of Mars. National Research Council of the National Academies. Washington, D.C.: The National ACADEMIES PRESS; 2006

[20] Kminek G, Rummel JD, Cockell CS, Atlas R, Barlow N, Beaty D, Boynton W, Carr M, Clifford S, Conley CA, Davila AF, Debus A, Doran P, Hecht M, Heldmann J, Helbert J, Hipkin V, Horneck G, Kieft TL, Klingelhoefer A, Meyer M, Newsom H, Ori GG, Parnell J, Prieur D, Raulin F, Schulze-Makuch D, Spry JA, Stabekis PE, Stackebrandt E, Vago J, Viso M, Voytek M, Wells L, Westall F. Report of the COSPAR mars special regions colloquium. Advances in Space Research. 2010;**46**:811-829

[21] Rummel JD. Special regions in Mars exploration: Problems and potential. Acta Astronautica. 2009;**64**:1293-1297

[22] Valentine DL, Reeburgh WS. New perspectives on anaerobic methane oxidation. Environmental Microbiology. 2000;**2**(5):477-484

[23] Pikuta EV, Hoover RB. Microbial extremophiles at the limits of life. Critical Reviews in Microbiology. 2007;**33**:183-209

[24] Morozova D, Möhlmann D, Wagner D. Survival of Methanogenic Archaea from Siberian permafrost under simulated Martian thermal conditions. Origins of Life and Evolution of the Biosphere. 2007;**37**:189-200

[25] Nicholson WL, Krivushon K, Gilichinsky D, Schuerger AC. Growth of Carnobacterium spp. from permafrost under low pressure, temperature, and anoxic atmosphere has implications for Earth microbes on Mars. PNAS. 2013;**110**(2):666-671

[26] Nicholson WL, Munakata N, Horneck G, Melosh HJ, Setlow P. Resistance of *Bacillus* endospores to extreme terrestrial and extraterrestrial environments. Microbiology and Molecular Biology Reviews. 2000;**64**(3):548-572

[27] Mykytczuk NCS, Foote SJ, Omelon CR, Southam G, Greer CW, Whyte LG. Bacterial growth at −15° C; molecular insights from the permafrost bacterium *Planococcus halocryophilus* Or1. ISME Journal. 2013;**7**(6):1211-1126

[28] Kuhn E. Toward understanding life under subzero conditions: The significance of exploring "cold-shock" proteins. Astrobiology. 2012;**12**(11):1078-1086

[29] D'Amico S, Collins T, Marx J-C, Feller G, Gerday C. Psychrophilic microorganisms: Challenges for life. EMBO Reports. 2006;**7**(4):385-389

[30] Hallsworth JE, Yakimov MM, Golyshin PN, Gillion JLM, D'Auria G, deLimaAlves F, La Cono V, Genovese M, McKew BA, Hayes SL, Harris G, Giuliano L, Timmis KM, McGenity TJ. Limits of life in $MgCl_2$-containing environments: Chaotropicity defines the window. Environmental Microbiology. 2007;**9**(3):801-813

[31] Smith SA, Benardini III JN, Strap JL , Crawford RL. (2009) Diversity of aerobic and facultative alkalitolerant and halotolerant endospore formers in soil from the Alvord Basin, Oregon.Systematic and Applied Microbiology 32:233-244

[32] Van der Wielen PWJJ, Bolhuis SB, Daffonchio D, Corselli C, Giuliano L, D'Auria G, de Lange GJ, Huebner A, Varnavas SP, Thomson J, Tamburini C, Marty D, McGenity TJ, Timmis KN, BioDeep Scientific Party. The enigma of prokaryotic life in deep hypersaline anoxic basins. Science. 2005;**307**:121-123

[33] Oren A. Thermodynamics limits to microbial life at high salt concentrations. Environmental Microbiology. 2011;**13**(8):1908-1923

[34] Tehei M, Franzetti B, Maurel M-C, Vergne J, Hountondji C, Zaccai G. The search for traces of life: The protective effect of salt on biological macromolecules. Extremophiles. 2002;**6**:427-430

[35] Lopalco P, Angelini R, Lobasso S, Köcher S, Thompson M, Müller V, Corcelli A. Adjusting membrane lipids under salt stress: The case of the moderate halophilic organism *Halobacillus halophilus*. Environmental Microbiology. 2013;**15**(4):1078-1087

[36] Krulwich TA, Sachs G, Padan E. Molecular aspects of bacterial pH sensing and homeostasis. Nature Reviews Microbiology. 2011;**9**:330-343

[37] Kubota H, Mitani A, Niwano Y, Takeuchi K, Tanaka A, Yamaguchi N, Kawamura Y, Hitomi J. *Moraxella* species are primarily responsible for generating malodor in laundry. Applied and Environmental Microbiology. 2012;**78**(9):3317-3324

[38] Chaibenjawong P, Foster SJ. Desiccation tolerance in *Staphylococcus aureus*. Archives of Microbiology. 2011;**193**:125-135

[39] Horneck G, Klaus DM, Mancinelli RL. Space microbiology. MMBR. 2010;**74**(1):121-156

[40] La Duc MT, Benardini JN, Kempf MJ, Newcombe DA, Lubarsky M, Venkateswaran K.Microbial diversity of Indian Ocean hydrothermal vent plumes: Microbes tolerant of dessication, peroxide exposure and ultraviolet and γ-irradiation. Astrobiology. 2007;**7**:416-431

[41] Kral TA, Altheide ST. Methanogen survival following desiccation, low pressure and martian regolith analogs. Planetary and Space Science. 2013;**89**:167-171

[42] Conrad R, Ji Y, Noll M, Klose M, Claus P, Enrich-Prast A. Response of methanogenic microbial communities in Amazonian oxbow lake sediments to desiccation stress. Environmental Microbiology. 2013;**16**(6):1682-1694

[43] Kempf MJ, Chen F, Kern R, Venkateswaran K. Recurrent isolation of hydrogen peroxide-resistant spores of *Bacillus pumilus* from a spacecraft assembly facility. Astrobiology. 2005;**5**:391-405

[44] McCoy KB, Derecho I, Wong T, Tran HM, Huynh TD, La Duc MT, venkateswaran K, Mogul R. Insights into the extremotolerance of *Acinetobacter radioresistens* 50v1, a Gram-negative bacterium isolated from the Mars Odyssey spacecraft. Astrobiology. 2012;**12**(9):1-9

[45] De Goffau MC, Maarten van Dijl J, Harmsen HJM. Microbial growth on the edge of desiccation. Environmental Microbiology. 2011;**13**(8):2328-2335

[46] Dubbs JM, Mongkolsuk S. Peroxide-sensing transcriptional regulators in bacteria. Journal of Bacteriology. 2012;**194**(20):5495-5502

[47] Mols M, Abee T. Primary and secondary oxidative stress in bacillus. Environmental Microbiology. 2011;**13**(6):1387-1394

[48] Gioia J, Yerrapragada S, Qin X, Jiang H, Igboeli OC, Muzny D, Dugan-Rocha S, Ding Y, Hawes A, Liu W, Perez L, Kovar C, Dinh H, Lee S, Nazareth L, Blyth P, Holder M, Buhay C, Tirumalai MR, Liu Y, Dasgupta I, Bokhetache L, Fujita M, Karouia F, Moorthy PE, Siefert J, Uzman A, Buzumbo P, Verma A, Zwiya H, McWilliams BD, Olowu A, Clinkenbeard KD, Newcombe D, Golebiewski L, Petrosino JF, Nicholson WL, Fox GE, Venkateswaran K, Highlander SK, Weinstock GM. Paradoxical DNA repair and peroxide resistance gene conservation in Bacillus pumilus SAFR-032. PLoS ONE. 2007;**2**(9):e928

[49] Reder A, Höper D, Gerth U, Hecker M. Contributions of individual σ^B-dependent general stress genes to oxidative stress resistance of *Bacillus subtilis*. Journal of Bacteriology. 2012;**194**(14):3601-3610

[50] Checinska A, Burbank M, Paszczynski AJ. Protection of *Bacillus pumilus* spores by catalases. Applied and Environmental Microbiology. 2013;**78**(18):6413-6422

[51] Hassler DM, Zeitlin C, Wimmer-Schweingruber RF, Ehresmann B, Rafkin S, Eigenbrode JL, Brinza DE, Weigle G, Böttcher S, Böhm E, Burmeister S, Guo J, Köhler J, Martin C, Reitz G, Cucinotta FA, Kim M, Grinspoon D, Bullock MA, Posner A, Gómez-Elvira J, Vasavada A, Grotzinger JP, MSL Science Team. Mars' surface radiation environment measured with the Mars Science Laboratory's Curiosity Rover. Science. 2013;**343**(6169):1244797

[52] Dartnell LR, Hunter SJ, Lovell KV, Coates AJ, Ward JM. Low-temperature ionizing radiation resistance of Deinococcusradiodurans and Antarctic Dry Valley bacteria. Astrobiology. 2010;**10**(7):717-732

[53] Rontó G, Bérces A, Lammer H, Cockell CS, Molina-Cuberos GJ, Patel MR, Selsis F. Solar UV irradiation conditions on the surface of Mars. Photochemistry and Photobiology. 2003;**77**(1):34-40

[54] Wassman M, Moeller R, Rabbow E, Panitz C, Horneck G, Reitz G, Douki T, Cadet J, Stan-Lotter H, Cockell CS, Rettberg P. Survival of spores of the UV resistant *Bacillus subtilus* strain MW01 after exposure to low-Earth orbit and simulated Martian conditions: Data from the space experiment ADAPT on EXPOSE-E. Astrobiology. 2012;**12**(5):498-507

[55] Vaishampayan PA, Rabbow E, Horneck G, Venkateswaran KJ. Survival of *Bacillus pumilus* spores for a prolonged period of time in real space conditions. Astrobiology. 2012;**12**(5):487-497

[56] Link L, Sawyer J, Venkateswaran K, Nicholson W. Extreme spore UV resistance of *Bacillus pumilus* isolates obtained from an ultraclean spacecraft assembly facility. Microbial Ecology. 2004;**47**:159-163

[57] Tauscher C, Schuerger AC, Nicholson WL. Survival and germinability of *Bacillus subtilis* spores exposed to simulated Mars solar radiation: Implications for life detection and planetary protection. Astrobiology. 2006;**6**(4):592-605

[58] Moeller R, Raguse M, Reitz G, Okayasu R, Li Z, Klein S, Setlow P, Nicholson WL. Resistance of *Bacillus subtilis* spore DNA to lethal ionizing radiation damage relies primarily on spore core components and DNA repair, with minor effects of oxygen radical detoxification. Applied and Environmental Microbiology. 2014;**80**(1):104-109

[59] Montero-Calasanz Mdel C, Göker M, Broughton WJ, Cattaneo A, Favet J, Pötter G, Rohde M, Spröer C, Schumann P, Klenk H-P, Gorbushina AA. *Geodermatophilus tzadiensis* sp. nov., a UV radiation resistant bacterium isolated from sand of the Saharan desert. Systematic and Applied Microbiology. 2013;**36**:177-182

[60] Keller LC, Thompson TL, Maxcy RB. UV light induced survival response in a highly radiation resistant isolate of the Moraxella-Acinetobacter group. Applied and Environmental Microbiology. 1982;**43**(2):424-429

[61] Gayán E, García-Gonzalo D, Álvarez I, Condón S. Resistance of *Staphylococcus aureus* to UV-C light and combined UV-heat treatments at mild temperatures. International Journal of Food Microbiology. 2013;**172**:30-39

[62] Santos AL, Moreirinha C, Lopes D, Esteves AC, Henriques I, Almeida A, Domingues MRM, Delgadillo I, Correia A, Cunha A. Effects of UV radiation on the lipids and proteins of bacteria studies by mid-infrared spectroscopy. Environmental Science & Technology. 2013;**47**(12):6306-6315

[63] Smith MR, Mah RA. Growth and methanogenesis by *Methanosarcina* strain 227 on acetate and methanol. Applied and Environmental Microbiology. 1978;**36**(6):870-879

[64] Kral TA, Bekkum CR, McKay CP. Growth of methanogens on a Mars soil simulant. Origins of Life and Evolution of the Biosphere. 2004;**34**:615-626

[65] Caldwell SL, Laidler JR, Brewer EA, Eberly JO, Sandborgh SC, Colwell FS. Anaerobic oxidation of methane: Mechanisms, bioenergetics, and the ecology of associated microorganisms. Environmental Science & Technology. 2008;**42**(18):6791-6799

[66] Krasnopolsky VA, Feldman PD. Detection of molecular hydrogen in the atmosphere of Mars. Science. 2001;**294**:1914-1917

[67] Schulze-Makuch D, Fairen AG, Davila AF. The case for life on Mars. International Journal of Astrobiology. 2008;**7**(2):117-141

[68] Zhou J, He Q, Hemme CL, Mukhopadhyay A, Hillesland K, Zhou A, He Z, Van Nostrand JD, Hazen TC, Stahl DA, Wall JD, Arkin AP. How sulphate-reducing microorganisms cope with stress: Lessons from systems biology. Nature Reviews Microbiology. 2011;**9**:452-466

[69] Karr EA, Sattley WM, Rice MR, Jung DO, Madigan MT, Achenbach LA. Diversity and distribution of sulfate-reducing bacteria in permanently frozen Lake Fryxell, McMurdo Dry Valleys, Antarctica. Applied and Environmental Microbiology. 2005;**71**(10):6353-6359

[70] Williamson AJ, Morris K, Shaw S, Byrne JM, Boothman C, Llyod JR. Microbial reduction of Fe(III) under alkaline conditions relevant to geological disposal. Applied and Environmental Microbiology. 2013;**79**(11):3320-3326

[71] Hand T. Phoenix fades away (news). Nature. 2008;**456**(6):8-9

[72] Coates JD, Michaelidou U, Bruce RA, O'Connor M, Crespi JN, Achenbach LA. Ubiquity and diversity of dissimilatory (per)chlorate-reducing bacteria. Applied and Environmental Microbiology. 1999;**65**:5234-5241

[73] Coates JD, Achenbach LA. Microbial perchlorate reduction: Rocket-fueled metabolism. Nature Reviews Microbiology. 2004;**2**:569-580

[74] Ju X, Sierra-Alvarez R, Field JA, Byrnes DJ, Bentley H, Bentley R. Microbial perchlorate reduction with elemental sulfur and other inorganic electron donors. Chemosphere. 2008;**71**:114-122

Chapter 13

Exploring the Stratosphere: What We Missed by Shooting for the Moon

Laura Galdamez

Additional information is available at the end of the chapter

http://dx.doi.org/10.5772/intechopen.73602

Abstract

Similar to outer space, the stratosphere experiences freezing temperatures, with atmospheric pressures and oxygen levels far below the level required for human survival. Exposure to this environment causes unique injuries to the human body that can be deadly if the correct management is not promptly initiated. The preceding decades are filled with stories of deadly failures from such exposures and marked achievement as we began to explore this section of our outer atmosphere. Through advances in technology, we have developed pressure suits and vehicles used for high altitude and outer space that provide protection and allow us to not only survive, but also explore these dangerous environments. The recent high altitude missions are examples of the remarkable capability of human innovation and ingenuity. These missions have fostered an explosion of interest and wonder, creating new demand for a commercial space industry that was virtually nonexistent in the previous century. Though recent tragedies have temporarily delayed the travel of eager citizens into space, the boom of the commercial space industry is pushing forward with new promises of space exploration available to the next paying customer, anticipated in the next few years.

Keywords: high altitude, ebullism, stratosphere, StratEx, Stratos, pressure suit, commercial space industry

1. Introduction

In 1969, when Aldrin and Armstrong first stepped onto the Moon's surface, the world was ablaze with excitement. This event generated a spark that had nearly vanished among those on earth watching the repetitive orbital missions. People organized "moonwalk" parties while children's imaginations across the world came alive with the thought of being able to float and bounce above that gray, dusty surface they saw on the television [1]. Yet in our hurry to aim for the lunar surface, we flew right through an unexplored and unconquered area of our own planet: the higher atmosphere.

2. Stratosphere and human physiology

2.1. Stratosphere

The stratosphere is the second lowest layer of Earth's atmosphere, stretching from 12 to 55 km above sea level (**Figure 1**) [2]. For comparison, the Everest summit sits at 8.85 km, and jumbo jets cruise at an average altitude of 12–16 km [3, 4]. To optimize fuel efficiency, the typical commercial airliner cruises at 9–12 km (just above the troposphere in the lower reaches of the stratosphere), where temperatures and air density are lowest [5]. Temperatures in the stratosphere are stratified and, somewhat non-intuitively, increase with higher altitude secondary to ability of the ozone layers to absorb ultraviolet light [5]. Due to increased energy absorption at higher altitudes, the top of the stratosphere remains near 0°C while the tropopause (which occurs between the troposphere and stratosphere) exhibits temperatures of −46 to −57°C [2, 4].

Humans on Earth live well below the altitudes that represent our species' physical limitations. Although only 15.6% of inhabited land occurs below 100 m, in 1994 approximately 33.5% of the world's population lived within these elevations [6]. While human beings can adapt to higher altitudes, the Everest "death zone" (mountain's altitude above 8000 m) earned its name for good reason. When people who are not acclimatized are exposed to equivalent levels of ambient hypoxia that exist at altitudes over 8500 m, they lose consciousness within 2–3 min [3]. Why does this occur and what changes in the environment lead to this? At sea level the average atmospheric pressure is 101.325 kPa, but at higher altitudes air pressure decreases. The mean atmospheric pressure on the summit of Everest is only 33.7 kPa, and atmospheric pressure at the top of the stratosphere is 1/1000 that of sea level [2, 7]. According to Dalton's law, the total pressure of a mixture of gases is made up of the sum of the partial pressures of each individual gas [8]. Given this, it is easier to understand how the partial pressure of oxygen is also much lower at these higher altitudes, which also results in lower partial pressures of oxygen in our blood at these altitudes. A study of the mean partial pressure of arterial oxygen (PaO2) in the blood gas of Everest climbers taken at 8400 m found their average PaO2 to be 24.6 mmHg, while normal is considered to be >80 mmHg [3]. But what happens when the human body is exposed to even higher altitudes than we are able to achieve via these hiking expeditions, such as the stratosphere or outer space?

2.2. Physics and human physiology

To understand this question, it is important to understand Boyle's law and the effect of pressure on liquids and gases as well. According to Boyle's law, pressure and volume have an inverse relationship; as the pressure on a given volume of gas decreases, the volume of the gas will increase [8]. This means that there will be fewer molecules of gas occupying any given space. For liquid to transform to vapor, consider vaporization or boiling, molecules on the liquid's surface must be able to leave, which means they either need sufficient energy (added

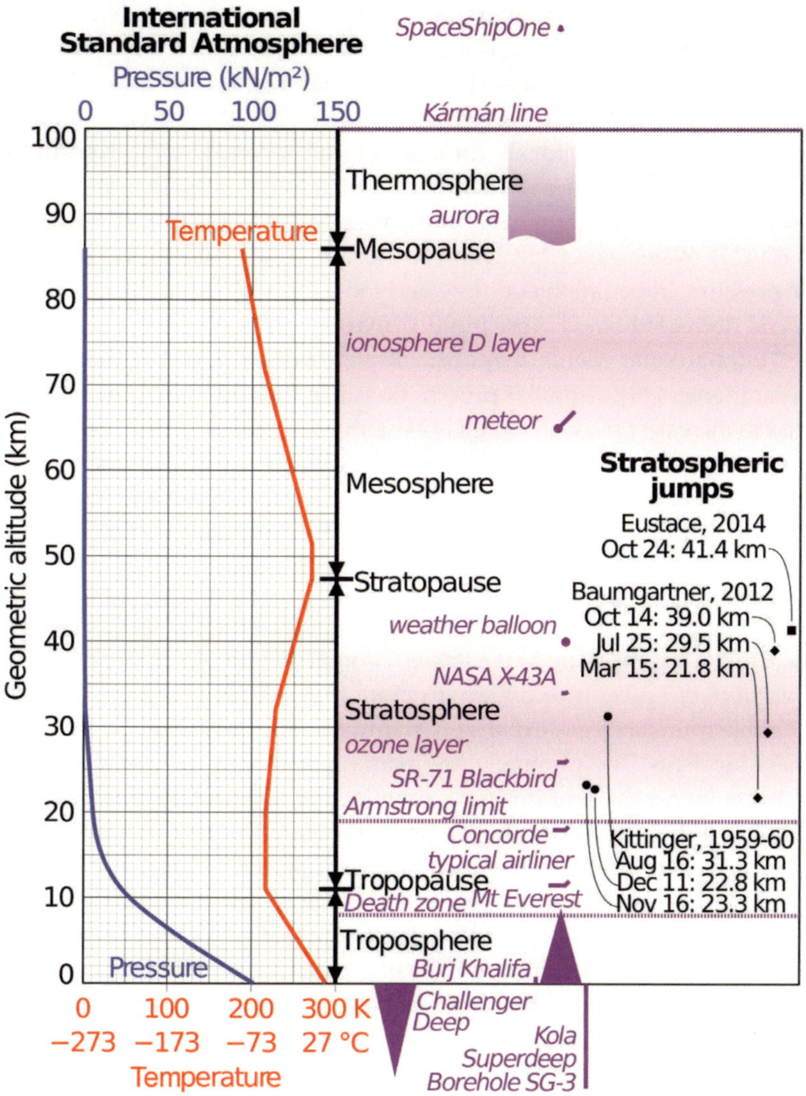

Figure 1. Comparison of international standard atmosphere, temperature and pressure at various altitudes including Armstrong limit and Mt. Everest death zone.

via heat energy water boils), or the number of molecules above the liquid (i.e. vapor pressure) needs to be sufficiently low to allow the surface molecules to escape (achieved by lowering the pressure above the liquid) [8]. Using these two concepts we better understand what happens at higher altitudes, both to the gas filled compartments and the water in our bodies.

2.2.1. Ebullism

A human's normal body temperature is 98.6°F or 37°C, well below the boiling point of water at our usual, livable altitudes. However, a man named of H.G. Armstrong defined the altitude at which the ambient pressure decreases to the point that spontaneous evolution of liquid water to vapor gas can occur at our body's temperature, a process called ebullism [9]. He coined this altitude "Armstrong's band", which occurs at approximately 18–19 km above sea level [9]. While survival at these altitudes without protection of a pressurized compartment is possible for several minutes, it can quickly lead to dangerous consequences [9]. The human body consists of >50% water, which exists ubiquitously throughout our bodies. During exposure to such low pressures, any water on our skin will vaporize, and resulting pressure in the extremity muscle compartments from the expanding gas can increase to the point that effective blood circulation may cease [10]. Water vapor forms in the thoracic cavity causing vapothorax, increasing intrathoracic pressure, impeding blood flow and inducing vagal stimulation which can lead to bradycardia and decreased blood pressures [9]. Investigations following human exposure have demonstrated intra-alveolar edema, hemorrhagic atelectasis, regional atelectasis, and simple or tension pneumothoraces [9]. A similar process occurs in the abdominal cavity and vapoperitoneum can lead to increased pressure on all the involved organs. In animal models, greater than 2 min of exposure to these simulated pressures led to vascular congestion and hemorrhage in the liver, spleen, kidneys and GI tract, as well as the brain. These findings were more pronounced if the exposure to these pressures was sudden and the decompression was explosive [9].

2.2.2. Anoxia and hypothermia

Tissue damage following exposure to the low environmental pressure above Armstrong's band, or the near vacuum of space, is secondary to more than simply ebullism. The combination of anoxia, gas bubble formation, gas expansion and rapid fluid loss underlie the array of injuries seen after these incidents. The extremely low partial pressure of oxygen at these altitudes can quickly lead to anoxic anoxia, which is most prominently demonstrated by cerebral hypoxia [9, 11]. The time of useful consciousness following exposure to this low pressure is approximately 9–11 s, at which time they lose voluntary control of their muscles, vomit, lose bowel and bladder control and collapse [9, 11]. Within 30 s severe neurologic manifestations ensue including tonic-clonic seizures which progress to spastic rigidity and eventually total flaccid paralysis [9, 11]. Hypothermia is also likely, and ensuing tissue damage depends on the length of exposure as well as the temperature [10]. Although actual tissue freezing is not commonly seen, the increased evaporative cooling on mucosal surfaces has been found to lead to ice formation in those areas [10].

2.2.3. Decompression illness

Gas bubbles form in three different ways: vaporization of water within tissues (previously discussed), evolution of dissolved gas in the vasculature (decompression illness), and direct injection of gas via ruptured alveoli, resulting in venous and arterial embolism [10]. Decompression illness (DCI) is related to diving, however a similar process can occurs any time the body experiences significantly decreased ambient pressures. In a manner similar to ebullism, pressure decreases to the point that dissolved gases in the blood spontaneously evolve out and form small bubbles that can become trapped within the smaller capillaries of the skin and organs including kidneys, brain and lungs [9]. Skin findings are more prominent in DCI with divers, but neurologic findings are more common in high altitude exposure [12]. DCI should be suspected in anyone with such exposure who displays altered mental status or abnormal neurologic findings that persist despite initial supportive treatment [9]. These bubbles can also have a significant effect on the heart. Irritation of the myocardium caused by these bubbles has been demonstrated in animal models to lead to various life threatening cardiac dysrhythmias such as ventricular fibrillation and heart block [9]. It was once believed that placing the exposed person on their left side with their lower body elevated above the heart was the optimal position, and would help prevent any arterial and venous gas emboli from entering cerebral circulation. Animal studies have demonstrated little utility to this approach. The key to treating neurologic DCI Is to provide 100% supplemental oxygen and repressurization via a hyperbaric chamber as soon as possible [9].

Decompression injury is most concerning for astronauts during their extra-vehicular activities, but fortunately is rare in space travel. There are only three astronaut deaths reportedly due to decompression from space exposure. On June 30, 1971, *Soyuz 11* was re-entering the atmosphere when the ship depressurized, leading to the death of the three crew members [13]. Two others experienced decompression during training missions in a chamber but for less than 5 min and survived without neurologic sequelae [14].

2.2.4. Barotrauma

Barotrauma occurs as decreased ambient pressures lead to increased volumes of gas within our gas-filled cavities, including the sinuses, ears and thoracic cavity. Usually, barotrauma to the sinuses and ears is less likely when the Eustachian tube and sinus passageways are functional; they act as connections between these cavities and the outside environment and normally allow any expanded volume of gas to exit before causing significant damage [9]. Typically, our airway acts in a similar fashion, but if a person were to hold their breath at the time of exposure, the closed glottis would not allow air to escape from the lungs, and air in the tiny alveoli of lung tissue would expand until their thin walls broke. The direct injection of a large volume of bubbles into the vasculature can result from the rupture of these alveoli, which directly exposes the nearby blood vessels [9]. This massive bubble load can proceed in a retrograde fashion to the right side of the heart and slow blood transport in the major veins, leading to congestion in the capillaries and subsequent tissue damage [9, 11]. As the bubbles prevent forward flow of blood, venous pressure increases and eventually equals that of arterial pressure, at which point cardiac output decreases and there is no effective circulation [11]. This can occur within 1 min of exposure and is a catastrophic condition known as 'vapor lock' [9, 11].

3. High altitude exploration

3.1. Pressure suits

The goal of the first high altitude explorers quickly became focused on finding a way to survive brief periods in this extreme environment. As stated earlier, a protective barrier is needed to survive the life-threatening pressures and temperatures, and so imaginations became focused on how to solve these challenges. Engineers designed pressurized suits which could prevent injuries such as ebullism; but the first suits were heavy, restricted movement and caused high thermal load [9]. Full pressure suits were such units, designed to be pressurized externally and then hold pressure for prolonged periods of time. They are typically are used for routine operations in hazardous environments [9]. In contrast, partial pressure suits have a low baseline profile and are worn overtop elastic undergarments which provide passive mechanical counter pressure. They are pneumatic and inflate when activated to provide near-instant pressurized conditions. They are less cumbersome than full pressure suits, but are not ideal for prolonged, routine activity, and are better used as a redundant protective measure in the event of a vehicle hull breach [9].

Pressure suit design is a very specialized field and the David Clark Company Incorporated (DCCI) has pioneered the field starting in the 1940s by developing partial pressure suits both for NASA as well as full pressure suits for the D558-2 and North American X-15 research aircraft [15]. DCCI created some of the first pressure suits used in high altitude missions, which then became the basis of the suits used by NASA for the space shuttle missions [15, 16]. The pressure suit designed for NASA, termed the Launch Entry Suit (LES), was modeled after the high-altitude protective outfit due to its combination of comfort and protection that could be provided in a short time. The LES was a counter-pressure garment with two separate bladders: one for acceleration, one for altitude protection [16]. The first worn by astronauts is made of athletic underwear, over which the counter-pressure garment fits. A nylon restraint layer is worn overtop which allows the counter-pressure garment to push inwards when inflated. The outer, waterproof, nomex cover is worn over this and is bright orange to assist in search and rescue operations [16]. The astronauts have a parachute harness, flotation devices and other supplies on the outside of their suit. The helmet is connected via a bearing assembly and gloves also strap on separately. The LES was used on 42 Space Shuttle missions from 1988 to 2001, at which time the new, full-pressure Advanced Crew Escape Suit was designed. It was lighter, less bulky, cooler and more comfortable with better physiologic protection and mobility [16]. This model was used through the discontinuation of the Shuttle program in 2011. It was developed in parallel with the S1034 PPA which became the standard pressure suit used by the United States Air Force and Department of Defense [15]. The S1034 and S1035 were the basis for the specialized suits developed for the latest high altitude missions, StratEx and Stratos [15].

3.2. Early near misses and failures

The modern quest to explore high altitudes began with Paul Bert in 1878, who developed the first altitude chamber complete with supplemental oxygen [17]. He was followed by Wiley Post, who in 1934 developed and demonstrated the effectiveness of the first pressure suit for high altitude flight [17]. New pressure suits that were thinner and more comfortable continued to feature improvements, and they became regular equipment for all high altitude pilots. In 1966, during a failed flight test of the Blackbird aircraft, one of these new pressure suits led to the safe landing and survival of its pilot, Bill Weaver. Weaver was piloting the Blackbird at Mach 3 and 75,000 ft when a structural failure caused the cockpit to detach from the fuselage [16, 18]. He was wearing the new S901 pressure suit, and after being violently ripped away from the disintegrating aircraft and ejection seat, he lost consciousness in free fall. His small drogue chute opened automatically, followed by his main chute at a lower altitude. He regained consciousness during the main parachute deployment, but his visor had completely frosted over and he could not see anything around him [15, 18]. After landing he noted that one of the two oxygen lines supplying his pressure suit had detached, and the second was barely connected. If it had detached, the suit would have depressurized [16]. It was then that he realized that this suit had protected him from the life-threatening low pressures and extreme cold. He referred to the suit later as his 'own escape capsule' [15].

However, as with most great feats, examples of exemplary achievement are likewise met with examples of great tragedy and failure. Pyotr Dolgov, a Soviet Air Force colonel, was part of a unit designing a pressure suit to allow for safe high altitude pilot bail-outs. In 1962, during field testing of this suit at an altitude just over 28.5 km, he attempted to exit the gondola when he accidently hit his helmet on part of the structure, cracking the visor which caused immediate suit depressurization. This led to catastrophic ebullism and hypoxia resulting in death [4, 12]. Another instance (1966), a Houston technician in a low-pressure chamber was testing a space suit when the suit suddenly lost pressurization. The technician was immediately exposed to atmospheric pressures equivalent to an altitude of 120,000 ft, and quickly lost consciousness. The chamber was repressurized and he awoke as the monitor read 14,000 ft. He reported recalling the feeling of saliva boiling off his tongue just before he passed out [17]. Amazingly, the technician had a quick recovery and did not require any hospitalization nor had any neurologic sequelae from the exposure. A similar incident occurred in 1982 when a technician was inside of a decompression chamber that unexpectedly began dropping pressure [17]. The technician experienced pressures correlating with altitudes greater than 74,000 ft for over 3 min, and was held at the maximum altitude for over 60 s. The chamber manager was forced to kick in the glass ionization gauge to repressurize the chamber by letting air in [17]. By the time the technician arrived at the hospital he was noted to be cyanotic, frothing at the mouth, bleeding from the lungs and had severe barotrauma in both ear drums. However, as serious as his presentation appeared, within 24 h he was awake and alert, and by day five he had been extubated. He, too, suffered no long term neurologic sequelae [17].

3.3. Nick Piantanida and Strato Jump I-III

Nick Piantanida, a truck driver from New Jersey with virtually no experience or training, became obsessed with the idea of high altitude jumps and breaking the world free fall record [12, 19]. He rallied a large amount of sponsor money and a group of volunteers to help him achieve this goal. However, his first attempt in 1965 with balloon ascent failed when wind shear tore off the balloon's top at 23,000 ft, forcing him to parachute into a nearby city dump [19]. During his second attempt, he successfully climbed to 123,500 ft but could not disconnect his oxygen hose from the gondola's oxygen supply in order to connect to a portable tank, and therefore he could not exit the capsule. Piantanida was then instructed by his ground crew to re-attach his seat belt and re-secure the belt across the capsule door, but his bulky gloves made this an impossible task [19]. He was forced to wedge himself inside as the gondola tipped forward at a 45° angle during its 15 s free-fall at 600 miles per hour, then brace inside the open gondola to avoid falling out when the cargo chute opened [20]. He reportedly said afterward: "If only I had a damned $1.25 wrench, gravity would have done the rest" [19]. His third and final attempt took place on May 1, 1966. Piantanida ascended uneventfully in a styrofoam-insulated gondola (**Figure 2**) when a loud 'whoosh' of rushing air was heard over the communications link by his ground crew, followed by the cut-off call of Nick's voice saying "Emerg-!" [19, 20]. The balloon was cut away and the mission immediately aborted. The gondola, with the pilot still inside, fell

Figure 2. Styrofoam-insulated gondola used by Nick Piantanida during the Strato Jump III mission. Now on exhibit at Smithsonian Institution's National Air and Space Museum.

for an agonizing 26 min descent from a height of 56,000 ft using the gondola's own emergency parachute. By the time the crew made it to the downed gondola Nick was outside the capsule, barely alive and conscious [19, 20]. Piantanida lapsed into a coma before arrival to the nearest hospital, and the doctors, with very little knowledge or training on high altitude pathology and management, did their best to help him [19, 20]. He had experienced massive tissue damage and brain injury secondary to emboli. He died 4 months later, never having come out of his coma [19]. While the exact circumstance that led to the incident may never be known, it is believed that Piantanida, who frequently reported discomfort from his pressure suit and had been known to quickly open and shut the faceplate to quickly depressurize the suit, had in fact opened his faceplate at that high altitude and experienced explosive decompression illness [19].

3.4. Joseph Kittinger and Project Excelsior

"There is a hostile sky above me. Man will never conquer space. He may live in it, but he will never conquer it. The sky above is void and very black and very hostile."

—Joseph Kittinger (12)

Colonel Joseph Kittinger is a pilot who personally understands the dangers of high altitude flight. He is a retired United States Airforce Colonel and, like Piantanida, desired to break

http://dx.doi.org/10.5772/intechopen.73602

the previous records for high altitude free fall [4]. During his first high altitude jump in late 1959 at just over 23 km, his drogue chute malfunctioned and deployed early causing him to go into a flat spin [12, 21]. A flat spin occurs in free fall when your body is horizontal and essentially spins in a cartwheel motion. A jumper can spin at a rate of up to 180–250 rotations per minute; the centrifugal force from a rapid spin creates negative G's that draw the blood to stagnate in the feet and head [4]. This can lead to headache, shortness of breath, vision failure, altered mentation, and loss of consciousness [4]. This is exactly what happened as Kittinger lost consciousness shortly into his flat spin, but was rescued by the automatic opening of his emergency parachute at 10,000 ft [12, 21]. Following this incident, Kittinger decided to have his small drogue chute open automatically after he jumped from the capsule, which helps to stabilize him in free fall and prevent flat spins. This slows freefall somewhat, but provides an extra layer of safety during that dangerous period [4]. His second jump proceeded uneventfully, however he did not achieve his desired altitude [22].

His third attempt on August 16, 1960, would be his final, and although this jump at 102,800 ft (**Figure 3**) was ultimately a success, it was not without problems [22]. During the ascent at approximately 43,000 ft, Kittinger noticed that his right hand began to feel strange. On inspection, he noticed the glove's airbladder was not inflated and realized that the pressurizing mechanism in his glove had malfunctioned. He decided not to notify ground control, knowing that he could still operate all of the necessary components of the gondola with minimal hand function since most of the controls were operated by a flick of a switch or nudge of the hand [21]. As he ascended higher his hand became increasingly swollen causing him extreme

Figure 3. Joseph Kittinger just after jumping from his capsule during the record-breaking skydive mission, Excelsior III.

pain and losing most of its circulation. By the time Kittinger made it to the ground, his right hand was nearly twice the size of his left due to swelling. However, 3 h after landing his hand had returned to normal size with no residual pain or deficits [21]. He would hold this record for many decades, until 2012 when another adventurous explorer would come along.

3.5. Felix Baumgartner and the Red Bull Stratos mission

On October 14, 2012, history was made as the Red Bull Stratos Capsule Jump set a precedent for high altitude exploration. The mission was no small feat; the goal was to safely ascend beyond Kittinger's previous altitude in a capsule and then free fall in a specially designed pressure suit, eventually using a parachute to descend the remaining elevation [22]. It was a huge success. The ascent was uneventful and the exit altitude for the pilot, Felix Baumgartner, was 127,852 ft (38.97 m). His maximum vertical speed was 843 miles per hour, at Mach 1.25 (377 m/s), making Baumgartner the first person to break the speed of sound in a freefall [22]. He was supersonic for 30 s of his 4 min 23 s freefall, during which he fell 119,431 ft (36,402 m). Other records broken during this jump include the largest balloon flown with a human aboard and the highest manned balloon ascent without a vehicle [22].

Unlike Joseph Kittinger's jump, the capsule (**Figure 4**) was pressurized, which allowed for continuous pre-breathing of oxygen throughout the entire ascent. Pre-breathing oxygen helps to decrease the risk of decompression illness. Increasing the amount of time the pilot can pre-breathe oxygen closer to the actual jump time increases the safety profile of the jump [22]. Using a pressurized capsule also prevents the discomfort and exertion of having to pressurize the suit prior to egress. The capsule provided thermoprotection; the lowest temperature

Figure 4. Gondola and pressure suit used by Felix Baumgartner during the Red Bull Stratos Mission. Now on exhibit in the National Air and Space Museum.

recorded inside the capsule was 13°F (−10.5°C), while the lowest recorded outside the capsule was −95.62°F (−70.9°C). The capsule consisted of fiberglass composite housing, with a hingeless acrylic door designed to pressurize at launch [22]. The capsule housing material selection was very important, as it would need to accommodate significant expansion and contraction expected with extreme variations of temperature. The hingeless door allowed for maximum range of movement within and outside the capsule, and was designed to maximize efficiency and ease of opening [22]. The capsule contained redundant life support systems including 10 h of breathable oxygen, glove and foot heaters, carbon dioxide and water scrubbers, among other systems. It would descend using a 100 foot diameter parachute; the typical landing shocks ranged from 4.5 to 8.0 G [22].

Baumgartner's pressure suit (**Figure 4**) was specifically designed to facilitate the capsule egress and exit. The major challenges of suit design were to provide adequate protection and life support during freefall yet allow the transition from sitting to standing, provide thermal protection, mitigate visor fogging, and create a system to sense and prevent the potential violent flat spin that trapped Kittinger. Baumgartner wore a baseline suit with standard undergarments which provided necessary thermal protection [15]. His chest pack contained three GPS units, an accelerometer, a mach speed indicator, a camera, a battery to heat the face plate and a power supply [22]. He also had an emergency cutaway knife in case the reserve chute opened at to high altitude, which would cause a slow descent and probably oxygen shortage. The helmet acted as an airtight gas container, impermeable to nitrogen and oxygen, yet breathable to allow water vapor to pass out and prevent fogging. The exterior of the suite included a fireproof cover [22]. Underneath he wore a fully integrated medical diagnostic tracking unit (Hidalgo Equivital) which provided continuous information on heart rate, respiratory rate and acceleration [22]. Taking another lesson from Kittinger's jumps, Baumgartner decided not to have a drogue chute deploy at the beginning of the jump, as this add unacceptable drag and he would not achieve his desired speed nor hope to achieve a new speed record [4]. Instead he wore a safety device on his right wrist which measured the amount of G-force throughout the entire mission. If he were to fall into a flat spin similar to Kittinger, and the device measured 3.5 G or higher for six continuous seconds, his drogue chute would automatically deploy, which would act to stabilize and pull him out of the spin [22].

The balloons used for these high altitude missions are also specially designed (**Figure 5**). They are made of thin plastic film, no more than 0.0009 inches (0.02 mm) thick [4]. As Piananida learned on his first failed ascent, the material needs to be thin and light enough to optimize the weight to lift ratio, but also needs to have a high drag limit (the point at which upward velocity creates drag strong enough to threaten damaging the balloon) to withstand the high winds of the upper altitudes [4, 21]. A lower drag limit could be mitigated by slowing the ascent, but only to the extent that the life support systems could allow. The Stratos balloon was made of 40 acres of polyethylene, at launch was twice as tall as a Saturn 5 rocket, and used 180,000 cubic feet of helium to launch [22]. Restraining fabric was placed around it for the initial ascent, holding the circumference to just under 17 ft. The balloon was released at 20,000 ft (6000 m) to allow its full expansion to 100 ft (30 m) diameter [22].

Figure 5. The balloon used for the StratEx mission being filled just as dawn breaks. The restraining fabric can be seen tied near the bottom, restricting the fill of the lower part of the balloon.

3.6. Gary Eustace and the StratEx mission

The StratEx (Stratospheric Explorer) mission (**Figure 6**) would take placed just over 2 years following Stratos, and though one goal was to break Baumgartner's record, a new dream was about to unfold. Unlike Stratos, a highly publicized project, StratEx was a privately financed scientific endeavor with focus on creating new technology to allow a less expensive, reusable way to explore the stratosphere without need for a constrictive and cumbersome capsule. Alan Eustace, the son of an aerospace engineer, worked as Google's senior vice president at the time of StratEx's development [23]. His vision was to use a balloon ascent system to transport a pilot, independent of a capsule, in a self-sustained pressure suit complete with all necessary life support systems for both ascent and descent [24].

The pressure suit designed for this mission required three distinct components: the actual pressure suit itself, an equipment module, and the parachute pack. The pressure suit was sandwiched between the equipment module in front, and the parachute pack in back. The pressure controller was housed inside the pressure suit, and the helmet was specially designed to force exhaled air down a valve and away from the facemask. The suit and helmet were essentially separate pressurized chambers, but the helmet had a suffocation valve that

Figure 6. Gary Eustace begins his ascent for the StratEx mission. He is barely visible in his special pressure suit secured to the bottom of the large balloon.

would allow for air passage into the helmet from the suit in an event of depressurization [25]. The equipment module included the life support and electronic support machinery which was primarily located in a large chest plate. The oxygen supply used a demand regulated system similar to SCUBA systems, where gas is supplied only when the pilot takes a breath and is not free flowing. Oxygen use was calculated beforehand based on suit pressure, suit temperature and pilot metabolic rate (breathing rate) during the various stages of the flight process. The oxygen cylinders used were modified standard aerospace composite cylinders [25]. The thermoregulatory system used a cartridge heater to heat a water-based fluid which was then pumped through a liquid thermal garment in order warm the pilot at high altitudes and cool the pilot while on the ground. This liquid cooling system was also used to help regulate the temperature of the electronic systems housed in the equipment module. This module also housed a power supply along with the voice and data communications hardware. The main parachute design was based on the Sigma tandem system, the most widely used tandem parachute [25]. These parachutes are normally built to support an instructor/student pair weighing up to 500 lbs, and fulfilled the needs of the StratEx system which weighed approximately 430 lbs (including the pilot). The parachute pack also acted as a suspension point during ascent, which provided the ideal 45° angle. The ascent balloon was similar to those used previously, however was much larger than Stratos, reaching 400 ft high at launch and spanned 275 ft in diameter at the maximum altitude [25].

A total of five test runs were performed, the second of which had to be aborted due to rising suit pressures caused by freezing with in the pressure control device [25]. The mission successfully concluded on October 24, 2014, when Eustace ascended to an altitude of 136,410 ft (41.5 km), and was released at 135,897 ft (41.4 km) to take the 9 min and 52 s ride back to earth. His free fall lasted 4 min and 27 s and spanned over 123,435 ft (37.6 m) [24].

3.7. Emergency medical planning for high altitude missions

The principles of medical coverage for this type of mission revolve around planning for the conventional expected injuries of a traumatic accident and well as those injuries specific to high altitude exposure. Medical conditions like those listed above are not routinely covered in medical training, so having a team of professionals familiar with the management of those specific injuries is paramount. Given the potential variability in landing location based on ascent time, wind speed and direction, as well as responding to emergency bail outs, having multiple recovery crews in a variety of vehicles is ideal. For instance, the StratEx mission utilized four vehicle chase teams, including two ground teams driving Suburbans, a helicopter and a fixed wing aircraft, which would deploy an additional parachutist to assist the StratEx pilot to identify a landing zone (limited sight due to the helmet design made clear ground views challenging) [24]. A central Mission Control coordinated these field teams, monitored all communications and worked to anticipate and mitigate any failures throughout the mission [26]. All chase teams included a combination of suit technicians, who specialized in rapid removal of the pressure suit, and volunteer medical personnel (mainly physicians and EMS) who were prepared with large selections of equipment to rapidly stabilize a downed pilot [24]. Each chase team had redundant means of communication including cell phones, laptops and tablets, in case of poor cellular signal [26]. Due to the well-supported utility of ultrasound in remote environments, two machines were carried along with the chase teams in order to perform a rapid, pre-specified, diagnostic and therapeutic assessment in the worst case scenario of an unstable, unconscious pilot [12]. One of the greatest challenges of these high altitude flights is the austere location of the launches in relation to medical facilities. The StratEx flight launched at a remote location in New Mexico, 500 miles from the closest hyperbaric chamber and 175 miles from the nearest level 1 trauma center [24]. Air transport was available in case of severe injury, but actual transport time was still daunting [26].

4. Commercial space industry

With the broadening scope of these high altitude missions, a new kind of space race has taken the Earth by storm, full of wealthy entrepreneurs who are now leading the commercial space industry [27]. In the 1980s two companies, Society Expeditions and Space Travel, proposed to NASA to begin offering passenger tickets aboard the space shuttle, but both were rejected. Through time, the once imagined concept of 'orbital space tourism' slowly shifted to 'suborbital space tourism', and the idea of experiencing longer time in lower earth orbit became a more feasible business strategy [28]. From that dream, companies emerged with the goal of providing wealthy ticketed passengers a few days of space training camp to then board a private spacecraft and float weightless for several minutes before returning to earth [27, 28].

In 2011, NASA began to lose much of United States government funding and was forced to shut down many programs and projects [29]. The new budget set forth by President Obama cut out the Constellation program, canceling the creation of the Orion spacecraft and Ares rockets. These crafts were supposed to replace NASA's three space shuttles which were set to retire that year. This effectively ended the shuttle missions, and NASA became more dependent on Russian colleagues for transport to the ISS, where they continued to perform many active research projects [30]. During this time the government has attempted to facilitate formation of the commercial space industry by providing only loose regulations, allowing an extended 'learning period' in order to support growth and practice of the companies [27].

4.1. Commercial space companies

4.1.1. Vulcan Aerospace

The first commercial aerospace vehicle, known as SpaceShip One, reached space in 2004. Developed by a Paul Allen (cofounder of Microsoft), in partnership with Burt Rutan, the project won the Ansari X contest which came with a 10 million dollar prize [27]. Their company, Vulcan Aerospace, has since begun building Stratolaunch, set to be the world's largest airplane with a wingspan wider than a football field, including the end zones. It is designed to carry a rocket tethered to its belly to an altitude of 35,000 ft, which then drops away and fires its engines to perform an 'air-launch' into orbit [27].

4.1.2. Space X

Space X, another growing commercial space company, was founded by Elon Musk, who made his first fortune on Zip2 and Paypal. However, Musk's dream extends beyond many other companies, with his main focus on eventual transport to Mars [27]. His goal is to build a 'Union Pacific' to Mars which would open entrepreneurial opportunities to anyone willing to make the journey. The initial plan of the company was to have its first unmanned flight to Mars by 2019, which would make Space X the first commercial space flight company to dare such a feat [27]. He also stated his next goal would be to execute a manned landing on Mars within the following 20 years [31]. Up to this point only 18 of the 43 robotic missions to Mars have been successful, including flybys [27].

Some of the newly developed technology from Space X includes the Dragon Spacecraft (**Figure 7**), selected as one of five competitors for NASA's commercial orbital transportation service to take astronauts to the International Space Station (ISS). With the NASA contract, Space X is granted the ability to use Kennedy Space Center's Launch Complex 39A, the same complex that launched Neil Armstrong and Buzz Aldrin to the moon in 1969 [27]. Following the discontinuation of NASA's shuttle program, US astronauts have become increasingly reliant on Russian flights for this transport service. The Dragon Spacecraft is approximately 9.5 ft in height, 11.8 ft in diameter, and weighs 9260 pounds. It contains a solar array, giving the

Figure 7. SpaceX's Dragon spacecraft pictured in Orbit.

aircraft a longer in-orbit duration which allows it to remain on the ISS for 1 week with astronauts aboard, or 1 year with only cargo aboard [31].

Space X has also developed a new rocket system. Their Falcon Heavy rockets (**Figure 8**) are designed to be twice as powerful as others currently in use, and can reach a maximum altitude of 75.8 miles. The goal for these rockets is to achieve high altitude and then arc parallel to the Earth's surface at 5 miles per second in order to stay aloft, which would allow them to circle the Earth in less time than it takes to watch a Star Wars movie [27]. However, their story has been laced with challenges. In 2016 one of the Falcon 9 rockets intended to transport cargo to the ISS incurred a malfunction which caused it to explode on the launchpad. That incident led the company to delay all further launches for 6 months, and created a backlog of over 70 missions, costing the company more than 10 billion dollars in revenue [27]. According to Musk, the company hopes to send two civilians around the moon in late 2018. Musk reportedly said this was "an important milestone as we work towards our ultimate goal of transporting humans to Mars" [32].

Figure 8. SpaceX's Falcon 9 v1.1 rocket being wheeled to the Cape Canaveral Space Launch Complex in preparation for the April 27, 2015 launch.

4.1.3. Blue Origin

Another giant of the commercial space industry is Blue Origin, founded by Jeff Bezos who made his fortune as founder, chairman and chief executive officer of Amazon.com and in 2013 purchased Washington Post. There appears to be some tension between Blue Origin and Space X, and Bezos and Musk have been known to make harsh comments about each other's accomplishments. The focus of Bezos' company is more to reduce the cost and increase the reliability of the commercial spacecraft, intended to open the opportunity for a suborbital experience to a greater consumer market [27]. His goal is to 'build a highway to lower orbit' so that contemporary infrastructure can be used by the next generation's entrepreneurs to further develop new technology and expand the space market. Blue Origin's rockets are designed to be fully reusable and can achieve a maximum altitude of 62.4 miles [27]. Their first unmanned test flight took place in April 2015, achieving an altitude of 93 km (57.8 miles) and speed of Mach 3. Since then, the same booster has been reflown four times during subsequent test flights from their secured launch space in Cape Canaveral [28]. However, Blue Origin has yet to send a rocket into space, and does not currently possess a rocket that is qualified to carry people [27].

4.1.4. Virgin Galactic

Virgin Galactic, a competitive entity in the commercial space industry, was developed by Richard Branson. Their spaceport is located in New Mexico, and Branson's stated goal is to be the first large volume commercial space line [27]. The company's main spacecraft is the SS2/WK2, a combination spacecraft and mothership design. The SS2 is an air-launched glider with capacity to carry six passengers and two pilots, and contains a rocket motor and extra systems for spaceflight [28]. Some 700 people jumped on the company's pre-sale tickets, some paying as much as $250,000. However, a tragic accident in 2014 involving one of their aircraft led to many delays in the planned flights [27, 28]. The incident occurred in California's Mojave Desert, involving a test pilot who was operating one of Virgin Galactic's newest spacecraft, SpaceShip Two [28]. According to the official report released by the National Transportation Safety Board, the pilot was believed to have unlocked the spaceship's 'feather system' prematurely, causing the vehicle to break apart in-flight, killing both the pilot and co-pilot [28, 33].

Three years from that tragic day, Virgin Galactic has yet to test any further powered spacecraft flights. Its newest vehicle, Spaceship Unity, has performed only a small number of glide flights. The company anticipates returning to unmanned, powered flights in late 2017 or early 2018. During the downtime, Virgin Galactic created a spin-off company called Virgin Orbit, which develops air-launched platforms for small satellites, with plans to begin launching 300 kg missions to Earth's lower orbit by 2018 [33]. The ultimate goal of Virgin Orbit is to distribute satellites around the solar system, starting with the low-Earth orbit constellation [33, 34]. This constellation will be part of the larger SpaceBelt satellite system, and will serve as a space-based data storage network. The company plans the first rocket test in 2018, using the Boeing 747-400 as part of its two-stage LauncherOne system, providing both an expendable and reusable air-launched platform [34].

4.1.5. Sierra Nevada Corporation, XCOR and other commercial space startups

There are a number of other smaller commercial space flight companies vying for position in this competitive market. Newer engineering including electronics miniaturization, advanced design of stronger and lighter materials, and new standards make it feasible for many of these developing companies to enter a market previously available only to those with the billionaires' backing. From 2007 to 2017, roughly 115 space-related companies were founded, with nearly 84 focusing on satellite technology [35]. The development of newer "microsatellites", weighing as little as 22–220 pounds, and "nanosatellites" which weigh less than 22 pounds, are sold by many of these companies. Roughly 2400 are projected to be launched within the next 6 years. The Cubesat is one satellite example, which weighs two pounds, is the size of a baseball and costs less than $100,000 to build [35].

Sierra Nevada Corporation (SNC) Space exploration a model company, with its Dream Chaser spacecraft (**Figure 9**) [28]. This winged spacecraft will reportedly allow for flexible, trustworthy, and affordable transport. This craft won NASA's Commercial Crew Integrated Capability award in 2012 as a potential spacecraft to provide transport of crews to the ISS, however SNC did not win NASA's commercial crew contract in 2014. The craft experienced some issues during a 2013 test flight in which the landing gear failed to deploy and sent the craft skidding off the runway after landing. Despite this setback, the company continued development of the a Dream Chaser cargo version and secured a Commercial Resupply Services-2 contract with NASA as one of three companies (including SpaceX's Dragon) to deliver cargo to and from the ISS from 2019 to 2024 [36]. SNC partnered with the United Launch Alliance, announcing in 2017 that they would employ the use of the Atlas 5 rocket, which includes 5 strap-on boosters and a twin engine upper stage, to send the first two Dream Chaser cargo aircraft to the ISS in 2020 and 2021. Utilizing these powerful rockets, the aircraft will be able to deliver nearly 12,000 pounds (5500 kg) of equipment and supplies on each non-piloted mission [37].

XCOR is a smaller commercial spaceflight company whose primary focus is a higher tempo flight operation. Their spacecraft, the Lynx Suborbital Vehicle, is a two seated, piloted transport with room for one pilot and one passenger or a specified payload [38]. XCOR plans to have a fast flight turnaround time and will prioritize low cost operations with minimal maintenance necessary for the fully reusable rockets between flights. This would potentially allow their goal of offering up to four flights per day. They are the first, and at the present time, the only, company to have successfully passed the Federal Aviation Administration's licens-

Figure 9. Dream Chaser spacecraft being lifted by an Erickson Air-Crane helicopter during a captive-carry flight test.

ing process for these aircraft. However, the actual two-seated rocket that XCOR plans to use remains under development in their Mojave, CA warehouse [38].

Naveen Jain, another visionary, plans to launch the Moon Express in late 2017 in an attempt to win Google's Lunar Xprize, a $20 million award for the first company to successfully land a robotic spacecraft on the moon and accomplish a variety of technical challenges. After landing, the Moon Express will need to extract iron ore, water, minerals and precious metals from the lunar soil, as well as capture nitrogen and hydrogen. Jain's vision is that the moon will one day become a fuel depot for spacecraft to dock and resupply before heading out on longer journey's [35].

Robert Bigelow, owner of Bigelow Aerospace which produces inflatable space habitats, has a similar vision. He even asked for government assistance to develop a 'lunar depot' that orbits the moon to allow for easier access to the lunar surface [39]. His company is currently testing a prototype, Bigelow Expandable Activity Module on the ISS, and has demonstrated free-flying prototypes in orbit. This could potentially give Bigelow a huge future advantage in the area of space tourism hotels.

Interorbital Systems, a small 12-person operation, is also based in the Mojave desert. Cofounders Roderick and Rnada Milliron started the company with the personal goal to eventually live on the moon. They also plan to compete for the Lunar XPrize. Currently, Interorbital Systems primarily sells satellites, with plans to launch more than 100 in the next year to provide revenue for equipment needed to achieve their moon landing goal [35].

4.2. Future of commercial space industry

Considering these commercial space industry startups, where will the next decades take us? Axiom Space, a Houston-based company, has partnered with Made In Space, a California company that specializes in 3D printing products. This partnership is intended to help facilitate Axiom's goal to develop a commercial version of the ISS, proposed as an outpost for private individuals and companies to conduct research and space exploration [40]. Its ultimate purpose is to help grow the space-tourism business. Made In Space's involvement is to expedite the logistics of creating an actual in-space factory that can produce equipment without the burden of transport from Earth's surface to the station [40]. Axiom has NASA backing; the current ISS has available funding currently to carry only through 2024. The full ISS expense cannot be overlooked; each day an astronaut is housed on the ISS costs roughly $7.5 million. The hope is that this deadline will be extended by at least 4 years, however there is a looming possibility that this extraordinary $100 billion structure will be brought down from orbit just over 25 years following its initial launch [35]. Axiom's current plan is to attach its first commercial module directly to the ISS in 2020. Upon ISS decommissioning, the module would detach and begin formation of the Axiom commercial space station [35].

Beyond a commercial ISS, plans to place orbital hotels have been the dream of many countries for some time. Consider the Space Hotel Berlin and Space Hotel Europe, which share a similar circular design with individual pods on the perimeter; each would provide accommodation for about 50 tourists [41]. A group from MIT won a NASA sponsored competition in early 2017 by designing a luxury space hotel purposed to help offset the cost of NASA research through commercial rental income [42]. The project, known as the Managed, Reconfigurable, In-space Nodal Assembly (MARINA), would be commercially owned and include a luxury hotel to serve as the primary anchor and a separate hub to serve as a temporarily co-anchor for NASA. The innovative aspect of MARINA is the external International Docking Adapter ports which allow modular service pods to connect to various points and, if standardized among space vehicles, would allow companies of all sizes to provide and request products and services from other companies in space [42].

5. Conclusion

Twenty-first century space exploration has transformed and taken on new meaning. What was once thought to be travel only to the moon or nearby planets now includes stratospheric exploration and commercial high atmosphere flights. Experiences available to only a select group of people with years of advanced training, are now close to being offered to a much wider group of eager customers.

Human ingenuity prevails yet again; we have developed technology to keep us safe in one of the most hostile environments of our home planet. As we continue to explore, we must never lose the sense of awe and respect for those visionary pilots and adventurers that helped us better understand and appreciate this aspect of our Earth. As Col. Joseph Kittinger said upon landing from his final mission: "Now that I am safely down, I realize once again how dependent upon the protection of the Almighty are all seekers of the unknown" [21].

Conflict of interest

There are no conflicts of interest to declare.

Author details

Laura Galdamez

Address all correspondence to: laura@outdoorem.com

Department of Emergency Medicine, Baylor College of Medicine, Houston, TX, United States of America

References

[1] Kirkpatrick R. 1969: The Year Everything Changed. New York: Skyhorse Publishing; 2011

[2] Zhu X. Radiative damping revisited: Parameterization of damping rate in the middle atmosphere. Journal of the Atmospheric Sciences [Internet]. 1993 Sep 1;**50**(17):3008-3021. Available from: https://doi.org/10.1175/1520-0469(1993)050%3C3008:RDRPOD%3E2.0.CO

[3] Grocott MPW, Martin DS, Levett DZH, McMorrow R, Windsor J, Montgomery HE Arterial blood gases and oxygen content in climbers on Mount Everest. The New England Journal of Medicine [Internet]. 2009 Jan 8;**360**(2):140-149. Available from: http://dx.doi.org/10.1056/NEJMoa0801581

[4] Mott N. Supersonic skydive's 5 biggest risks: Boiling blood, deadly spins and worse. National Geographic [Internet]. 2010 Oct. Available from: https://news.nationalgeographic.com/news/2012/10/121005-felix-baumgartner-skydive-science-sound-barrier-joseph-kittinger/

[5] The Great Leap Forward. Wired [Internet]. 2001 Aug. Available from: https://www.wired.com/2001/08/scale/

[6] Cohen JE, Small C. Hypsographic demography: The distribution of human population by altitude. Proceedings of the National Academy of Sciences of the United States of America [Internet]. 1998;**95**(24):14009-14014. Available from: http://www.pnas.org/content/95/24/14009.full.pdf

[7] West JB. Barometric pressures on Mt. Everest: New data and physiological significance. Journal of Applied Physiology. 1999 Mar;**86**(3):1062-1066

[8] Myers RL. The Basics of Physics. Westport, CT: Greenwood Press; 2006. pp. 113-120

[9] Murray DH, Pilmanis AA, Blue RS, Pattarini JM, Law J, Bayne CG, et al. Pathophysiology, prevention, and treatment of ebullism. Aviation, Space, and Environmental Medicine. 2013;**84**(2):89-96

[10] Stegmann B. Considerations for the Survival of Ebullism [Internet]. Fairborn, OH: Wright State University; 1989. Available from: http://www.geoffreylandis.com/ebullism.html

[11] Parker JF, West VR. Bioastronautics Data Book. 2nd ed. Arlington, VA: National Aeronautics and Space Administration; 1973

[12] Galdamez LA, Clark JB, Antonsen EL. Point-of-care ultrasound utility and potential for high altitude crew recovery missions. Aerosp Med Hum Perform. 2017 Feb;**88**(2):128-136

[13] Stewart LH, Trunkey D, Rebagliati GS. Emergency medicine in space. The Journal of Emergency Medicine [Internet]. 2007 Jan 1 [cited 2017 Dec 6];**32**(1):45-54. Available from: http://www.ncbi.nlm.nih.gov/pubmed/17239732

[14] Pilmanis AA, Sears WJ. Physiological hazards of flight at high altitude. Lancet (London, England) [Internet]. 2003 Dec 1 [cited 2017 Dec 10];**362**(Suppl):s16-s17. Available from: http://www.ncbi.nlm.nih.gov/pubmed/14698113

[15] Jacobs SE. Pressure suit design for commercial spaceflight: Lessons learned from Red Bull Stratos. In: 43rd International Conference on Environmental Systems [Internet]. Reston, Virginia: American Institute of Aeronautics and Astronautics; 2013 [cited 2017 Oct 23]. Available from: http://arc.aiaa.org/doi/10.2514/6.2013-3398

[16] Jenkins DR. Dressing for Altitude. Washington, DC: National Aeronautics and Space Administration; 2012. pp. 408-410

[17] Czarnik TR. Ebullism at 1 Million Feet: Surviving Rapid/Explosive Decompression. Fairborn, OH: Wright State University; 1999

[18] Graham RH. Technical features of the SR-71. In: SR-71 Revealed: The Inside Story. Minneapolis, MN: MBI Publishing Company; 1996. pp. 87-90

[19] Sherman T. A deadly fall: 46 years ago, a Jersey daredevil died while trying to set record. The Star Ledger [Internet]. 2012 Oct 12. Available from: http://www.nj.com/news/index.ssf/2012/10/a_deadly_fall_46_years_ago_a_j.html

[20] McKay B, McKay K. Skydiving from space part II: Nick Piantanida's "magnificent failure" [Internet]. The Art of Manliness. 2010. Available from: http://www.artofmanliness.com/2010/10/07/skydiving-from-space-part-ii-nick-piantanidas-magnificent-failure/

[21] Luce HR, editor. Fantastic catch in the sky. Life. 1960;**49**(9):20-24

[22] Walshe A, et. al. Red Bull Stratos Summit Report Final. Los Angeles; 2013

[23] Shetty S. How working at Google led this man to jump from the stratrosphere. CNBC [Internet]. 2016. Available from: https://www.cnbc.com/2016/11/18/how-working-at-google-led-this-man-to-jump-from-the-stratosphere.html

[24] Menon AS, Jourdan D, Nusbaum DM, Garbino A, Buckland DM, Norton S, et al. Crew recovery and contingency planning for a manned stratospheric balloon flight – The StratEx program. Prehospital and Disaster Medicine [Internet]. 2016 Oct 30 [cited 2017 Oct 23];**31**(5):524-531. Available from: http://www.journals.cambridge.org/abstract_S1049023X16000601

[25] Leidich J, Maccagnano Z, McFatter D, Lee GR, Hahn N. StratEx pressure suit assembly design and performance. In: 45th International Conference on Environmental Systems [Internet]; 2015. Available from: https://ttu-ir.tdl.org/ttu-ir/bitstream/handle/2346/64411/ICES_2015_submission_138.pdf?sequence=1&isAllowed=y

[26] Law J, Clark JB, Blue RS. Medical mission control planning and operations for a manned stratospheric flight test program. Space Operations Communicator. 2013;**10**(3):1-7

[27] Davenport C. The inside story of how billionaires are racing to take you to outer space. The Washington Post. 2016 Aug 19

[28] Chang Y, Chern J. Ups and downs of space tourism development in 60 years from moon register to spaceshiptwo CRASH. Acta Astronautica [Internet]. 2016 Oct 1 [cited 2017 Oct 23];**127**:533-541. Available from: http://www.sciencedirect.com/science/article/pii/S0094576516300066

[29] Sykes MV. The Obama legacy in planetary exploration (Op-Ed) [Internet]. Space.com. 2014. Available from: https://www.space.com/24157-obama-legacy-in-planetary-exploration.html

[30] Malik T. NASA grieves over canceled program [Internet]. NBC News. 2010. Available from: http://www.nbcnews.com/id/35209628/ns/technology_and_science-space/t/nasa-grieves-over-canceled-program/#.Wi1iCEqnHD4

[31] Szondy D. SpaceX Dragon's ultimate mission is Mars colonization. New Atlas. 2012

[32] Mosher D. Elon Musk: SpaceX is going to launch 2 space tourists "beyond the moon" [Internet]. Business Insider. 2017. Available from: http://www.businessinsider.com/spacex-moon-mission-elon-musk-2017-2

[33] Sheetz M. Virgin Galactic is returning to powered flights, CEO says, in a crucial next step for the spaceship company [Internet]. CNBC. 2017. Available from: https://www.cnbc.com/2017/09/07/virgin-galactic-ceo-company-is-returning-to-powered-flights.html

[34] Sheetz M. Virgin Orbit announces contract to launch a dozen Cloud Constellation satellites. CNBC [Internet]. 2017. Available from: https://www.cnbc.com/2017/09/14/virgin-orbit-announces-contract-to-launch-a-dozen-cloud-constellation-satellites.html

[35] Alsever J. Space startups are booming in the Mojave Desert. Fortune [Internet]. 2017. Available from: http://fortune.com/2017/02/20/space-startups-travel-satellites/

[36] Forest J. Sierra Nevada Corp. prepares for next round of Dream Chaser tests. Space News [Internet]. 2016 Jul. Available from: http://spacenews.com/sierra-nevada-corp-prepares-for-next-round-of-dream-chaser-tests/

[37] Clark S. Sierra Nevada confirms ULA will launch first two Dream Chaser cargo missions. Spaceflight Now [Internet]. 2017 Jul. Available from: https://spaceflightnow.com/2017/07/22/sierra-nevada-awaiting-direction-from-nasa-confirms-ula-will-launch-first-two-dream-chaser-cargo-missions/

[38] Chang Y-W. The first decade of commercial space tourism. Acta Astronautica [Internet]. 2015 Mar 1 [cited 2017 Oct 23];**108**:79-91. Available from: http://www.sciencedirect.com/science/article/pii/S0094576514005013

[39] King L. The commercial space industry wants updated regulations, more aid. USA Today [Internet]. 2017 Apr 27. Available from: https://www.usatoday.com/story/news/politics/2017/04/27/commecial-space-industry-wants-updated-regulations-more-aid/100948378/

[40] David L. Private space station coming soon? Company aiming for 2020 launch [Internet]. Space.com. 2017. Available from: https://www.space.com/35488-private-space-station-2020-axiom-space.html

[41] Reichert M. The future of space tourism. In: 50th International Astronautical Congress. Amsterdam: American Institute of Aeronautics and Astronautics; 1999

[42] Malone D. This space hotel design from MIT won NASA's graduate design competition. Building Design and Construction [Internet]. 2017. Available from: https://www.bdcnetwork.com/space-hotel-design-mit-won-nasa's-graduate-design-competition

Chapter 14

The Mortality of Space Explorers

Robert J. Reynolds and Steven M. Day

Additional information is available at the end of the chapter

http://dx.doi.org/10.5772/intechopen.73603

Abstract

Outer space exploration poses unique risks to human survival. Here, we review the current literature on United States astronauts and Soviet and Russian cosmonauts and provide updated and original research findings. As in previous research, both astronauts and cosmonauts are shown to have reduced risk of death by natural causes, particularly from chronic diseases such as cardiovascular disease and cancer, compared with appropriately matched general populations. Simultaneously, space explorers are at increased risk of death by external forces, particularly accidents such as plane crashes and spacecraft accidents. In total, both astronauts and cosmonauts are at reduced risk of all-cause mortality in comparison to the general populations of the United States and Russia. However, in comparison to astronauts, cosmonauts have been at equal risk of accidental death, but increased risk of death by chronic disease. We conjecture that the lack of risk from chronic disease may be due to the excellent health and medical monitoring of space explorers coupled with the deliberate attempts to limit their radiation exposure levels below those that would be detrimental. The differences in the astronaut and cosmonaut mortality experiences are likely due to lifestyle factors and the background rates of mortality in the two nations.

Keywords: astronauts, cosmonauts, mortality, cancer, cardiovascular

1. Introduction

For nearly 60 years, space exploration has captured our imagination and advanced human knowledge. Yet, in many ways, our understanding of space exploration is in its infancy. This is particularly true regarding possible long-term health consequences of living and working in space. For example, it is unknown whether humans will be able to safely explore deep space, colonize other planets, or live indefinitely on space stations even within Earth's orbit.

Data that will shed light on such questions have been, and continue to be, collected, and ongoing epidemiological analyses of those data will be required to determine the long-term hazards of space travel and to test whether countermeasures designed to mitigate those hazards are effective [1].

Measures of injury and morbidity, including incidence rates of acute physiological side effects or injuries related to space travel, and incidence rates or prevalence of chronic disease are important indicators of the relative safety of space exploration. Measures of mortality provide one kind of summary measure of such injury and morbidity outcomes that can potentially answer a fundamental question: is space exploration altering the lifespan of those who participate in it?

Exposures accrued during space exploration may elevate the risk of some causes of death, including those related to equipment failures or other accidents and cancers that may be related to exposure to radiation in space. Such elevated risks may be counterbalanced by the rigorous physical fitness and other health requirements of programs that have traditionally provided humans the opportunity for space travel, which may protect against some causes of death. In this chapter, we carefully examine the current evidence on mortality rates of space explorers, including cause-specific rates, and how they compare to those of age-, calendar year-, geography- and sex-matched general populations. We discuss the implications of this evidence and consider future steps in ongoing surveillance of the mortality and longevity of space explorers. Because space exploration to date has been largely confined to government-sponsored corps of astronauts from the United States and cosmonauts from Russia (via the former Soviet Union and current Russian Federation), our review will focus on these cohorts. As we shall see, evidence compiled and analyzed to date, as well as updated data and current analyses reported here for the first time, demonstrate that mortality rates for astronauts and cosmonauts differ in important ways from those of the general population and from each other.

2. Astronaut and cosmonaut cohorts

2.1. Demographic characteristics

The United States and Soviet manned space programs started at roughly the same time, with the selection of the first National Aeronautics and Space Administration (NASA) astronaut class in April 1959, the first group of United States Air Force (USAF) astronauts in June 1959, and the first Soviet cosmonauts in March 1960. The USAF eventually relinquished all astronaut training and manned space activities to NASA in 1969. The Soviet Space Agency is considered to have operated from 1957 to its official dissolution in 1991, after which its successor Roscosmos has continued space exploration for Russia. For purposes of analysis, we consider the "Soviet era" to be from the selection of the first Soviet cosmonaut class (March 15, 1960) until approximately the time of the fall of the Berlin Wall, which marked the beginning of the dissolution of the Soviet Union (December 31, 1989). The demographics of astronauts and cosmonauts are displayed in **Table 1**.

Characteristic	Astronauts, n (%)				Cosmonauts, n (%)				ALL	
	NASA		USAF		USSR		Russia			
Total cohort size	338	(100)	22	(100)	194	(100)	68	(100)	622	(100)
Sex										
Female	50	(14.8)	0	(0)	14	(7.2)	4	(5.9)	68	(10.9)
Male	288	(85.2)	22	(100)	180	(92.8)	64	(94.1)	554	(89.1)
Race/ethnicity										
Asian/E. Indian	4	(1.2)	0	(0)	0	(0)	4	(5.9)	8	(1.3)
Black	17	(5)	2	(9.1)	0	(0)	0	(0)	19	(3.1)
Hispanic	12	(3.6)	0	(0)	0	(0)	0	(0)	12	(1.9)
White	304	(90.0)	20	(90.9)	194	(100)	64	(94.1)	581	(93.4)
Other	1	(0.0)	0	(0.0)	0	(0.0)	0	(0.0)	1	(0.0)
Military										
Yes	237	(70.1)	22	(100)	127	(65.5)	36	(52.9)	422	(67.8)
No	101	(29.9)	0	(0)	67	(34.5)	32	(47.1)	200	(32.2)
Education										
High School	0	(0)	2	(9.1)	0	(0)	0	(0)	2	(0.3)
Bachelor	55	(16.3)	10	(45.5)	130	(67)	51	(75)	246	(39.5)
Master	173	(51.2)	9	(40.9)	6	(3.1)	6	(8.8)	194	(31.2)
Doctoral	110	(32.5)	1	(4.5)	58	(29.9)	11	(16.2)	180	(28.9)

Table 1. Demographic characteristics of astronauts and cosmonauts, 1959–2017.

In spite of the Soviets sending the first female to space in 1963, the U.S. Astronaut Corps has selected a larger percentage of females (14% vs. 7%) over the entirety of the follow-up period. The NASA group is also more ethnically diverse, with about 10% of the astronauts selected having non-White race/ethnicity.

Table 2 shows the average ages for the cohorts, including age at selection, average age at death for those who died, and average age of survivors as of 31 October 2017. Cosmonauts were slightly younger at the time of selection, on average, at 31.3 years versus 34.4 years for astronauts. There were no significant differences in the age at death or the mean age at the end of the study; thus, the cosmonauts were followed for 2.6 years longer, on average.

2.2. Actuarial characteristics

Table 3 lists selected actuarial characteristics of the astronaut and cosmonaut cohorts as of 31 October 2017, including counts of astronauts and cosmonauts, total amount of follow-up time

Event	Astronauts		Cosmonauts		p > t
	Mean	**(sd)**	**Mean**	**(sd)**	
Age (year)					
Selection	34.4	(3.6)	31.3	(5.4)	<0.0001
Death	59.6	(17.8)	61.4	(14.2)	0.4762
End of study	64.0	(11.7)	63.2	(14.7)	0.5523
Follow-up (year)	28.6	(14.1)	31.2	(14.8)	0.0269

Table 2. Average ages and follow-up time for astronauts and cosmonauts, 1959–2017.

(in person-years), and counts of deaths. Crude death rates, being in each case the ratio of total deaths to total person-years lived by a respective group, are also reported. Crude rates are highly dependent on the age structure of each cohort, and differences in crude rates may be due to differences in one or more cause-specific rates, or due to age differences in the cohorts, or both. To better understand whether and to what extent the rates in the various cohorts may differ, a more careful examination of these issues is required.

There have been 622 individuals selected and trained as astronauts or cosmonauts between April 1959 and October 2017. These men and women have contributed a total of 18462.8 person-years of observation time and 176 deaths, for an overall crude mortality rate of 9.53 deaths per 1000 person-years of observation.

The United States space programs account for 360 astronauts, contributing 10,291.8 person-years of follow-up and 80 deaths, yielding a crude mortality rate of 7.77 deaths per 1000

Source	Count	Exposure*	Deaths	Crude rate[†]
All astronauts and cosmonauts	622	18462.9	176	9.53
U.S. astronauts-all sources	360	10291.8	80	7.77
NASA Astronaut Corps	338	9565.7	62	6.48
USAF programs	22	726.1	18	24.79
X-15	7	230.7	7	30.34
X-20	5	160.6	5	31.13
MOL	10	334.7	6	17.93
Cosmonauts-all sources	262	8171.1	96	11.75
Soviet (1960–1989)	194	7089.2	91	12.84
Russian (1990–2017)	68	1081.9	5	4.62

*Pooled observation time since selection, expressed as person-years.

[†]Crude death rate, expressed as deaths per 1000 person-years.

Table 3. Actuarial characteristics of astronauts and cosmonauts, 1959–2017.

person-years. The vast majority of astronauts have been selected by NASA: 338 versus just 22 from the USAF. The last of the USAF astronauts were selected in 1967, while NASA astronauts have been periodically selected across the follow-up period. This makes the USAF astronauts some of the oldest in the overall cohort; this is reflected in their large crude mortality rate in comparison to that of NASA astronauts.

Data on cosmonauts are shown stratified by era of selection: the era of the Soviet Space Program (1960–1989) and the post-Soviet era of the Russian Space Program under the Russian Space Agency, Roscosmos. The 262 Soviet and Russian cosmonauts have accrued 8171.1 person-years of follow-up and 96 deaths, yielding a crude mortality rate of 11.75 deaths per 1000 person-years (**Table 3**).

3. Causes of death and comparative mortality for United States astronauts

We focus here and throughout this chapter on underlying cause of death as reported in official NASA astronaut biographies and in the news media. In this way, astronaut and cosmonaut deaths are categorized according to a single underlying cause of death.

Though there are many ways to quantify the mortality experience of groups such as astronauts and cosmonauts, here, we will focus on the Standardized Mortality Ratio (SMR). SMR is a risk ratio; it is computed by dividing the observed number of deaths in a group by the number of deaths that would be counterfactually "expected" were the group subject to a set of death rates from a reference population. (By convention, the resulting ratio is multiplied by 100.) Thus, SMRs of 100 represent equal risk between the group under study and the reference population, SMRs above 100 represent increased risk for the group under study, and SMRs below 100 represent decreased risk for the same.

3.1. Numbers and causes of death

Figure 1 shows the distribution by cause of the 80 astronaut deaths recorded through 31 October 2017. More than half of all deaths (46/80) were due to natural causes. Among natural causes, most deaths have been due to cancer (41.3%), followed by cardiovascular disease (CVD) (23.9%).

The 34 deaths from external (i.e., not natural) causes are dominated by 33 accidental deaths. Plane crashes and space craft accidents account for 29 of these deaths, with vehicular accidents accounting for the other 4. The only nonaccidental externally caused death was attributed to suicide (**Figure 1**).

For comparisons, all-cause United States general population mortality rates were taken from the Human Mortality Database for years 1960 through 2015 [2] and cause-specific rates from the CDC WONDER database for 1970 to 2015 [3–5]. The 2015 rates were used as the comparison rates for astronaut data from 2016 and 2017.

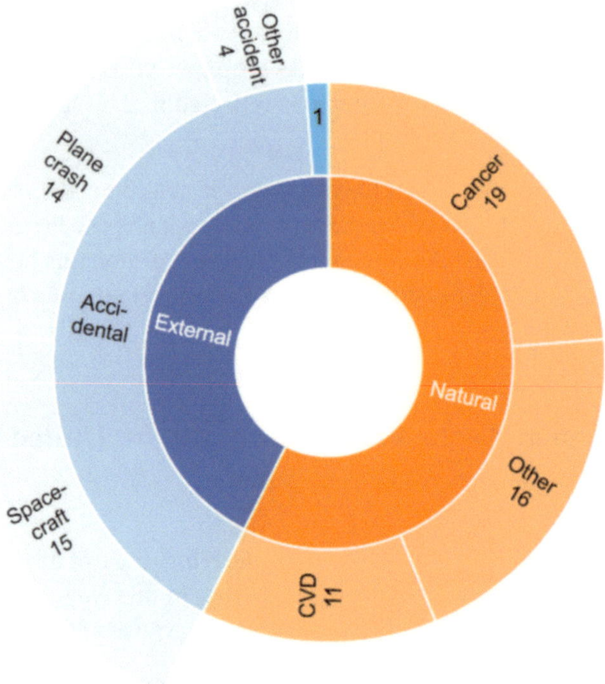

Figure 1. Causes of death among United States astronauts, 1960–2017.

3.2. All-cause mortality

In perhaps the first systematic analysis of astronaut mortality, based on data from 15 April 1959 to 30 September 1991, a nearly two-fold increased risk of death was reported compared to age- and gender-matched general population rates (SMR = 181, 95% CI = 110–279) [6]. The result was surprising, as it was expected that astronauts might experience lower all-cause mortality rates than the general population, thanks to their high levels of physical fitness, socio-economic status, and free access to presumably top-quality healthcare (a phenomenon often referred to as the healthy worker effect (HWE)) [7]. Another analysis of the same data compared astronauts with ground-based employees of the Johnson Space Center (JSC) in Houston, TX and found astronauts to be at more than 5 times the risk of death from all causes (hazard ratio = 5.07; 95% CI = 2.46–10.41), adjusting for sex, education, marital status at selection, and smoking history [8]. Comparing astronauts to a similar occupational cohort seemed to suggest that astronauts really were at greater risk of death by virtue of their status as space travelers. But if so, why?

By 2009, the picture had become clearer. After 1980, all-cause SMRs began declining: from a statistically insignificant 115 (95% CI = 53–219) in the 1980s, to 61 (95% CI = 29–112) in the 1990s, to finally a statistically significant 43 (95% CI = 23–74) in the first decade of the 2000s [9]. The overall SMR for 1980–2009 was also significant, at 59 (95% CI = 40–83) [9]. It appeared, then, that astronauts were at lower overall risk of death than the general population as a whole.

As it turned out, external causes-accidental deaths in particular-were the main drivers of the observed increased mortality risk in the early years, mainly due to deaths that occurred in the 1960s. A drop in the accidental death rate over time lowered SMRs from the year 1980 onward, though astronauts were still at a significantly higher risk of accidental death than the general population throughout the study period. Eventually, as the cohort of astronauts aged, the near absence of death by chronic diseases reduced astronaut all-cause mortality rates to levels significantly below those in the general population.

Figure 2 displays all-cause decade-specific SMRs for astronauts for 1960 to 2017, as well as a summary SMR for the entire 1960 to 2017 period, based on the latest available data. It is immediately apparent in **Figure 2** that 1960–1969 was a period of high risk for astronauts, as they were more than 8 times as likely to die during this period as were age- and gender-matched members of the United States general population. The pattern across the decades is consistent

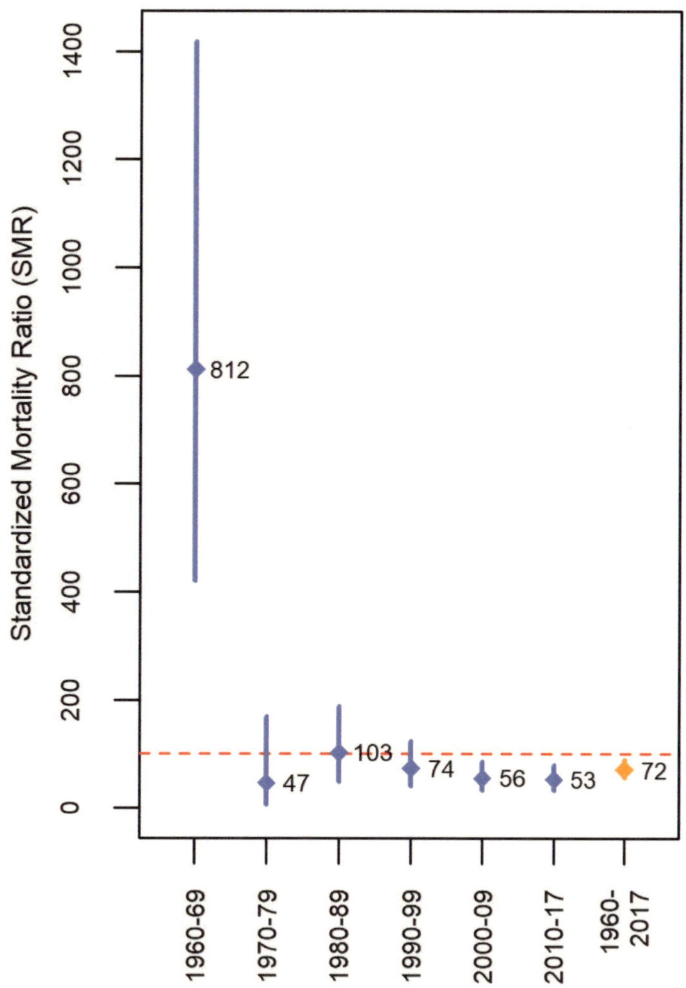

Figure 2. SMRs for all causes of death among United States astronauts, 1960–2017.

with previously published results, with higher astronaut mortality in the 1960s, then falling rates up to the current decade, and an overall lower than expected rate of mortality for astronauts from 1960 to 2017.

SMR of 53 (95% CI = 34–80) for 2010–2017 is reported here for the first time, based on all currently available data. This is broadly consistent with SMR from 2000 to 2009, suggesting a sustained reduction in risk for astronauts and a possible plateauing of the protective effect. If trends in specific causes of death continue as they have, this trend in all-cause SMR will continue as well.

3.3. External causes

Causes of death are subdivided at the most basic level into internal and external causes. Examples of external causes include drowning, electrocution, poisoning, burns, and trauma. For purposes of analysis, we consider external causes to be any death with primary cause code of E800–E899 in the International Classification of Diseases code set, Eighth Edition (ICD-8), E800–E899 in the ICD-9 code set, and V01–Y89 in the ICD-10 code set.

3.3.1. All external causes of death

As we have noted, accidental deaths accounted for all external deaths in the astronaut cohort for many years, and no published study bothered to report on nonaccidental external causes for this reason. Current data include a fair number of deaths due to external causes other than accidents, and we present in **Figure 3**, for the first time, SMRs for all external causes combined, as well as SMRs for accidental causes.

Across all decades, United States astronauts have been at approximately 250% risk of the general population of death due to external causes. This excess risk continues to be driven almost entirely by accidental deaths, some of which occurred in catastrophic accidents that many readers will recall, which took the lives of multiple astronauts in single events.

In the 1980s and 2000s, astronauts were at significantly increased risk of death from external causes, and this is unsurprising. These two decades each saw the destruction of a space shuttle, with the death of multiple astronauts in each: the space shuttle Challenger explosion in January 1986, which claimed the lives of five astronauts, and the Space Shuttle Columbia reentry disintegration in February 2003, which killed 6 astronauts.[1] These deaths pushed SMRs significantly high, even as deaths from other external causes-particularly other accidental sources-were on the decline. These deaths are largely responsible for the overall 2.5-fold increased risk of death due to external causes for the entire follow-up period. Indeed, without these deaths, overall SMR in the 1980s would be approximately 160, and not statistically significant. Likewise, SMR for the 2000s would be approximately 115, and not statistically significant.

[1]Three additional people were killed in these two disasters, all of were whom Payload Specialists, i.e., civilians trained for single missions only. As they are not considered a part of the NASA Astronaut Corps, those individuals were not included in either prior research or the updated analyses presented here.

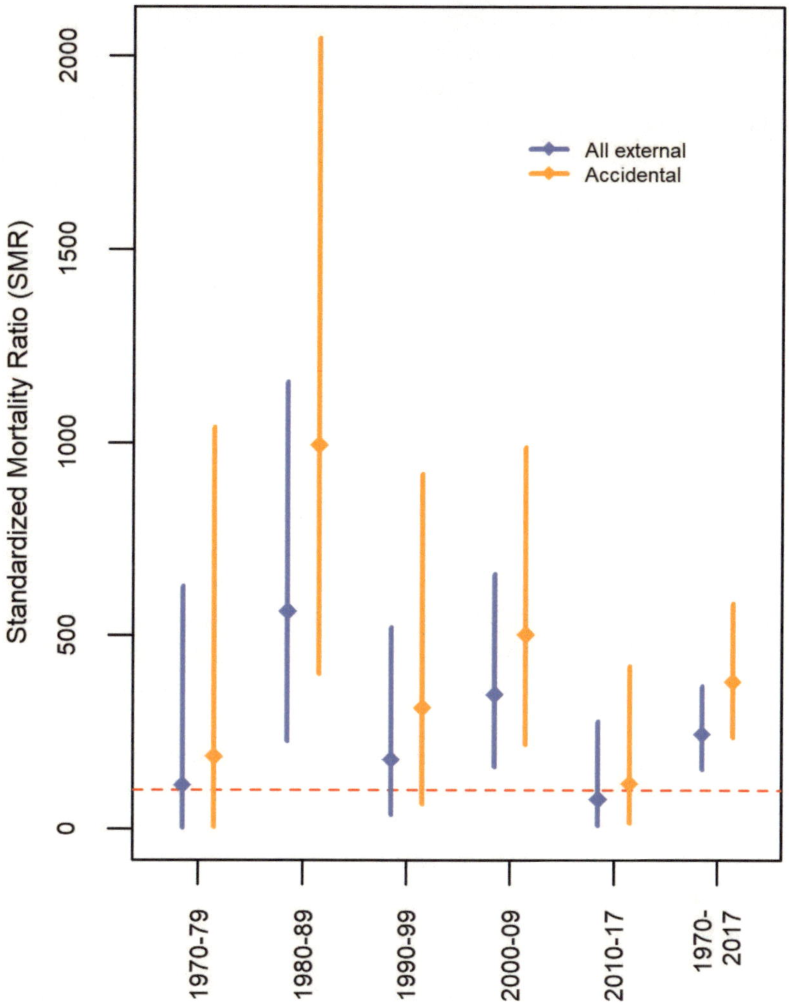

Figure 3. SMRs for all external causes and accidental causes of death among United States astronauts, 1970–2017.

3.3.2. Accidental death

The high risk of accidental death was anticipated from the beginning of the United States space program, when then U.S. President Dwight Eisenhower called for astronauts to be selected from the ranks of military test pilots, because, among other reasons, they were accustomed to high-risk vocational activities [10]. This proved to be prescient: by the end of 1991, 16 of the 20 recorded astronaut deaths were accidental, with half due to space craft accidents, and several of which were duty-related plane crashes. As we have noted, the rate of accidental deaths for this cohort is significantly greater than expected in an age- and sex-matched cohort from the general population.

Accidental death SMRs for all astronauts based on data through 2017 are also given in **Figure 3** as the orange points and lines. The progression of risk for accidental death is the main driver of the risk for death by all external causes. As such, SMRs follow much the same pattern as those from all external causes. The main difference is that SMRs for accidental death are higher than those for all external causes due to the smaller number of expected deaths in the general population from this causal subset.

3.4. Natural causes

Natural causes of death are causes originating within the body (albeit possibly as a result of an unseen and perhaps unfelt external force, such as cancer precipitated by radiation exposure), rather than a cause related to an obvious external force or object. In the universe of all possible causes, natural causes are the complement of external causes, and thus, in terms of ICD-8, −9, and −10 codes, these would include all codes other than those noted above for external causes in Section 3.3. Natural causes are also a major concern for astronauts in relation to space travel: does time in space equate to a higher mortality risk due to disease?

Natural causes of death include cancer, cardiovascular disease (CVD), and myriad other less common diseases. Testing hypotheses related to large numbers of potential causes as a group can boost statistical power, which can be helpful when analyzing rare events that may have long latency periods (e.g., cancers related to radiation exposure). However, studying natural causes as a single group has its drawbacks, too. The rigorous physical fitness requirements, high-quality health care and careful follow-up of health, and relative high socio-economic status of astronauts may all be protective of some diseases. Thus, there may be two forces at work, each of which may mask the effects of the other relative to all natural-cause mortality: exposure to space on the one hand, which may elevate the risk of death due to some causes, and a healthy lifestyle on the other hand, which may mitigate the risk of other (or even some of the same) causes. We will explore these issues in some depth now.

3.4.1. Any natural cause of death

As far as we are aware, analysis of natural-cause mortality of astronauts has only previously been reported in the context of a dissertation project [11]. In that project, the authors fit a Poisson regression model to explore possible differences in natural-cause mortality rates of astronauts above and below the median in estimated space radiation exposure. Unfortunately, the risk of misclassification of exposure and demonstrably low statistical power made the results inconclusive [11]. We present SMRs for death by natural causes, CVD, and cancer in **Figure 4**.

The darker blue points in **Figure 4** suggest that astronauts have been at a consistently reduced mortality risk due to natural causes and significantly so since the year 2000. For the entire 1970 to 2017 period, astronauts were at less than half the risk of death from natural causes as the general population. As with SMRs for external causes, SMRs in **Figure 4** are based on only a few deaths through the end of the 1990s and thus do not reach statistical significance. From the year 2000 onward, the aging astronaut cohort resulted in dramatic increases in the number of expected deaths, and observed astronaut deaths did not keep up, resulting in SMRs statistically significantly less than 100 for the latest two periods.

http://dx.doi.org/10.5772/intechopen.73603

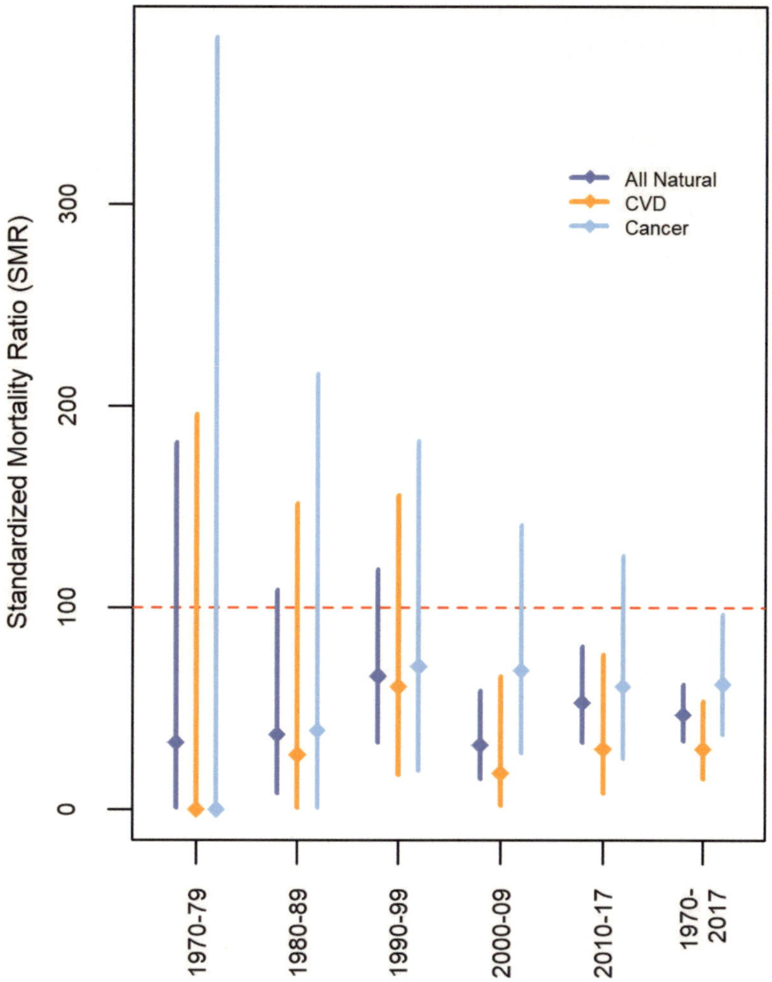

Figure 4. SMRs for natural-cause, CVD, and cancer mortality among United States astronauts, 1970–2017.

3.4.2. Cardiovascular disease

Cardiovascular disease includes ischemic heart disease, heart failure, stroke, and any other circulatory disease captured by ICD-8 codes 390–458, ICD-9 codes 390–459, and ICD-10 codes I00–I99.

Death from CVD is an outcome that has been tracked since the first published research on astronaut mortality in 1993. Even though CVD has been one of the most common causes of death for astronauts, SMR for CVD through 1991 was reported to be a statistically insignificant 47 (95% CI = 5–168), suggesting (inconclusively) that astronauts may be at less than half the risk of death from CVD in the general population [6]. An analysis comparing the same astronaut mortality rates to those of ground-based controls from the JSC found astronauts to have an insignificant elevation in risk (HR = 1.20; 95% CI = 0.27–5.28) [8]. Considering the wide range of possible effect sizes those two studies suggested were possible (and the lack of statistical significance in both of them), no solid conclusions could be drawn about CVD mortality from the 1991 data set.

The 2010 study of astronaut mortality by Reynolds and Day [9] reexamined CVD mortality and found that, in comparison to the general population for the 1980 to 2009 period, CVD SMR was a statistically significant 27 (95% CI = 9–63), validating the findings from Peterson et al. [6].

Again, it has been speculated that HWE is responsible for the reduction in risk of CVD mortality even in the face of potentially heavy smoking by the earliest groups of (mostly military) astronauts [9, 12]. Prior reports of astronaut biometric measurements have shown astronauts to be at or below suggested normal limits for blood pressure, cholesterol, and body mass index, all important risk factors for cardiovascular disease [8, 13].

Figure 4 shows updated SMRs for CVD among astronauts (orange points and lines), once again adding in the USAF astronauts and extending follow-up to the end of 2017. Thin data in the early decades give statistically insignificant SMRs through 1999. However, in the last two periods, SMRs show significant reductions in CVD mortality rates for astronauts, as does the overall 1970–2017 SMR. For the overall period, astronauts are at less than one third the risk of death from CVD as is the general population of the United States (SMR = 30; 95% CI = 15–54).

In total, the evidence thus far suggests that astronauts are at a greatly reduced risk of death from CVD in comparison to the general population.

3.4.3. Lunar astronaut mortality from CVD

In 2016, a published study investigated the risk of CVD mortality for astronauts who had either walked on the moon or who completed circumlunar flights as part of the Apollo missions (so-called lunar astronauts) [14]. The study compared lunar astronauts to astronauts who only completed missions on low Earth orbit (LEO), or to astronauts who had not flown to space at all ("nonflight" astronauts). Dividing the lunar astronaut percentage by the percentage of deaths from CVD in the comparison groups (up to the end of 2015), the authors computed proportionate mortality ratios (PMRs). PMRs attempt to look for differences in the proportion of deaths by a specific cause between two groups. In this case, PMRs demonstrated that a higher proportion of lunar astronauts died of CVD than either of the comparison groups, leading the authors to conclude that lunar astronauts were at greater risk of death by CVD than astronauts who had never left low Earth orbit or never flown into space at all. Conflating these findings with a model of how radiation may damage vascular endothelial tissue in rats, the authors proposed a potential biological mechanism to match their purported epidemiologic findings [14].

Other authors were quick to point out flaws in the study. Questions arose regarding the data set used, the analytic method, and the potential imprecision of the disease outcome definition [15]. Several comments on the online version of the paper pointed to methodological flaws in the PMR analysis, particularly the potential biases related to competing risks or confounding by age when the age structure is markedly different between groups [16]. As it turns out, these two issues were, in fact, driving the PMRs to be misleading.

A reanalysis of the astronaut data set used by Delp et al. [14] revealed that lunar astronauts were significantly older than the nonflight and LEO-only astronauts at the start and end of the study, making the potential for age confounding in the PMR high [17]. Using SMRs instead of PMRs demonstrated that when the age structure was taken into account (and all the available information, including follow-up time, used), there were no significant differences between any of the astronaut groups in risk of death from CVD [17].

In comparison to nonflight astronauts, lunar astronauts had no significant increase in risk, with an SMR of 117 (95% CI = 24–343) for the 1968–2015 period. In comparison to LEO astronauts in the same period, SMR was 67 (95% CI = 14–197), and, in comparison to the combined group of LEO and nonflight astronauts (nonlunar astronauts) in the same period, the SMR was 77 (95% CI = 17–237) [17]. These results offered no plausible evidence of an increase in risk of death from CVD for Apollo lunar astronauts and showed that prior research on the topic was still entirely tenable: there is simply no evidence to support the hypothesis that lunar astronauts are at elevated risk of CVD mortality compared to nonlunar astronauts [17].

3.4.4. *Cancer mortality among astronauts*

Since the beginning of the space program in the United States, there has been concern that radiation exposure in space may lead to greater mortality from cancer, particularly from the unique radiation sources found in outer space [18, 19]. Because of this, cancer incidence has been under regular surveillance by NASA, and cancer mortality has been reported on repeatedly over the years [6, 8, 12].

Cancer deaths are those with a primary cause ICD-8 code of 140–239, ICD-9 code of 140–239, or ICD-10 code of C00–D48. The first study of astronaut mortality from 1993 did not report any mortality measures related to cancer and for good reason: at that point in time, there had only been one death due to cancer among astronauts. The first reported analysis of cancer mortality among astronauts was from the 1998 Longitudinal Study of Astronaut Health (LSAH) [12]. The study compared the three observed astronaut deaths from cancer to the number expected in two comparison groups: the general population of Public Health Region 6 of Texas and ground-based controls from Johnson Space Center (JSC) [12]. In comparison to the general population, astronauts were found to be at less than half the risk of dying from cancer (SMR = 47; 95% CI = 10–105). In comparison to the LSAH controls, however, astronauts were at almost three and a half times the risk of death from cancer (SMR = 345; 95% CI = 66–756) [12]. These results suggested three possibilities for astronauts: (1) they are at greater risk of cancer due to exposure to space travel, (2) the JSC ground-based controls are healthier than astronauts in ways that protect against cancer, or (3) these results represent a statistical anomaly of this relatively small sample of data. In any case, the small number of observed deaths provides for low statistical power, making results inconclusive.

The LSAH reexamined cancer mortality in a study published in 2000 [8]. In a proportional hazards model, again comparing astronauts to the LSAH controls, astronauts were found to have a hazard ratio of 3.19 (95% CI = 0.93–21.85), adjusted for sex, education, marital status at selection, and smoking history [8]. This similar result is not surprising, as the analysis was based on essentially the same data set as the prior study: there had been only one additional cancer death among astronauts since the 1998 study. Once again, the small number of deaths plus lingering doubt about the suitability of the controls yielded an underpowered and, ultimately, inconclusive analysis.

In the 2010 study of astronaut mortality, cancer mortality was analyzed by decades between 1980 and 2009 [9]. By the end of 2009, the number of astronaut deaths from cancer had risen to a total of seven. For the three decades 1980–2009, SMRs for cancer in comparison to the general population were consistently below 100, but still, small numbers of cases made SMRs nonsignificant. Pooling the data from the entire 30 years led to SMR of 47 (95% CI = 19–97) [9]. This was the first time that the observed reduction in cancer mortality among astronauts reached statistical significance and was consistent with prior general population comparisons.

Updated findings through 2017 are displayed as the lighter blue points and lines in **Figure 4**. With the additional follow-up time and incorporation of USAF astronauts, we can see that astronauts are still estimated to be at about half the risk of cancer mortality as the general population between 2010 and 2017, though the results are again not significant (**Figure 4**). SMR for the period 1970 to 2017 is significant, however, with SMR of 62 (95% CI = 37–97).

3.4.5. The effect of competing risks on rates of death by natural causes

As we have seen, astronauts have at times been at elevated risk of death due to external causes (primarily accidents) and have been and continue to be at reduced risk of death due to natural causes. External and internal (or natural) causes are mutually exclusive, competing causes of death (at least in terms of an underlying cause of death). That this may explain the low natural-cause SMRs that have been observed has been suggested in recent literature [17, 20]. Given that a significant number of astronauts have died (some quite young) of external causes, this could alter natural-cause mortality rates at older ages.

Such altered natural-cause mortality rates would in turn affect SMRs for natural causes, and this could explain the low natural-cause SMRs that have been observed for astronauts. We can attempt to quantify the potential effect of this via sensitivity analyses, whereby we compute SMRs under extreme hypothetical alternative scenarios regarding the reassignment of observed deaths due to external causes. These alternative SMRs are shown in lighter blue (left-most SMR in each cluster) and darker blue (right-most SMR in each cluster) in **Figure 5**, along with the observed natural-cause SMRs in orange.

Specifically, the lighter blue and darker blue point estimates in each period form an upper and lower bound to our natural-cause SMRs under two differing counterfactual assumptions regarding the occurrence of deaths due to external causes: (1) assuming the external deaths did not happen at all and the astronauts are still alive as of the end of the follow-up period (lighter blue); (2) assuming all external deaths occurred at the observed times, but were due to natural causes instead (darker blue). The former is the most optimistic, biasing SMRs downward by inflating only the denominator of the natural-cause SMR. As a result, all of these point estimates are lower than their corresponding observed (orange) point estimates. The latter scenario is the most pessimistic possible scenario, biasing SMRs for natural causes upward by inflating the number of natural-cause deaths without changing the observation time. This

http://dx.doi.org/10.5772/intechopen.73603

leads to all darker blue point estimates being higher than their corresponding orange point estimates. The true (unbiased) estimates, and estimates for all other potential scenarios, must be somewhere in between. Thus, the light blue zone surrounding these sets of SMRs show the total range of possibility for the true SMR (with 95% CI) in each period.

In the observed SMRs for natural causes (**Figure 5**), no SMRs are statistically significant until the 2000–2009 period. However, SMRs in the optimistic scenario (**Figure 5**) reach statistical significance in the 1980s and remain significant through the end of the observation period. This suggests that early deaths due to external causes may in fact be biasing SMRs for natural causes upward. Had those astronauts lived, SMRs for natural causes could be even lower than observed.

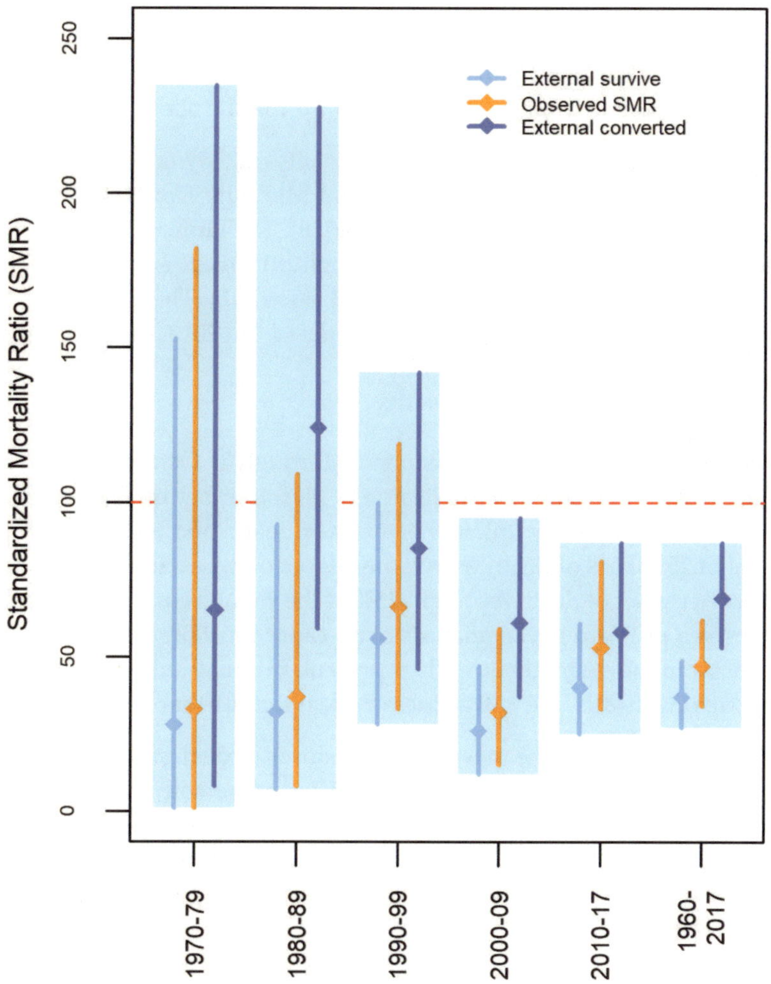

Figure 5. Alternative SMRs for natural-cause mortality among United States astronauts, 1970–2017.

4. Causes of death and comparative mortality for Soviet and Russian cosmonauts

In comparison to astronaut mortality, there has been comparatively little research on cosmonaut mortality. To date, only three studies have been published on this topic, all since 2014 [21–23]. From this research, we know that the patterns of mortality risk of cosmonauts are quite similar to those of astronauts in terms of how they compare to Soviet and Russian general population controls.

Cosmonauts tend to have the same major causes of death as astronauts and, like astronauts, tend to have lower mortality rates for those causes than the general population. This is perhaps unsurprising given the similar vocational backgrounds, similar (and now joint) training and physical readiness criteria required to be either a cosmonaut or astronaut, and the similar levels of biomedical monitoring of cosmonauts throughout their careers [24, 25]. That cosmonauts would be at elevated risk of death due to accidents, especially in the early years of the space programs, might also be expected, and as we shall see, this is indeed true. We refer the reader again to **Tables 1–3** for demographic and actuarial information about the cosmonaut corps.

For comparisons, Russian general population mortality rates were taken from the Human Mortality Database [2] and the Russian Fertility and Mortality Database from the Center for Demographic Studies at the New Economic School [26]. The Human Mortality Database supplied all-cause mortality rates for the years 1960 through 2015; the Russian Fertility and Mortality Database supplied cause-specific rates for the years 1960 to 2014. The latest rates available (2014 or 2015) were used for comparison with cosmonaut data from 2015, 2016, and 2017 as needed.

4.1. Numbers and causes of cosmonaut deaths

Figure 6 shows the causes of death for cosmonauts through 31 October 2017. What should be immediately apparent in **Figure 6** is the relatively high number of deaths due to unknown causes. There were 24 completely unknown causes of death, and a single unknown external cause, for a total of 25 causes of death with some degree of uncertainty as to their causes. The 24 totally unknown causes of death represent 25% of the 96 total cosmonaut deaths. Unknown causes do not pose a problem for an analysis of all-cause mortality. However, depending on which causes these unknown might actually represent, this may substantially alter cause-specific mortality rates and SMRs. We will address these issues further in the sections that follow.

Similar to the astronaut cohort, the most common causes of death for cosmonauts were CVD (25 deaths) and cancer (20 deaths), and accidents (19 deaths), with a small number of deaths by other natural causes (7 deaths). However, external causes of death have accounted for a smaller share of cosmonaut deaths (21%), compared to the astronaut cohort (42%).

4.2. All-cause mortality

Much like United States astronauts, cosmonauts have been at decreased risk of all-cause mortality when compared to the general population [21–23]. This has been shown to be true for the cosmonaut cohort as a whole [21], as well as the subset of cosmonauts who went to space [22, 23].

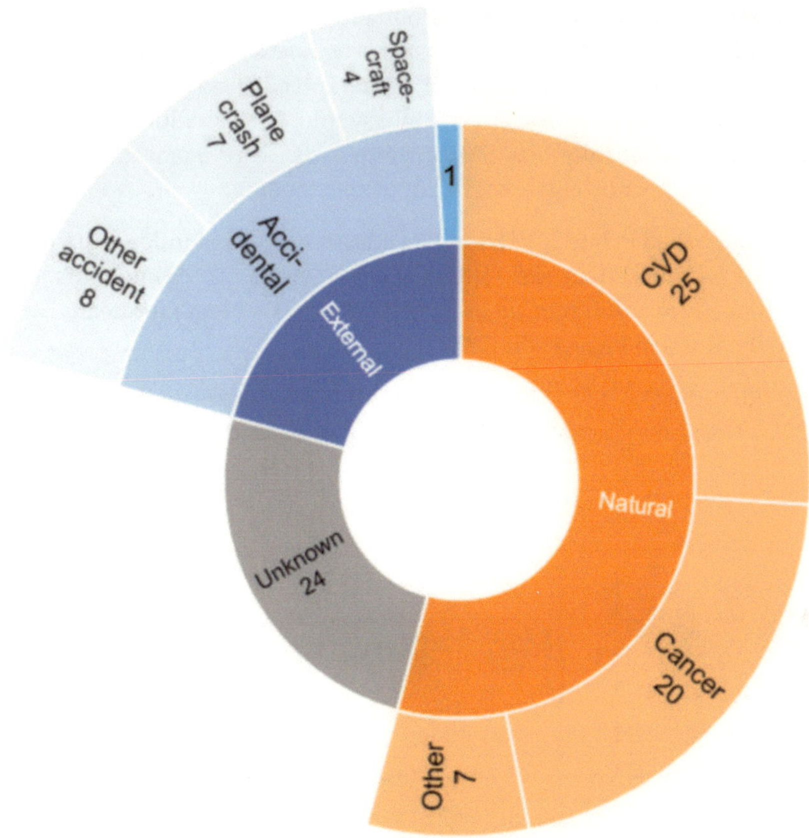

Figure 6. Causes of death for Soviet and Russian cosmonauts, 1960–2017.

The first study of cosmonaut mortality, published in 2014, studied the entire cohort of people who were not only selected for but also completed cosmonaut training [21]. The study reported a dramatic (and statistically significant) reduction in risk of death from all causes. However, SMRs reported in this study were incorrect, as SMRs were computed using probabilities of death for the astronaut cohort rather than mortality rates. While the difference between age-specific probabilities of death and mortality rates may be negligible at young ages, as age increases mortality rates become substantially higher than probabilities of death. This made SMRs and the corresponding bounds on their 95% confidence intervals too low. Because of this, no conclusions can be drawn from the all-cause SMR results from that study alone.

Two additional studies on cosmonaut mortality have been published after the 2014 Reynolds et al. study. These two studies are highly similar to each other in that they use only cosmonauts who went to space. The first, published in 2016 in the Russian journal *Air-Ecosystem and Environmental Medicine,* studied cosmonaut mortality within the cohort of 114 cosmonauts who flew to space at least once [22]. The authors compared the mortality experience of these cosmonauts to that of the general populations of the Moscow region and the Russian Federation as a whole, with follow-up through the end of 2013. Unfortunately, the study counted observation time as starting with enrollment in the cosmonaut corps, meaning SMRs were incorrect (biased downward) due to an *immortal time bias* [27]. In this instance, the time between selection and the first space mission is guaranteed survival time, as the cosmonauts who may have died before their first space flight would not be included in the analysis. The effect of this is to systematically bias SMRs downward. Once again, this error renders the study inconclusive on the mortality experience of cosmonauts.

A second study published in late 2017 added an additional year of follow-up to the 2016 study and corrected the immortal time bias. This study confirmed that cosmonauts are at reduced risk of death from all causes, reporting SMR of 40 (95% CI = 27–61) for death by all causes in comparison to the Russian Federation [23].

A reanalysis (with correction) of the Reynolds et al. data [21] is presented in **Figure 7**, updated to 31 October 2017. This shows the trend in SMRs for all-cause mortality for cosmonauts over time.

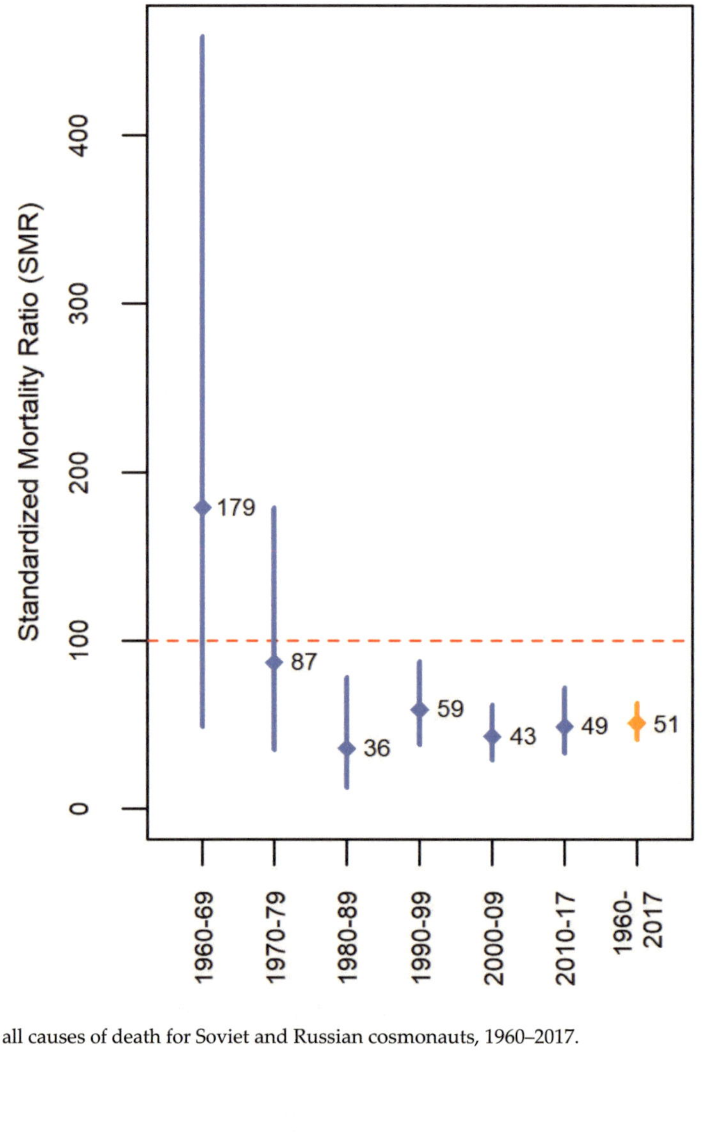

Figure 7. SMRs for all causes of death for Soviet and Russian cosmonauts, 1960–2017.

Though mortality for cosmonauts in the 1960s was 179% of that expected, the SMR was not statistically significant. Cosmonauts experienced 87% of the mortality expected in the 1970s, also not significant. From the 1980s onward, the mortality risk for cosmonauts has been statistically significantly *reduced* in comparison to the general population, with SMRs ranging from a low of 36 to a high of 59, and an overall SMR for the 1960–2017 period of 51 (95% CI = 41–63). These long-term and correctly-computed findings solidify the result that cosmonauts are at an overall lower risk of mortality than the Russian general population. They also largely agree with the prior results from the 2016 study by Ushakov et al. [22] even with slightly different cohort definitions. From this, we can conclude that, overall, cosmonauts have been at a lower risk of death from all causes, between 1960 and 2017, than the year-, age- and sex-matched general population of Russia.

4.3. External causes

4.3.1. All external causes of death

As noted above, (known) external causes of death represent just over one-fifth of cosmonaut deaths. Yet, the only previously reported mortality measure in relation to external causes came from the 2016 study, which, as previously mentioned, was biased [22]. Nevertheless, the authors reported that, for the period 1960–2013, cosmonauts who had flown on at least one mission to space had SMR for external causes of death of 42 (95% CI = 16–107) in comparison to the general population of the Russian Federation [22].

Figure 8 shows updated SMRs for all external and accidental causes of death for all trained cosmonauts by decade between 1960 and 2017. The 1960s saw three times the number of deaths from external causes as expected, but this quickly tapered off to cosmonauts having fewer deaths than expected.

Results were generally not statistically significant by decade, but over the entire period from 1960 to 2017, cosmonauts were at significantly lower mortality than the general population, with SMR of 47 (95% CI = 29–72). This overall SMR is quite close to that published by Ushakov et al. [22], in spite of the bias in that article and the differing cohort definitions. Overall, the evidence shows that cosmonauts are at lower risk of death from external causes than is the Russian general population.

4.3.2. Accidental causes of death

Similar to astronauts, the majority of deaths due to external causes (19 of 20 for which causes were known) for cosmonauts have been accidental. Given this, it is not surprising that the pattern of SMRs for accidental deaths (**Figure 8**) is very similar to that for all external causes combined.

SMR of 62 (95% CI = 38–98) for the entire 1960–2017 period is close to the SMR of 52 (95% CI = 19–139) reported by Ushakov et al. [23] The wider confidence interval on the Ushakov SMR is a direct result of the more limited data used in that study (only male cosmonauts who had been to space, with follow-up to the end of 2014).

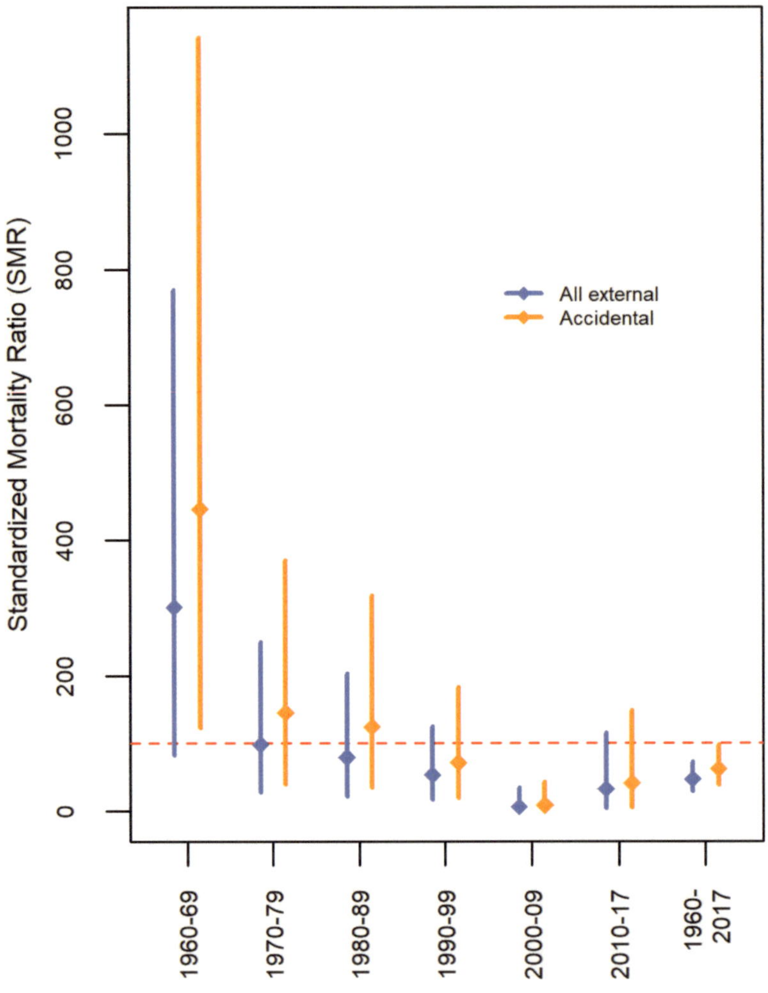

Figure 8. SMRs for all external causes and accidental causes of death for Soviet and Russian cosmonauts, 1960–2017.

4.4. Natural causes

Natural causes of death are responsible for more than 50 cosmonaut deaths to date (**Figure 6**). Primary among them have been cancer and cardiovascular disease, with a small number of assorted other natural causes.

No prior research has reported measures of mortality for all natural-cause deaths for cosmonauts. **Figure 9** shows SMR for natural causes (darker blue), CVD (orange), and cancer (lighter blue) for 1960 to 2017.

SMRs for all natural causes of death in all decades (darker blue) were below 100, though they only reached statistical significance in the 1980–1989 period and remained significant thereafter. SMRs ranged between 0 and 76, with SMR of 33 (95% CI = 25–43) for the entire 1960–2017 period, indicating that cosmonauts were at one-third the risk of death by natural causes as the general population of Russia.

http://dx.doi.org/10.5772/intechopen.73603

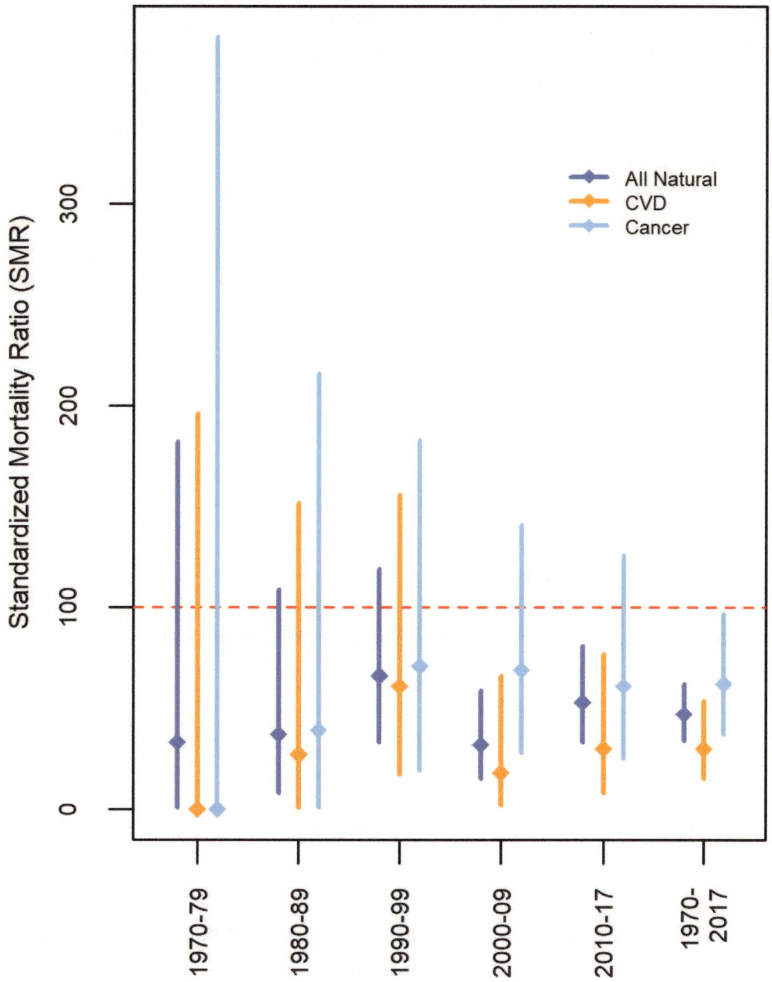

Figure 9. SMRs for all natural-cause, CVD, and cancer mortality among Soviet and Russian cosmonauts, 1960–2017.

4.5. Cardiovascular disease

One of the major components of death by natural causes, CVD has been responsible for 25 out of 52 (48%) of the known natural-cause deaths among cosmonauts. Prior research reported that cosmonauts who have been to space at least once had between 35% and 40% risk of death by CVD compared to the general population through the end of 2014 [22, 23].

Looking to **Figure 9** once again, we can see SMRs for the trained cosmonaut cohort (orange). Interestingly, there were no deaths from CVD in either the 1960s (not shown) or 1970s. From 1970 to 2017, SMRs show decreased risk of death from CVD, with SMRs ranging from 0 to 62. Overall, the SMR from 1960 to 2017 is 28 (95% CI = 18–41).

The trend here may be due in some part to missing information on CVD deaths in the decades since 1990. In those decades, there have been 25 deaths from unknown causes. Depending on how many of those deaths were due to CVD, this could be enough to push some of these

SMRs to reflect significantly increased risk for cosmonauts. (See Section 4.7 below for further discussion and analysis of this point.) Nevertheless, based on **Figure 9** and the prior research, it appears that cosmonauts are at decreased risk of CVD mortality compared to the Russian general population.

4.6. Cancer

The other major component of natural causes of death for cosmonauts is cancers of various types. Previous studies of cosmonaut mortality reported cosmonauts who had been to space as having an SMR of approximately 75 for death by cancer [22, 23].

Figure 9 also shows SMRs for cancer in the wider cosmonaut cohort (lighter blue points and lines). While most SMRs over time are not statistically significant, the SMR of 60 for the entire period from 1960 to 2017 did reach significance (95% CI 36–92). As with natural causes and CVD, SMRs for cancer from 1990 onward could be influenced by the number of unknown causes of death that might rightly be attributed to cancer. Section 4.7 explores this possibility.

4.7. The effects of unknown causes

One potential limitation of the cause-specific analyses presented for cosmonauts is the 24 deaths due to unknown causes. Depending on the distribution of the true causes of these deaths, they could dramatically alter the cause-specific SMRs reported here. In order to explore this possibility, we recomputed SMRs under various assumptions about the distribution of the unknown causes of death.

We assumed for these analysis that the unknown causes were cancer deaths, CVD deaths, or deaths due to other natural causes. This allows us to more deeply explore the question of whether space travel is shortening the longevity of space explorers through increased rates of death by chronic disease, a question of primary concern (that accidents related to space travel will shorten the lives of some astronauts is accepted).

As reflected in **Figure 10**, the 1990–1999 period had 3 unknown deaths, the 2000–2009 period had 5 deaths with unknown causes, and the 2010–2017 period had 16 deaths with unknown causes.

To see the effect these deaths may have on natural-cause SMRs, we recomputed SMRs assuming that all deaths due to unknown causes in a period were due to each respective cause in turn (natural causes overall, CVD or cancer). The result is **Figure 11**.

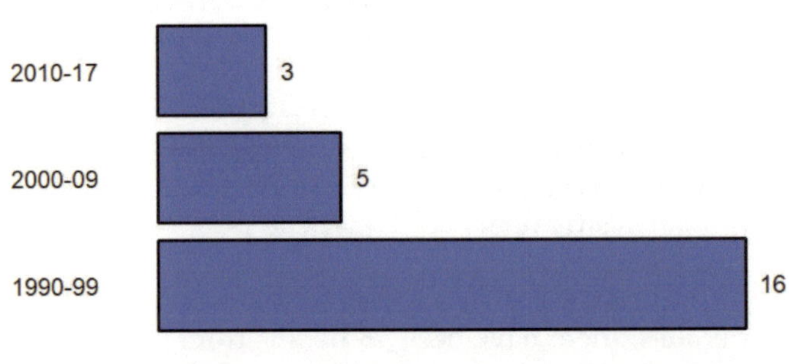

Figure 10. Distribution over time of deaths by unknown causes for Soviet and Russian cosmonauts.

http://dx.doi.org/10.5772/intechopen.73603

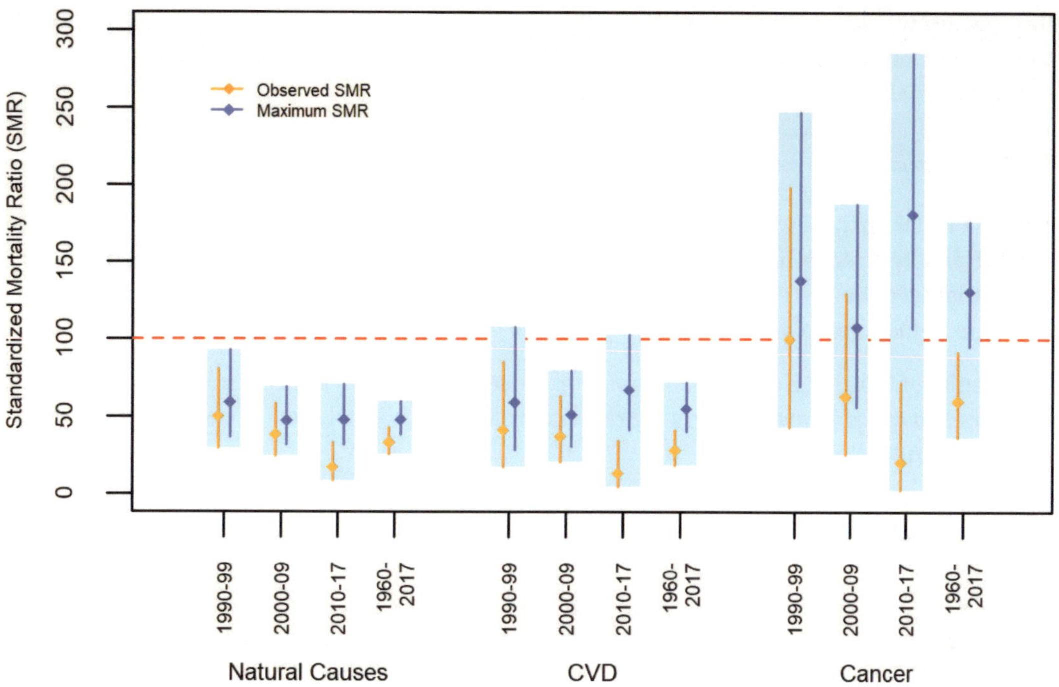

Figure 11. Alternative SMRs for Soviet and Russian cosmonauts, 1990–2017.

In **Figure 11**, the orange points and lines represent observed SMRs and 95% confidence intervals for natural causes, CVD and cancer (exactly as in **Figure 9**), omitting from the analyses all unknown causes. The blue points and lines in **Figure 11** are SMRs and 95% confidence intervals calculated by counting all deaths due to unknown causes as deaths due to the respective cause (natural causes overall, CVD or cancer). As in **Figure 6**, the light blue rectangular zone surrounding these sets of SMRs shows the total range of possibility for true SMR (with 95% CI) by cause. As in the prior SMR figures, the red dashed line is drawn across 100, the point of parity with the general population.

4.7.1. Cosmonaut natural-cause SMRs revisited

Observed SMRs for death by all natural causes combined in **Figure 11** show significantly reduced risk for cosmonauts in comparison to the general population, even with the inclusion of the unknown causes within this causal category. The light blue interval for the entire study period 1960 to 2017 shows that the true value of SMR for the entire study period of 1960 to 2017 is likely to be between 25 and 60, whatever the truth about the unknown causes of death may be. In total, we conclude that, though the value of SMRs reported in Section 4.4 above may be too low due to misclassification of some deaths, the overall conclusion remains the same: cosmonauts are at reduced risk of death from natural causes.

4.7.2. Cosmonaut CVD SMRs revisited

The implications of categorizing all unknown causes of death as CVD deaths have similar implications to the overall natural-cause death analysis, although in this case SMRs for two of the decades would no longer be statistically significant (**Figure 11**). The overall conclusion for the period from 1960 to 2017 also does not change: cosmonauts are at significantly lower risk of CVD mortality than the Russian general population.

4.7.3. Cosmonaut cancer SMRs revisited

Figure 11 demonstrates that the impact of reassigning deaths with unknown causes has the most dramatic effect when all are counted as cancer deaths. In this case, all SMR point estimates that were below 100 are now above 100, though in most cases the confidence intervals continue to include 100 (thus failing to reach the level of statistical significance).

The results of these exploratory analyses for cancer suggest that cosmonauts likely have little difference in cancer-specific mortality rates compared to those of the Russian general population between 1960 and 2017.

5. Astronauts vs. cosmonauts

Having examined the mortality experience of astronauts and cosmonauts separately, we can conclude that both groups have similar patterns of mortality in comparison to general population control groups: lower mortality rates overall, with higher rates of accidental deaths (more so for astronauts) and much lower rates of death from chronic diseases. We now turn our attention to how astronauts and cosmonauts compare directly to one another. Given the similarity in selection criteria, background, training, and career duties, we might expect to find similar mortality rates for astronauts and cosmonauts over the last 60 years. However, given that mortality rates for the Russian general population are known to be higher than those in the United States, we may find some differences.

To explore these possibilities, we computed SMRs using the observed age-, sex-, and period-specific mortality rates among United States astronauts to generate expected numbers of deaths for cosmonauts based on their corresponding age-, sex-, and period-specific exposure times. The ratios of observed cosmonaut deaths to expected deaths determined in this way thus provided SMRs for cosmonauts compared to astronauts.

5.1. All-cause mortality

Figure 12 displays SMRs for cosmonauts in comparison to astronauts for death by all causes. In all but two decades, cosmonauts were at significantly greater risk of death than astronauts; only in the 1960s were cosmonauts at reduced risk of death, and only in the 1980s was there no significant difference between the two groups of space explorers.

What is perhaps most striking about **Figure 12** is that from 1990 to 2017 cosmonauts were more than twice as likely to die as astronauts. Confidence intervals for individual decades are wide, but over the entire period from 1960 to 2017, cosmonauts experienced a nearly doubling of risk compared to astronauts (SMR = 186, 95% CI = 150–228).

http://dx.doi.org/10.5772/intechopen.73603

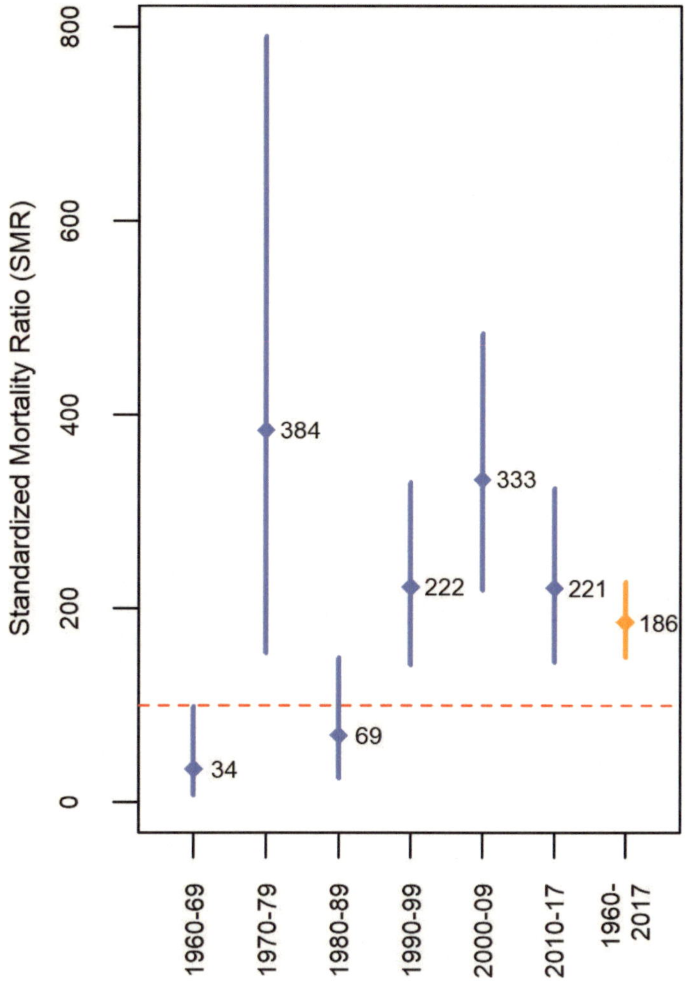

Figure 12. SMRs for all causes of death among Soviet and Russian cosmonauts compared to United States astronauts, 1960–2017.

5.2. External causes

5.2.1. All external causes

No measures of mortality have been previously reported comparing rates of death from all external causes between cosmonauts and astronauts. We report cosmonaut to astronaut SMRs for all external causes here for the first time.

In **Figure 13**, we see that the external-cause SMRs for periods before the year 2000 are similar to those for death by all causes, which might be expected given that astronauts and cosmonauts were relatively young in those years and most deaths observed were externally caused deaths. In the case of the 1960s, SMR for all causes and external causes is identical, as the only causes of death to both astronauts and cosmonauts in the 1960s were accidental, a subset of external causes.

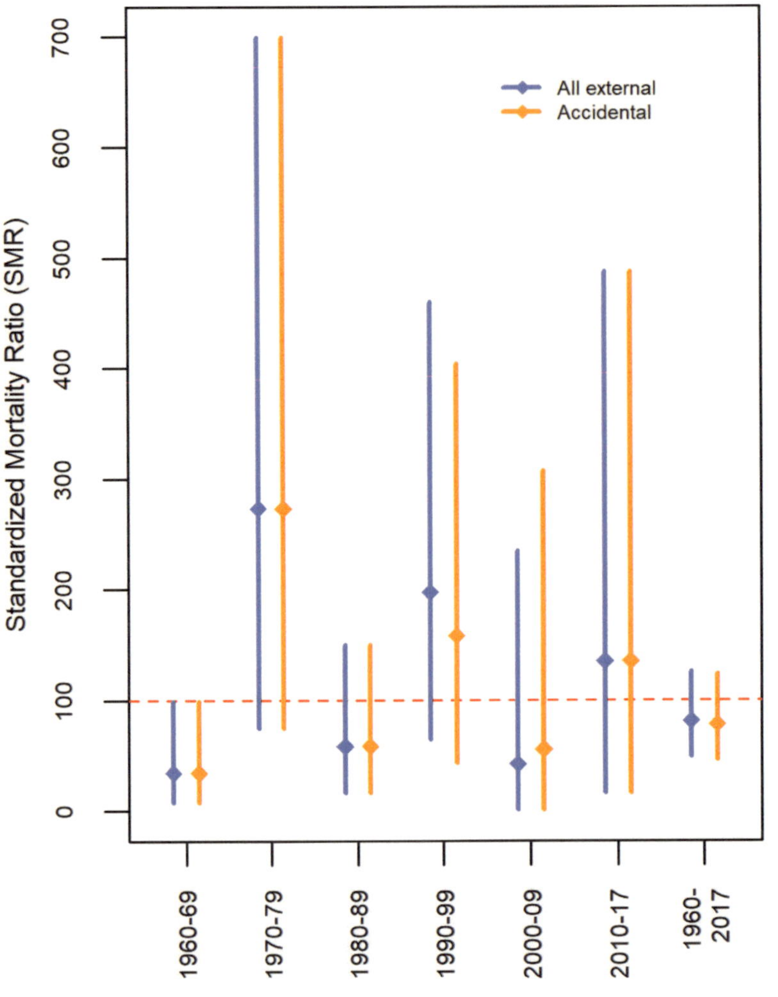

Figure 13. SMRs for all external and accidental causes of death among Soviet and Russian cosmonauts compared to United States astronauts, 1960–2017.

From 2000 onward, the all-cause and external-cause SMRs diverge, as the cohorts' age and other causes of death are observed more frequently among astronauts and cosmonauts. None of the decade-specific results are statistically significant, nor is the overall SMR of 81 (suggesting a somewhat lower risk of externally caused deaths for cosmonauts, but with 95% CI = 49–126).

5.2.2. Accidents

SMRs comparing rates of accidental death among cosmonauts to those of astronauts were reported in the first published study on cosmonaut mortality [21]. SMRs showed an insignificant reduction in risk in the Soviet era, an insignificant increase in risk in the Roscosmos era, and an insignificant reduction in risk for the overall 1960–2013 period (SMR = 88; 95% CI = 54–136) [21].

Our updated analysis comparing rates of accidental deaths in the two groups decade-by-decade is also given in **Figure 13**. SMRs show us that cosmonauts were at particularly lower risk of accidental death in the 1960s (owing to several astronaut plane crash deaths and three astronaut deaths in the Apollo 1 fire), and then again in the two decades that experienced space shuttle disasters (1980–1989 and 2000–2009). As with all external causes, the results here do not reach a level of statistical significance (aside from 1960 to 1969, which barely reaches significance).

Most noticeable in **Figure 13** is the fact that SMRs for all external causes are largely identical to those for accidental causes, since most of the deaths by external causes in both cohorts are accidental in nature. Only in the 1990s and the 2000s are SMRs for the two different, and even then only slightly.

From this, we might conclude that the occupation of cosmonauts and astronauts demands of them that they lead comparably risky lives and that the estimates of relative risk for them within particular periods have more to do with chance timing than systematic differences in risk of accidents.

5.3. Natural causes

5.3.1. All natural causes

Figure 14 displays cosmonaut/astronaut SMRs for natural causes of death (darker blue), CVD mortality (orange), and cancer mortality (lighter blue). Since there were no deaths to astronauts or cosmonauts by natural causes in the 1960s, it is impossible to define an SMR for that period. Few natural-cause deaths in the 1970s and 1980s result in wide confidence intervals for those periods, with no significant evidence of excess mortality for either cosmonauts or astronauts.

In general, cosmonauts have been at higher risk of death by natural causes after 1989. The exception is the 2010–2017 period, when there was essentially no difference between the two cohorts. However, this result is highly suspect, as this period contains 16 deaths in the cosmonaut cohort that are of unknown cause. If we examine the possible range of SMRs as we did in Section 4.7, adding as few as 9 deaths to the cosmonaut death count would render significant SMR of 175 (95% CI = 104–276). If all 16 deaths from unknown causes were actually deaths from natural causes, the SMR would be 243 (95% CI = 157–358). Either value certainly seems plausible, and the net effect is that we should not rule out the possibility of a significant increase in mortality for cosmonauts in this period.

There were cosmonaut deaths from unknown causes in the 1990–1999 and 2000–2009 periods as well. However, since the observed SMRs in **Figure 14** already show statistically significant increases in mortality risk for cosmonauts, adding more deaths from natural causes would only further increase SMR values. For example, assuming all three deaths from unknown causes in the 1990s were truly from natural causes would raise SMR to 229 (95% CI = 138–357) up from the current 193 (95% CI = 110–313). Assuming all five deaths from unknown causes in 2000–2009 were from natural causes would raise SMR to 453 (95% CI = 296–664).

Given the SMRs (and hypothetical SMRs from assumptions about the distribution of causes of death among unknowns), we conclude that cosmonauts are at higher risk of death by natural causes than are astronauts from 1990 through 2017.

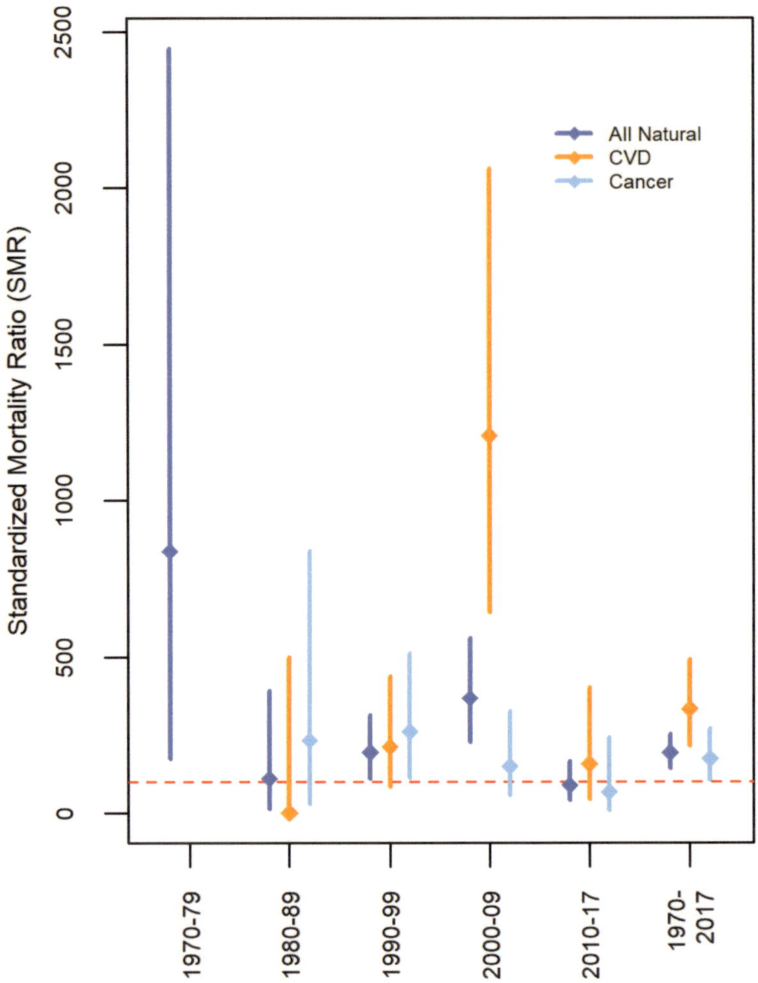

Figure 14. SMRs for all natural-cause, CVD, and cancer mortality among Soviet and Russian cosmonauts compared to United States astronauts, 1960–2017.

5.3.2. Cardiovascular disease

Cardiovascular disease is a cause of death for which both astronauts and cosmonauts have greatly reduced mortality risk in comparison to the general populations of the United States and Russia, respectively. The evidence for this relative to cosmonauts may not be as convincing given the high numbers of deaths of unknown causes, however (see Section 4.7 above). In the prior cosmonauts-to-astronauts comparison, cosmonauts were noted to have a significant increase in mortality due to CVD between 1960 and 2013 (SMR = 364, 95% CI = 225–557) [21].

Figure 14 only includes SMRs for CVD starting with the 1980–1989 period since there were no astronaut deaths from CVD between 1960 and 1979, again making SMRs impossible to define for those two decades (orange points and lines). Cosmonauts experienced no deaths from CVD in the 1960s, but did experience one CVD death in the 1970s.

Three out of the four decade-specific SMRs for CVD in **Figure 14** are not statistically significant. Only the SMR for 2000–2009 was large and statistically significant (SMR 1206, 95% CI = 642–2062). The large confidence interval on this estimated SMR is evidence of the small number of deaths in the astronaut cohort (from which the comparison mortality rates were derived). This small number of deaths led to low mortality rates and thus a low expected number of deaths. This extreme SMR would grow larger if any of the unknown causes of deaths among cosmonauts in that period were in fact deaths due to CVD. However, the very low number of observed deaths among astronauts in this period makes the estimate somewhat unstable.

Like in the prior study [21], overall SMR is statistically significant at 332 (95% CI = 215–491), heavily influenced by the 2000–2009 period. As noted in prior sections, the number of deaths from unknown causes could change the results of recent SMRs. If even one additional death were added to the tally for the 1990–1999 period, SMR would be 243 (95% CI = 105–478), a statistically significant result. Similarly, a reassignment of some unknown causes of death in 2000–2009 and in 2010–2017 could easily raise these to a level of statistically significant elevated risk of CVD mortality for cosmonauts. From the observed data and hypothetical SMRs under various assumptions about unknown causes of death, we can conclude that cosmonauts have been at greater risk of dying from CVD since the 1990s.

5.3.3. Cancers

Cosmonauts have previously been reported to be at elevated risk of cancer mortality in comparison to United States astronauts [21]. Though SMRs for the Soviet and Russian periods were not significant separately, the overall 1960–2013 SMR was significant at 177 (95% CI = 108–274) [21].

Updated SMRs for cancer mortality for Soviet and Russian cosmonauts in comparison to United States astronauts are shown as light blue points and lines in **Figure 14**. Periods for which SMR could not be calculated include 1960–1969 or 1970–1979; there were no astronaut deaths due to cancer in either of those periods.

SMRs for cancer show that in comparison to United States astronauts, Soviet and Russian cosmonauts have largely been at increased risk of death from cancer, though most SMRs are not statistically significant. The exceptions are 1990–1999 and the overall 1970–2017, both of which demonstrate a statistically significant increased risk for cosmonauts. SMR of 67 for 2010–2017 (95% CI = 8–241) is the only SMR that shows a reduction in risk, and not coincidentally, this is in the period in which there are comparatively many observed cosmonaut deaths due to unknown causes.

If we were to add three more deaths to the count of cancer deaths for cosmonauts in the 1990–1999 period, SMR would rise to 356 (95% CI = 178–638). As few as three extra deaths in the 2000–2009 period would drive SMR to significance at 224 (95% CI = 102–425), and six additional deaths in the 2010–2017 period would yield significantly increased SMR of 267 (95% CI = 115–527). Finally, assuming all 24 deaths from unknown causes were due to cancers would increase the overall 1980 to 2017 SMR to 392 (95% CI = 284–529). Even without these potential extra deaths, SMR seen here is highly similar to that reported previously, at 173 (95% CI = 104–271) vs. 177 (95% CI = 108–274) [21]. As with mortality due to CVD and natural causes, we conclude that cosmonauts are at increased risk of death due to cancer in comparison to United States astronauts.

6. Chapter summary

In this chapter, we have examined the mortality of astronauts and cosmonauts in comparison with the general populations of their respective nations, with specially selected controls, and in comparison with one another. The results of prior research and the new analyses presented here indicate that both astronauts and cosmonauts have much lower rates of death by chronic disease (such as CVD and cancers) than do their respective general populations. However, the mortality rate from plane crashes and spacecraft accidents over the years has made both groups of space explorers more likely to die from external causes in general and accidental death in particular, than is expected in the general population. The net effect is that all-cause mortality risk for space explorers is still lower than that of the general population.

Careful interpretation is needed for the reduced risk of chronic disease among astronauts and cosmonauts. Space agencies to date have intentionally tried to limit the potential harmful exposures from space radiation. This has included both projecting and measuring the lifetime dose of radiation for individual astronauts [28]. The evidence gathered thus far seems to indicate that few to no astronauts or cosmonauts have received detrimental doses of space radiation and that we will only begin to understand the mortality risks space radiation can bring once humans start performing longer missions, such as to Mars or beyond.

When comparing cosmonauts to astronauts, we see that astronauts tend to have a slightly higher risk of accidental death compared to cosmonauts, but a significant reduction in the risk of CVD and cancer. The net difference places cosmonauts at an overall greater all-cause mortality risk than United States astronauts.

The lower death rate among cosmonauts from accidental causes is due to fewer plane crashes among cosmonauts, fewer spacecraft accidents, as well as fewer deaths per spacecraft accident (owing to the smaller, 3-person crews in Soviet spacecraft as compared to 7-person crews on United States space shuttles). The reduction in risk due to accidental causes was most pronounced in the 1960s and 1970s, periods of relatively many accidental deaths for both nations. The combined loss of 11 United States astronauts in space shuttles in 1986 and 2003 coincided with no Russian spacecraft accidents in the same period. From this perspective, we could rate the Soviet and Russian space programs as "safer" than the United States program.

The reason behind the greater rate of death by CVD and cancer among cosmonauts is unclear. The most obvious explanation would be lifestyle differences between the United States and Russia, as reflected in the greater mortality risk for these diseases in the general population death rates between Russia and the United States. This could be most salient after retirement from active duty of astronauts and cosmonauts, as the mortality rate from CVD and cancers begin to climb steeply after age 50, in both the United States and Russia [4, 26]. Differences in diet and greater alcohol consumption and tobacco use in Russia/Ukraine than in the United States may explain the risk differences between the groups, especially if those habits were consistent over a period of years [29].

Still another possibility, though less likely, is differing occupational exposures between the cohorts, particularly radiation dose while in space. We are aware of no published work to date that has examined the relationship between time in space or radiation dose and cosmonaut death rates from cancers or cardiovascular disease. However, given the similarity of the Soviet/Russian and the United States space programs over the years, and their explicit partnership over the last 17 years collaborating on the International Space Station, the overall dose per person-day in space in recent years is likely equivalent.

Finally, the quality of medical care could be an explanation for the differential mortality rate due to chronic diseases. Even among equal rates of incidence, differences in the accessibility or effectiveness of treatment for CVD and cancers could lead to a higher case-fatality rate in cosmonauts compared to astronauts. This could result in cosmonauts dying younger of the same diseases afflicting astronauts, driving up SMRs.

The balance of evidence accumulated to date regarding mortality of space explorers suggests that they are, overall, at less risk of death on an age- and gender-matched basis than the baseline risks in their respective countries of origin. As humans continue to explore space, and in particular as they engage in longer trips deeper into space, it may be inevitable that the unique exposures they will face will ultimately lead to some increased risk of mortality due to at least some particular causes (tragic accidents and radiation-related cancers perhaps chief among them). However, it may be worth bearing in mind that as a profession, being an astronaut or cosmonaut is not a terribly risky business. Based on the research reviewed here, as well as the original research presented for the first time, the job of space explorer should not make any top 10 lists of the world's deadliest jobs.

Author details

Robert J. Reynolds* and Steven M. Day

*Address all correspondence to: rreynolds@mortalityresearch.com

Mortality Research and Consulting, Inc., USA

References

[1] Nicogossian AE, Williams RS, Huntoon CL, Doarn CR. Living and working in space: An overview of physiological adaptation, performance, and health risks. In: Nicogossian AE, Williams RS, Huntoon CL, Doarn CR, Polk JD, Schneider VS, editors. Space Physiology and Medicine. 4th ed. New York: Springer; 2016. p. 95-134

[2] Human Mortality Database. University of California, Berkeley (USA), and Max Planck Institute for Demographic Research (Germany). Available from: www.mortality.org or www.humanmortality.de [Accessed: Oct 27, 2017]

[3] Centers for Disease Control and Prevention, National Center for Health Statistics. Compressed Mortality File 1968-1978. CDC WONDER Online Database, compiled from Compressed Mortality File CMF 1968-1988, Series 20, No. 2A, 2000

[4] Centers for Disease Control and Prevention, National Center for Health Statistics. Compressed Mortality File 1979-1998. CDC WONDER On-line Database, compiled from Compressed Mortality File CMF 1968-1988, Series 20, No. 2A, 2000 and CMF 1989-1998, Series 20, No. 2E, 2003

[5] Centers for Disease Control and Prevention, National Center for Health Statistics. Compressed Mortality File 1999-2015 on CDC WONDER Online Database, released December 2016. Data are from the Compressed Mortality File 1999-2015 Series 20 No. 2U, 2016, as compiled from data provided by the 57 vital statistics jurisdictions through the Vital Statistics Cooperative Program

[6] Peterson LE, Pepper LJ, Hamm PB, Gilbert SL. Longitudinal study of astronaut health: Mortality in the years 1959-1991. Radiation Research. 1993;**133**(2):257-264

[7] Fox AJ, Collier FF. Low mortality rates in industrial cohort studies due to selection for work and survival in the industry. British Journal of Preventive & Social Medicine. 1976;**30**:225-230

[8] Hamm PB, Nicogossian AE, Pool SL, Wear ML, Billica RD. Design and current status of the longitudinal study of astronaut health. Aviation, Space, and Environmental Medicine. 2000;**71**(6):564-570

[9] Reynolds RJ, Day SM. Mortality among U.S. astronauts: 1980-2009. Aviation, Space, and Environmental Medicine. 2010;**81**(11):1024-1027

[10] National Aeronautics and Space Administration. Where No Man Has Gone Before, CH5-2. Webpage. Available from: https://www.hq.nasa.gov/pao/History/SP-4214/ch5-2.html [Accessed: Oct 25, 2017]

[11] Reynolds RJ. Development of a location exposure matrix for ionizing radiation in extraterrestrial environments and its application in the study of mortality for U.S. astronauts. Texas Medical Center Dissertations (via ProQuest). Paper AAI3604553. 2013. Available from: http://digitalcommons.library.tmc.edu/dissertations/AAI3604553. [Accessed: Oct 25, 2017]

[12] Hamm PB, Billica RD, Johnson GS, Wear ML, Pool SL. Risk of cancer mortality among the Longitudinal Study of Astronaut Health (LSAH) participants. Aviation, Space, and Environmental Medicine. 1998;**69**:142-144

[13] Ade CJ, Broxterman RM, Charvat JM, Barstow TJ. Incidence rate of cardiovascular disease end points in the national aeronautics and space administration astronaut corps. Journal of the American Heart Association. Aug 7, 2017;**6**(8)

[14] Delp MD, Charvat JM, Limoli CL, Globus RK, Ghosh P. Apollo lunar astronauts show higher cardiovascular disease mortality: Possible deep space radiation effects on the vascular endothelium. Scientific Reports. 2016;**6**:29901

[15] Cucinotta FA, Hamada N, Little MP. No evidence for an increase in circulatory disease mortality in astronauts following space radiation exposures. Life Sciences and Space Research. 2016 Aug;**10**:53-56

[16] Comments on "Apollo lunar astronauts show higher cardiovascular disease mortality: Possible deep space radiation effects on the vascular endothelium." Available from: https://www.nature.com/articles/srep29901#article-comments [Accessed: Nov 14, 2017]

[17] Reynolds RJ, Day SM. Mortality due to cardiovascular disease among Apollo lunar astronauts. Aerospace Medicine and Human Performance. 2017;**88**(5):492-496

[18] Bacal K, Romano J. Radiation health and protection. In: Nicogossian AE, Williams RS, Huntoon CL, Doarn CR, Polk JD, Schneider VS, editors. Space Physiology and Medicine. 4th ed. New York: Springer; 2016. pp. 197-224

[19] Chancellor J, Scott G, Sutton J. Space radiation: The number one risk to astronaut health beyond low earth orbit. Life. 2014;**4**:491

[20] National Aeronautics and Space Administration. Risk of Cardiac Rhythm Problems During Spaceflight. Available from: https://humanresearchroadmap.nasa.gov/evidence/reports/Arrhythmia.pdf [Accessed: Nov 14, 2017]

[21] Reynolds RJ, Day SM, Nurgalieva ZZ. Mortality among Soviet and Russian cosmonauts: 1960-2013. Aviation, Space, and Environmental Medicine. 2014;**85**(7):750-754

[22] Ushakov IB, Voronkov YI., Bukhtiyarov IV, Tikhonov GI, Gorkachova TY. Retrospective analysis of status of cosmonauts' health after participation in space flights. Air-ecological and Environmental Medicine. 2016;**50**(2):14-20. [Russian]

[23] Ushakov IB, Voronkov YI, Buhtiyarov IV, Tikhonova GI, Gorchakova TY, Breyleva MS. A cohort mortality study among Soviet and Russian Cosmonauts, 1961. Aerospace Medicine & Human Performance. 2017;**88**(12):1060-1065

[24] Bugrov SA, Voronia LI, Voronokov YI, Korotayev MM, Senkevich YA. Selection and biomedical training of cosmonauts. Advances in Space Research. 1992;**12**:347-350

[25] Bogomolov VV, Castrucci F, Comtois J, Damann V, Davis JR, Duncan JM, Johnston SL, Gray GW, Grigoriev AI, Koike Y, Kuklinski P, Matveyev VP, Morgun VV, Pochuev VI, Sargsyan AE, Shimada K, Straube U, Tachibana S, Voronkov YV, Williams RS. International space station medical standards and certification for space flight participants. Aviation, Space, and Environmental Medicine. 2007;**78**(12):1162-1169

[26] Center for Demographic Studies at The New Economic School. Russian Fertility and Mortality Database. Available from: http://demogr.nes.ru/index.php/en/demogr_indicat/data_description

[27] Lévesque LE, Hanley JA, Kezouh A, Suissa S. Problem of immortal time bias in cohort studies: Example using statins for preventing progression of diabetes. British Medical Journal. Mar 12, 2010;**340**:b5087

[28] National Aeronautics and Space Administration. How Do We Protect the Astronauts From Space Radiation? Webpage. Available form: https://srag.jsc.nasa.gov/SpaceRadiation/How/How.cfm. [Accessed Nov 14, 2017]

[29] Shkolnikov VM, Cornia GA, Leon DA, Mesle F. Causes of the Russian mortality crisis: Evidence and interpretations. World Development. 1998;**26**:1995-2011